The Avid® Handbook

FOURTH EDITION

The Avid® Handbook

INTERMEDIATE TECHNIQUES, STRATEGIES, AND SURVIVAL INFORMATION FOR AVID® EDITING SYSTEMS

FOURTH EDITION

Steve Bayes

ELSEVIER

AMSTERDAM • BOSTON • HEIDELBERG • LONDON • NEW YORK • OXFORD
PARIS • SAN DIEGO • SAN FRANCISCO • SINGAPORE • SYDNEY • TOKYO

Focal Press is an imprint of Elsevier

Focal Press is an imprint of Elsevier
200 Wheeler Road, Burlington, MA 01803, USA
Linacre House, Jordan Hill, Oxford OX2 8DP, UK

 Recognizing the importance of preserving what has been written, Elsevier
prints its books on acid-free paper whenever possible.

Library of Congress Cataloging-in-Publication Data

Bayes, Steve, 1959-
    The Avid handbook : intermediate techniques, strategies, and survival
information for Avid editing systems / Steve Bayes.— 4th ed.
      p. cm.
    Includes index.
    ISBN 0–240–80553–4 (pbk. : alk. paper)
      1. Video tapes—Editing—Data processing. 2. Motion pictures—
Editing–Data processing. 3. Avid Xpress. 4. Media composer.
I. Title

   TR899.B37 2004
   778.59'3–dc22

                                         2003058304

British Library Cataloguing-in-Publication Data
A catalogue record for this book is available from the British Library.

ISBN: 0–240–80553–4

For information on all Focal Press publications
visit our website at www.focalpress.com

04 05 06 07 08 09 10   9 8 7 6 5 4 3 2 1
Printed in the United States of America

To Mary Andress, Marjorie Bayes, and Jane
MacFarlane for all the birthdays

Thanks to:

Roger Shufflebottom
David East
Jeff Greenberg
Johnathon Amayo
Jay Lasoff
Peter Bos
Joe Doyle
Matthew Feury
Jaime Fowler
Mark Geffen
Elliott Kaplan
Stephen Hullfish
Dave Meichsner
Jim McKenna
Don Nelsen
Paul Pearman
Greg Staten
Tim Vandawalker
Paul Sampson
Wes Plate
Jeannie Munro
James McKenna
Julia Miller
Michael Phillips
Doug Hansel

"It is circumstances and proper timing that give an action its character and make it either good or bad."

— Agesilaus, 444–400 B.C.

"A globe-full of people, and not one is ignorant of the worth of twenty minutes, each minute is a Pearl, let slip, one after the next, into Oblivion's Gulfs."

— Thomas Pynchon

"The fact would seem to be, if in my situation one may speak of facts, not only that I shall have to speak of things of which I cannot speak, but also, which is even more interesting, but also that I, which is if possible even more interesting, that I shall have to, I forget, no matter."

— Samuel Beckett,
The Unnamable

Table of Contents

Preface

The world has changed in many ways since the first edition of *The Avid® Handbook*. This version reflects the changes in the industry and in the technology since then. Amazing breakthroughs have been made with faster computers using optimized software for more realtime streams than ever. Drives have increased in size and speed while plummeting in price. Many techniques that were common sense have become archaic and amusing. But there is still a core of basic information and ways of using the gifts of progress that are embedded in this text. Our creative challenge is in how to use the new tools to tell stories as best we can—a challenge that must be taken up every day.

The *Avid® Handbook* is a book about nonlinear editing that can be read in a linear fashion, but it can also be kept handy to refresh your memory before you start a big job. It will give you the confidence you need after taking an introductory Avid course to deal with the more complex situations that occur outside the classroom.

Writing about a series of keystrokes does not represent a coherent view of the machine or its capabilities. Many people who are looking for a tips book are striving for a deeper understanding of ways to make things go faster and smoother to give them an edge over their competition. You won't get that by memorizing ever more arcane variations on secret handshakes. You will gain the power of technology as part of the new production *procedures*, which is the way this book is focused.

Avid editing systems are complex because they deal with complex, ever-changing, and unpredictable situations. If you are a professional and are learning a high-end system, I recommend that you take a class or find a good teacher. Spend the time with an experienced human being to get going and learn the basics. Having a good teacher is clearly worth the money. It's your career—start this part right!

This book is really intended for overworked editors, assistants, or postproducers who find themselves needing to know more than they really have any chance of learning through their limited experience. They are at a stage

bordering on intermediate, but are being asked to make predictions and esti-
mates based on procedures they have never seen and functions they have never
used. Everyone in the industry should have a healthy respect for the vast
amount of things that can go wrong and increasing levels of complexity in *any*
production. A complete overview of procedures will show you everyday savings, shortcuts, and strategies. Display *The Avid® Handbook* proudly in your
Avid suite and reread it while you are digitizing and rendering!

1

Talking to the Machine

"Anything may, with strict propriety, be called perfect, which perfectly answers the purpose for which it was designed."

—AMERICAN SHAKER PROCLAMATION, 1823

The Shakers, a utopian sect that lived in complete racial and sexual equality during the late 1700s, attempted with their numerous inventions to create "heaven on earth." Their burst of creativity during this social experiment lasted 125 years, and when the Shakers built, refined, and used their own tools they were directly attempting to improve their lives. This concept—that the refinement of tools could create earthly paradise—has been drowned out by our industrial culture. But in the postindustrial age we look again to models that can guide us to an ideal for usability. Simple on the surface, deep underneath; power when you need it, but not before; and the anticipation of earthly needs through the encoding of experience are all areas to which our modern software now aspires.

For artists and craftspeople working today in the communications and entertainment industries, the main characteristic that will distinguish this time period from others will be the speed of change. Tools have always evolved, and clever workers have always adapted their existing tools to the job at hand. Once a tool was refined to a certain point, it may have stayed in that particular state of grace for centuries. No such luck these days. Coming to grips with new techniques and features is a constant and never-ending process. We now face an everyday choice of whether to exchange the power of the newest and the fastest for the comfort of the familiar.

But maybe it's not really so bad. Maybe there is a thrill to the new and different—a pleasure in finding the best way to work for you. The secret pleasure of a complex system is simplifying the tools to meet your needs, streamlining them so that you are in complete control. You don't need to buy new software for a new job; just get out the manual and find out what is hidden inside this one giant piece of software. It's a treasure hunt to find the extra bit of power that can propel you through a tedious task and finish while your client is still on the phone. Or like a magic trick, you do have a few things up your sleeve, capabilities your clients have never seen before, things you have developed for yourself that can astonish and amaze.

That's the positive side to all this change. You don't need to know everything. There is no way you can know everything, but you should know how to find what you need. The time you spend memorizing arcane computer jargon and seldom-used techniques is time away from the creative process and time away from friends and family.

This book is meant to be the next step after formal training. I strongly recommend taking a formal beginner's course. I helped write the courseware (along with experienced technical writers and teachers), and I used to certify the instructors. No matter how experienced you are, if you do not start with formal training you will have gaps in your knowledge. You are forced to guess and improvise when someone has already thought it out for you. For example, after viewing a demo reel filled with beautiful, multilayered graphics, I asked the editor to send me his sequence. "How do you do that?" he asked. An odd gap, but one that is holding him back.

This handbook is meant to fill in those gaps. It provides the extra level of depth that you can't get in a class full of techniques; the type of experience and insight that comes from watching beginners make mistakes and advanced people fly; the type of experience that takes a long time to sink in. Along with this handbook you should experiment, read the user documentation, take an advanced course, read the Internet, and never stop learning. By necessity, training for adults must be short and intense. Unfortunately, it is hard to retain all the information taught in a short course. When I taught trainers how to teach the Media Composer, my mantra was, "Learn what to leave out." In two, three, or five days you can do more damage trying to overload your students than you can by presenting less. Make them feel confident and competent with the basics. You will not be "an Avid Editor" after a short course, but you will be able to start and complete your first job. Editing is a job that takes some technical knowledge but an even larger amount of creativity and diplomacy. You won't learn the critical aspects of being an editor by taking a course. Sometimes I feel an evil joy in telling my students that I am always learning something new on the Avid, after 12 years. They don't think that's funny.

A very important translation goes on when you take a vague idea in your head and translate it into pull-down menus. I have found a little trick that allows the creative side to cohabitate with the technical side of my brain. The trick stems from the understanding that the computer is doing only what you ask it to do. The computer is interpreting those commands through the filter of the minds of the people who programmed it. If the designers of editing software don't understand professional postproduction, then you will have to fight and trick the machine at almost every step beyond the most basic tasks. When you are talking to the machine the result should be a mirror of your demands.

When you sit in front of an Avid interface you trust that the Avid designers understand what you want to do and, if they have not written something to accomplish exactly that, at least they haven't held you back with arbitrary restrictions or complex workarounds. You have some freedom to improvise and customize. Many of the techniques described in this book have been developed because the system is so flexible. Over the many years of development more direct functionality has been added to the software, much of it through my design work and that of a team of dedicated engineers.

The writing of this handbook has directly affected my design ideas and manifested itself in minimizing complexity without reducing functionality. My favorite design icon is a Thermos™ container. It keeps hot things hot and cold things cold—it just knows. I wish all software could be that adaptable and simple to use!

Before I became a principal designer, I taught Avid editing around the world for five years. What was really interesting was that I was learning as much as my students. I learned about sit-coms, army VIP videos, promotion departments, and reality television. I saw this same piece of software used every day in many completely different ways. I understood very clearly that there was no right way to do most things, and what is a minor problem for some is a catastrophe for others. Features that you have never seen anyone use will regularly save entire productions. I had to throw away a lot of preconceptions and realize that the more I learned the less I knew.

The value for me in teaching Avid in so many different environments is that I experienced different needs and ways of doing things. The purpose of this handbook is to synthesize all of these experiences into an easy format for you to apply immediately. While using the system for the last 12 years, I have seen hundreds of situations where just a little advantage is enough to keep a client and make the difference between losing money, breaking even, or making a profit.

This handbook describes procedures that go beyond the basic functions of the system and adapt to changing situations and demands. Knowing that the procedures are available will allow you to plan more efficient strategies. The more you understand the capabilities for organizing and the database power of Avid's methods, the more you will be able to plan for and stay calm during the inevitable chaos. There is depth to this software that extends far beyond the basic interface, and that is hard to show in a demo or a side-by-side comparison. It will save you if you know how to use it.

So what should you focus on to learn and master the Avid editing systems? Although each model has a different market, there are some very important similarities both in the software and in the way it is used. This handbook covers, but does not does not focus on, the individual menus or buttons, although they must be covered to get any substantial hands-on value. Buttons and menus change and go out of date so fast that standard printed books no longer can really keep up when you get too specific with professional software (OK, this is the fourth edition in six years, but you know what I mean). The ideas and procedures in this book should help you for some time to come because they deal with the entire process and where the Avid fits in.

The Avid® Handbook was first written during the beta testing of version 7.0/2.0. Today, the era of the Meridien system is slowly being replaced by much more sophisticated, programmable hardware at the high end and host-based, software-only systems at the low and midrange. The curse of interesting times is certainly alive and well, but for some, being in the middle of yet another technological upheaval is part of why the film and video industry is challenging and rewarding. Many times a technique for Media Composer or Xpress will apply to Symphony or Xpress DV, so models are not specifically mentioned most of the time. You can count on just about anything that does not have to do with platform differences as being the same across all the models. With so many versions (both recent and legacy), two platforms, and four different video boards, it is hard to make sweeping statements on features and functionality. For convenience sake, I have included both Windows and Macintosh modifier keys like this: Ctl/Cmd or Alt/Option. Even though there is a Control key on the Macintosh, it is not used very often, so when you see the Ctl/Cmd mentioned you should assume the Ctl is on Windows since that is the modifier that maps most closely to the Macintosh Command key. This edition will remove some of the information on older systems like the NuVista and reduce the detail of ABVB-based systems. There is so much new material to cover! If you are confused about whether some information applies to your system, just give the technique a try anyway — you might be surprised!

There will be times when a particular function does not exist in Xpress DV that is standard in Media Composer, but for the most part a technique described in this handbook will make sense anyway and may give you some new ideas about how to use Xpress DV. Even if you are using your Xpress DV only once a week, it is still worthwhile to learn these techniques and polish your skills in order to be more proficient.

This handbook makes some assumptions about basic understanding of the computer's operating system; otherwise this book would take on the proportions of Avid, Macintosh, and Windows documentation combined. This handbook is not a replacement for reading those fine manuals, but it will point you very quickly to the most useful areas of the software and focus on explaining some of the most confusing. There will be some complex techniques to help you deal with complex situations. I like to think of them as recipes, the kind of lists of instructions you deal with every day. Follow the steps, understand the big picture of what you are trying to accomplish, and you will find yourself at the next level of understanding as you work your way through the book.

I hope this handbook helps you avoid down time and maximize your creative time. Get past the bells and whistles to the really important stuff and have confidence that with better tools you can become a better editor. You can't use a shot if you can't find it, and getting to a level of technical proficiency will let you spend time on creating your perfect vision. You're working with the best system in the world for telling stories with pictures and sound. Good luck and let's go!

2

Workflow of a Nonlinear Project

Workflow is where many beginning editors and producers can take advantage of new opportunities and improve their efficiency by looking at the whole process. The main steps to any nonlinear digital project are input, editing, and output. This chapter describes in-depth strategies for these three basic steps, and how to adapt to fit them easily into your production methods. You must get a clear picture of how this technology will change your scheduling and budget. You must plan how you are going to organize and finish a project before you begin or you will lose yourself in the many choices available to you at each step.

CHANGE YOUR MIND WITHOUT LOSING IT

First let's deal with the most commonly repeated myth of the benefits of this technology: "It's faster." Well, yes, the computer does arithmetic better than you, but it can't decide between green or blue for a title or whether a logo is too small. You still need to make all the decisions. The computer makes it easier and faster to implement those decisions and to see the many subsequent versions for the client's new vice president, but let's face it, humans don't make decisions faster just because they are using a computer. In fact, some things take longer using the computer. Once you have a clear idea of where you want to go, the mechanics of getting there are accelerated.

You prudently would have stopped at the third or fourth version of a sequence using more traditional methods from the sheer exhaustion of creating them all or from the loudly ticking clock of the expensive online suite. Now you are forced to defend your cut for ethereal reasons of timing, sensibility, and art—reasons that can take much longer to justify. It is also easier to lose that argument since "the client is always right" (or at least they are paying the bills). The freedom to experiment and explore can be a liability unless you understand how to take control of a situation, fight for what you think is best, and know when you are making the scene worse. Multiple versions mean multiple screenings for various groups of people with conflicting and passionate opinions. Lots of discussion ensues with arm twisting, insults, shameless self-promotion, and occasionally a decision. This is progress, but it's not fast.

Another reason that your work may not go faster is that since all of these changes are supposed to be so easy and painless, you will get them at the last possible moment. Before nonlinear editing, there was a considerable penalty for changing your mind at the last minute, and this put pressure on producers to be better prepared before the final stage of the project. It was easier to charge for the overages because everyone understood how hard those changes were.

Some changes are easy with a nonlinear system. Making Part 3 into Part 2 takes just a few minutes, but the difference between an easy fix and a much more complicated one may not be obvious. Some changes have always been difficult; for instance, changing layer three in the middle of a 20-layer effect. Those familiar with traditional layering methods actually will be surprised at how easy it is now to make that kind of change. What they aren't prepared for are all the little changes that ripple throughout the entire show, all the new details that add up to extend a project in unexpected ways. In the workflow of a nonlinear project, there are places where the work will go amazingly fast, but occasionally, the technology and the added capabilities will add some requirements to slow you down.

Venturing into a method of working where thousands of decisions and many hours of diplomacy are integral to finishing on time and on budget is daunting. You may want to restrict the choices offered to your indecisive client by offering only a few of the directions you think are best. Clients must learn to trust you and your judgment—once they do you can lead them through the areas that you think are the most important without getting distracted or fixated on details that can be worked out later. Instead of asking, "What do you think we should do here?" you could think of offering direction A or B and maybe C. Use the power of the system to show the choices quickly to the client and help them pick the best before moving on.

INPUT

Let's look at the first step: capture. Some Avid systems refer to this as digitizing or recording, but the current term used is *capture* because it takes into consideration that some media are already digital or are already on disk. Some people are shooting straight to disk these days and using tape only as a backup. This presents new challenges to keep track of everything as it goes flying by. If you synchronize your watch to the timecode generator that is producing the time-of-day timecode for all the camera feeds, you will be much happier when backtracking through your logs if you are going directly to disk. Even if you are not capturing, you still need to incorporate preproduction planning and project organization in a way that will influence the shooting and logging. Take into consideration that a shoot may go past midnight; then all the timecodes based on time of day will be lower numbers than the material shot first. You may want to sort material in the bin based on multiple criteria, like date and timecode hour, which we will discuss later.

Logging: Film and Video

Logging is your link between production and postproduction, so don't treat it lightly! If you have strategies in place to minimize hand entry or eliminate duplication of effort to input data to the Avid system, you will have instant savings. Logging and analyzing as much of the footage as possible before the editing begins are two of the best ways to improve the speed and efficiency of capturing.

If you are working on a film-based project, much of the initial logging will be taken care of for you during the film-to-tape transfer process. The transfer house will create a transfer log file or database, such as a FLExfile. Use the transfer log file from the telecine session and convert it to an ALE file (Avid Log Exchange) for importing into your project as a bin. Use the free Avid ALE program to do this. Do not skip this step to save a little money! The bin created from the ALE file will show you where the 2:3 cadence begins on an NTSC (National Television Standards Committee) project; usually this is an A frame, and the keycode relationship with the original film (see Figure 2.1).

During the capture process the Avid system will strip out the extra pull-down frames and leave you with the original 24 fps. If you do not have the ALE you will have to identify the A frame by eye by jogging through the tape frame by frame, then manually entering the keycode for that film frame into the Avid bin. Current Avid versions can detect the 2:3 cadence "on the fly" so you no longer need an ALE file just for capturing, but the other important function of the ALE is to track what video frames correspond to the original

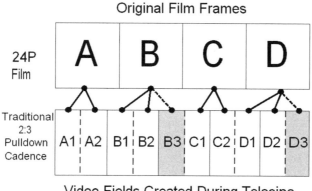

Figure 2.1   Traditional NTSC 2:3 Pulldown Showing the Frames Removed during Capture

film frames. This is done by tracking the keycode, the frame numbers burned onto the edges of the original film and recorded in the log file during the telecine session. You need to know how the keycode matches the video time-code of the film-to-tape transfer master tape. If you do not have the keycode information entered in your bin, you will not be able to make a negative cut list or be able to cut the negative accurately (unless you match the negative to the video output by eye).

If you do not have the keycode tracked by the telecine operator, then you will have to make dubs of your telecine transfer and burn in the keycode to the video frame (which has been recorded in the auxiliary lines Vertical Interval Time Code [VITC] during the telecine session) so it can be viewed on the screen. Capture the burned-in dubs to the Avid film project. Go to the first frame of every master clip and enter the keycode into the keycode bin column by hand. You may want to create burn-in dubs with the keycode anyway and check the head of every master clip, even if you have a proper ALE, just to check that the keycode in the ALE file and the keycode on the telecine master match up. Make sure no mistakes were made in the film transfer to avoid disaster when it finally comes time to cut the negative. And you should absolutely always get the telecine log file from the film transfer session.

There is a different kind of NTSC cadence that records extra frames in the field while shooting 24p in DV. This is called advanced cadence pulldown and is characterized by a 2:3:3:2 pulldown pattern (see Figure 2.2). There are 24 fps recorded on location and then the extra pulldown frames are added by the camera to pad the rate to a true 29.97 fps. These extra frames are eliminated easily while capturing over firewire on Xpress Pro or Media Composer Adrenaline. This means that when you are shooting advanced cadence on location you can

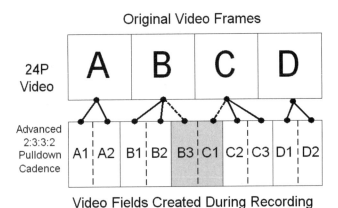

Figure 2.2 Advanced Pulldown Showing the Frames Removed during Firewire Capture to a 23.976p NTSC Project

capture a true 24p media, edit at 24p, and output to DVD or film with true 24p. The further benefits of staying 24p are addressed in Chapter 12.

Some telecine facilities are transferring directly to hard disk. Using *chunking*, the ability to use a shot before it is finished capturing, you can start to edit before the telecine transfer is done! If you have your film transferred directly to disk (with a master tape backup, of course) you can connect that drive directly to your Avid system and skip the entire capture process. It is more likely that the telecine house will transfer to their Avid Unity MediaNetwork and then copy the files to an inexpensive firewire drive for transport to your editing location. You can then choose to copy those files to your faster SCSI or Fibre drives and immediately get to work. By skipping the capture stage, you can work on a film project at 24p on an Xpress Pro because you no longer need the Avid hardware to remove the 2:3 pulldown. The pulldown has been removed during the telecine transfer to disk stage and you are left with the original 24 fps. Don't forget to get the telecine log file anyway! Editors working on video-based projects must spend more time preparing for the capturing stage.

Although it may seem tedious at this stage to think deeply about logging, you will find that the more thought you put into the creation of an edit session at the beginning of the process, the more efficient you will be for the rest of the job. If you are going to cut the material together then you should be looking at everything during the logging process. If the first time you look at the material is in the session with the client, then you will not be providing the service they expected. This chapter contains a lot of detail about logging, even though it seems unimportant to workflow compared to the editing. Ignoring the importance of keeping track of tape names and timecodes is a mistake you make only once!

MediaLog, the logging software that ships with every system, creates Avid bins on almost any computer without any special hardware. A deck control cable comes with MediaLog and connects from the tape deck to the serial port or com port of the computer. This connection can be used for an external modem or a printer, so if you are using MediaLog you may want to look into a USB or firewire connection for these other peripherals. If you can control a video deck with the computer while using MediaLog, you can mark start and end points while the tape is playing. MediaLog will read the timecode that is on the address track of the tape. You then can bring the removable media containing the logged project to the Avid suite and copy the bins and project to the Avid system. The capturing is now an automated process controlled by the computer, which asks for each tape as it is needed.

A limited range of professional decks can be controlled only with extra deck control hardware like a virtual local area network (VLAN) or a VLX. These are external deck control devices that can be downloaded with the proper deck control protocols for your deck. These are becoming increasingly rare, but make sure the deck you want to use is on the Avid list of supported decks in the user documentation. More likely you will have a Rosetta Stone or an extra device that will convert the control signal from the computer to standard deck control protocol. Avid uses standard Sony protocol, RS-422, but strangely, many popular decks don't comply with this standard. This is especially true with the less expensive DV25 decks, and if you choose to use your camera as a deck will find some control limitations. Take care that you log onto the same version of Media-Log as the one you will use for the actual edit. MediaLog 11.x will have problems sending bins back to Media Composer 7.1. Forward is good, backward is problematic. If you must work with mixed versions of the software (a bad idea), make sure that the version of MediaLog is earlier than the version of Avid editing software. Here is an important tip: If you are having problems getting your bins into the Avid system, you can display the most important headings in the bin and export from MediaLog as a shot log. Importing is much more forgiving using a text file instead of an unrecognizable version of a bin.

Timecode Breaks

Videotape shot in the field generally will have breaks in the timecode whenever the deck is completely stopped, not just paused. Occasionally, the deck operator actually will be making backspaced edits in the field to eliminate the timecode breaks, but this is unlikely from a camera crew on the run. However, if the crew is shooting a time-of-day (TOD) timecode, you will have a break in the timecode every time the deck is paused. The TOD timecode generator continues to increment when the deck is paused and, when the tape starts rolling again, you have a timecode that reflects the new time of day. The clock keeps ticking!

TOD timecode (sometimes called freerun) is especially problematic when logging since someone must be alert to every break. TOD is recommended when you are shooting an event where there is continuous action, long takes, or when multiple cameras are running (like multicam). It is also the most popular timecode for newsgathering. The many quick timecode breaks created when using are also a pain if there are a lot of quick shots. New Avid features allow you to preroll over timecode breaks on the source tape during capturing, but this won't help if you missed the timecode break when logging and logged it incorrectly.

Avid systems can create a new master clip whenever they capture across timecode breaks. You can just pop a tape in, set the Capture setting to Capture across timecode breaks, but still keep an eye on it. Depending on the tape format, DV25 especially, the system may miss a timecode break occasionally and keep capturing. The Avid software should sense every time there is a break in the timecode, make a new master clip, roll forward to come up to speed for a few seconds, and then start capturing again. If there is plenty of time between the field deck starting and the beginning of the action, this works out fine.

However, chances are that at least some of the takes will not come up to speed completely before the action begins and the beginning of the action is not captured. This may result in the beginning of a take being cut off. Although the reaction time of the Avid has gotten better over the years, you may occasionally have to modify and recapture a clip where the action started too quickly after the tape started to roll.

There is still a known issue with missing the single very small (a couple of frames) timecode breaks that come from the Sony Digital Betacam cameras. If the cameraperson is not shooting TOD and decides not to preroll the tape before recording each take, occasionally there may be one-frame breaks on a source tape. These one-frame discrepancies may seem minor, but they can add up over the course of a 30 minute reel if there are lots of short shots. Avid systems today need about six frames in order to catch a timecode break because anything shorter happens quite often during common tape playback. The system would abort capturing constantly, just because of basic tape transport issues which in the long run would be a much bigger problem. As computers get faster and capturing is not such a strain on the bandwidth of the system, eventually this problem will be solved. However, in the meantime, be extra vigilant for these one-frame breaks. Either use TOD timecode or carefully prelog tapes that come from a Sony Digital Betacam camera before capturing.

Preroll and postroll are keys to good logging. You must have a minimum preroll of several seconds after the timecode stabilizes and the deck comes up to speed. If you are working with Betacam SP, then you need about three seconds. U-Matic and cheaper DV decks struggle with anything less than four or five seconds. Don't log the beginning of a shot until this amount of time has elapsed after the field deck has started recording. Otherwise the deck will rock back and forth during capturing as it tries to find the proper preroll point—those three to five seconds before the logged inpoint. If the Avid can't find the proper preroll point, it stops and gives you the error message "failed to find preroll coincidence point." The expected preroll point isn't there.

The feature called Adaptive Preroll will try to capture a shot several times before giving up. The first time the system senses a timecode break it will attempt to preroll using control track. Failing that, it will attempt to count backward from the last known good timecode. If you are going to be conforming this project on a system without Adaptive Preroll (pre-version 10 Media Composer or version 2 Symphony), you may have difficulties since the older system will not automatically retry to capture the clips with short preroll. However, this setting is optimized for speed and should be used if possible. Adaptive Preroll, combined with the Capture Setting "Log errors to the Console and Continue Capturing" has significantly reduced the reasons a capture session will stop for a problem. I will discuss a few more important capture speed settings later.

Set the preroll setting for the playback deck to be as short as possible. This will speed up capturing and cause fewer problems by reducing the occasional

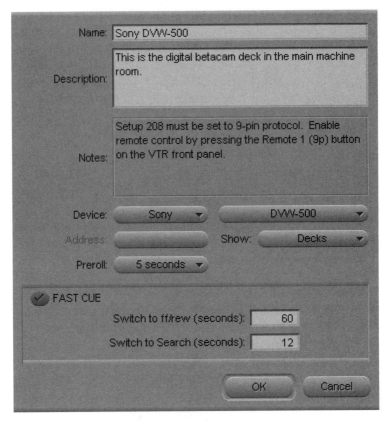

Figure 2.3 Deck Settings Dialog Box

prerolls across a break in the timecode. You can change the deck preroll under the Deck Configuration setting, so that the preroll is shorter than the default. The default preroll time changes depending on the type of deck you choose. If you set a very short preroll time, you run the risk of the first few frames of the captured shot being unstable. If you are considering taking the final online to a linear tape suite you run the risk that the edit will abort in the online suite. The online editor will have to change the preroll time for that shot, which may stop the auto assembly process, prompting the question: "Who logged this?" But finding the shortest preroll time your deck can handle will help optimize the capturing process. If all this fails and you are absolutely positive you will not go to a linear tape suite, you can capture on the fly by letting the deck roll over the break and quickly clicking on the red capture button to start the capture without preroll. However, if you must make an edit decision list (EDL) to go to a linear tape suite or recapture this entire master clip when finishing, you are better off dubbing this shot to another tape and then capturing the dub. If this is an analog tape, dub it to a digital source to prevent the loss of a full generation.

DV Scene Extraction

Under the Bin menu, there is a great time-saving feature in Xpress DV called DV Scene Extraction. This setting also can be found in the Record Settings in Xpress DV so that you can have the scene extraction done automatically as you record. When you are starting and stopping a DV camera in the field there are signals being recorded on the tape that can be read and used to break a master clip into smaller subclips. You capture the entire masterclip, then select it in the bin and use the DV Scene extraction to break it into subclips. Considering the poor reaction time of many DV decks, this ability to break things apart after capturing is something you will wish all tape formats could do.

Using Modify

One of the best tools you have for fixing a capture problem is to modify the logged clip and change the inpoint. Highlight the clip and click on the start time in the bin. You can change that number to anything. Alternately, you can go under the Clip menu and pull down Modify. Modify allows you to enter a specific number to increment or decrement the starting timecode of the master clip. It also allows you to modify all the shots at once in a particular bin. In the case of a preroll problem, either increment the timecode by enough to get the start time away from the control track or timecode break, or decrement the timecode so you don't chop the first word of the action and take your chances with a short preroll.

Modify is a very useful command that will be used many times throughout this book. Be aware that when you modify an inpoint on a master clip, the duration of the clip stays the same—the outpoint will change by the same amount as the inpoint. You may have just chopped off the end of the shot! As a second step you must change the outpoint of the shot if you want the duration to change after you modify an inpoint.

The Avid does not allow you to change duration of a shot if it has already been completely captured. This protects you from messing up every potential sequence where this shot was used. Any change to a master clip will ripple across your entire project and affect every sequence and every bin that contains that shot. All the sequences will inherit the change to the master clip. If you have already been editing with a shot you probably don't want to modify it! To change the duration of a master clip after it has been captured you must first "uncapture," which is really an "unlink." This allows you to change the inpoint or the duration, and unlinks the master clip from the media and from the sequences. This book will discuss linking and unlinking in more detail later.

Postroll presents its own challenges. If you log the outpoint of a particular clip after a break in timecode, you will not discover the error until after you have almost finished capturing the master clip. You may have to wait until the end of a five-minute clip before you get the message that there is a discontinuity in the timecode and the entire material so far must be discarded. Then you must modify the outpoint and determine where the timecode really ends. It pays to double-check any written logs you are handed for these exact reasons. Tell your logging personnel not to jog to the absolute last field and not to add a frame to the outpoint just to be complete. A break in the timecode during capture outweighs any extra few frames you might get by being extra persnickety. After you type in the handwritten field logs, you should always "go to" both the ins and outs (using the Go To buttons) to check that the numbers really exist. Using the Go To buttons is a great way to check whether you have enough preroll and not too much postroll.

As you begin to use MediaLog more, you will discover that there are a few shortcuts. For instance, you should use the mouse as little as possible. You can tab through the various entry fields and use the B key (older systems) or F4 (current systems) to actually enter the log (make sure the Caps Lock key is off). You then can type the name of the clip and use Control/Command-7 to get back to the logging interface and tab through again—all without using the mouse! If you really want to move quickly then you should disable the setting Activate Bin window after Digitize in the Capture setting. This will keep you focused on the Capture Tool when logging and you can go back and rename all the clips later (perhaps after they are captured). Since logging is inherently repetitive, even these small shortcuts will seem like lifesavers later on.

Tape Wear and Dubbing

Some producers are concerned about tape wear and the possibility of creating dropouts or creasing the tapes by playing them too often. This is a serious concern if there is much logging to do. Dropouts occur when small parts of the tape's magnetic coating, the oxide, flake off and stick to the playback heads. This is more noticeable with the smaller formats like Hi-8 or DV since a small amount of oxide on a small format makes a more visible dropout. Although improvements in tape stock are continuing constantly, "prosumer" formats do not find it cost effective to use the more expensive materials or to use better quality control when manufacturing these tapes. You may find yourself dubbing to a higher format like IMX or Digital Betacam and then using those tapes throughout the production while the originals are carefully packed away for a worst-case scenario. If an IMX dub of your Hi-8 is lost or damaged, you have to go back to the original field tapes and redub them. If there is timecode on the original tapes, then carry it over to the dubs (called jam-sync). This way you can make another set of dubs to match what you have already captured. Many DV cameras reset back to one-hour timecode for many reasons. If the cameraperson does not catch this then there will be many source tapes with identical timecode. To counteract this, some producers will have the timecode on the dubs increment correctly or correspond to the number of the tape. If there is no time-code on the original tapes or the dubs have different timecode from the originals, you may be forced to re-edit by eye if the dubs are damaged.

In situations where tape wear, a long postproduction schedule, and cost are factors, even with a high-end source tape, you may want to dub all of your tapes to a lower quality format like 3/4-inch U-Matic, S-VHS, or even DV. Dubbing from a higher format like Digital Betacam or HDCAM to DV or Betacam SP saves money by reducing the cost of renting an expensive deck for the offline. Transfer the source timecode to the dubs (maybe even burning the timecode into a visible window on the image) so you will have a perfect match back to the original sources when you recapture for the finishing stage or when you make an EDL for the online linear assembly. Evaluate your cost savings and safety factor by the amount of tapes shot for the project and the extra time for dubbing. The cost savings are quickly undone if there are dozens of tapes to dub, the offline has a short completion time, and dubbing is a cost that must go out of house.

Logging by Hand

To err is human, and logging by hand is certainly no exception. The main mistake users make when logging in NTSC without a deck attached to their computer is to forget to change the user setting to assume drop-frame or non-drop-frame timecode. The default setting (except on NewsCutter) is non-drop-frame. There is more information on this NTSC distinction in the next section. In PAL they may forget to change the format from the default NTSC setting and log the tape accidentally as NTSC.

Logging by hand makes sense if you want to reduce costs and tape wear even more. Dub all the source material to a VHS tape with a timecode window burned into either the upper or lower part of the video frame. You can use this "burned in" timecode as a visual reference and something you can easily show the client for comments. You can even make a QuickTime movie of this footage and still have a timecode reference if you want to exchange footage electronically. There is no actual timecode on this tape or QuickTime export, just the display of timecode, so you can't actually use the VHS dubs to capture or the QuickTime to edit. The producer, director, or editor can view the material at home (or some other place where they don't get charged by the hour) and control the tape with a standard remote control on a VHS deck with a good, clear freeze frame. Unfortunately, this method requires that someone type by hand the timecode numbers for the start and end of each shot into MediaLog. This option is faster if you purchase the number keypad attachment that connects to your laptop's peripheral port (where the mouse goes) and have a VHS deck with a search knob for finer control.

Some people have VHS decks that can read VITC, a timecode recorded in the video signal that can be controlled by the Sony deck protocol, but this is rare. If you have a VHS deck that can read VITC, you can use MediaLog to control the deck and log using the Mark In and Mark Out buttons on the Avid interface. This should reduce pilot error when entering the timecodes. One product, EasyReader™ from Telcom Research, actually can record VITC onto a VHS tape during the shoot or dubbing stage, which then can be read by the Avid if you capture using the same EasyReader device.

Drop-Frame vs Non-Drop-Frame

What kind of timecode has been recorded on the source tapes: drop-frame (DF) or non-drop-frame (NDF)? PAL people don't laugh; someday you will be forced to work in NTSC, too! How can you tell the difference and, um, what is the difference?

Here is a quick primer on drop-frame and non-drop-frame. NTSC videotape runs at 29.97 frames per second (fps), not 30 fps. If you count frames at 30 fps, then you have an interesting problem. A show that uses timecode incrementing at 30 fps actually will have a running time longer than one hour even though the timecode says it is exactly 1:00:00:00 in duration. Because of the slight difference between 29.97 fps (the true playback frame rate) and the more convenient 30 fps, at the end of one hour a non-drop-frame master is too long by exactly 108 frames. This is enough to cut off the credits, the syndicator's logo, or worse, interfere with the commercial coming next!

The two frame rates do not change the speed of the tape, just the way that the frames are counted. In fact, both drop-frame and non-drop-frame run at 29.97 fps, but non-drop-frame counts the frames at 30 fps. To keep the frame count even with the correct running time or duration, certain frame numbers are skipped in drop-frame timecode. Frames of the recorded image are not actually dropped, but two frame numbers per minute, except for the tenth minute, are not used. This means that when using drop-frame and the timecode says one hour, the real-time playback of the show will take exactly one hour, not one hour, three seconds, and 18 frames.

Since accurate timing is most important for broadcast purposes, drop-frame is traditionally used in broadcast. Non-drop-frame is used in corporate work, film transfers for cutting at 24 fps, animation, and other nonbroadcast uses like laser discs. It also is used for advertisement since they generally do not run longer than a minute, and putting multiple commercials on a reel can get confusing when you have to take into account missing frame numbers. Really, once you choose one format or the other, the choice is no longer apparent until it is time to output.

How does the computer know what kind of timecode is on your tapes? When you pop a tape into a deck that is connected to the Avid application, the computer plays a few seconds of that tape automatically. The software reads what kind of code is on the address track before it allows you to name a tape.

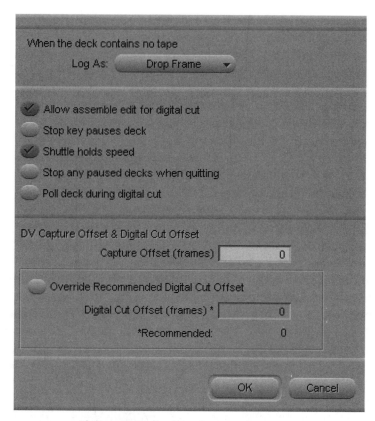

Figure 2.4 Deck Preferences Dialog Box

In earlier versions, you must play the tape yourself for a second in the deck before naming it.

What if you are not controlling the deck with a computer? You must enter the type of timecode by hand. Look at the burn-in timecode in the small window display on the VHS dub. Look at the characters between the last two digits, the frames, and the next two digits, the seconds. If the character is a colon (:), you are working with non-drop-frame. If it is a semicolon (;) (which you could say "drops down"), you are working with drop-frame. There is a Deck Preferences setting that says, "If there is no tape in the deck log as" and then a choice between DF and NDF. Well, there is a tape in the deck! But, you are not controlling the deck (it is not physically connected to the computer), so this setting applies. Choose the setting that matches your burn-in timecode and make sure that this type of timecode does not change! If the timecode type in your timecode window changes, you have to go back to the Deck Preferences setting and change the type of timecode there.

There is a difference between drop-frame on the field tapes and drop-frame on the finished sequence. You can use any kind of timecode on the field tapes, and you can mix and match as you create the finished master sequence. The master sequence will have one type of timecode throughout, and the source tape does not influence this. It is a good idea for the timecode type in your project to be consistent, but sometimes this is not possible because of the use of historical or stock footage. Camera people who always shoot one type of timecode may be unwilling or unfamiliar with how to change the type of timecode recorded on their field deck. One freelance crew may shoot non-drop-frame and the next may shoot drop-frame. This isn't critical, just annoying. To be fair, on some field decks you need to pull out a circuit board to make the change. If you have mixed DF and NDF in your project, you will need to catch the timecode change when you log different tapes.

The Importance of Tape Names

When logging in either MediaLog or Avid editing software, start by creating a new tape in the project and give that tape a simple and unique tape name. Tape name is very important at this stage. Short names are good, and a name consisting of all numbers is even better, especially if you are planning to create an EDL. When you create a new tape name, you are assigning properties to this tape. Attached to the tape name is an indication whether this tape is drop-frame or non-drop-frame. (This is not an issue in PAL.) The second property attached to the tape name is the name of the project. Tape 001 in Project X is a completely different tape than Tape 001 in Project Y. Keep this in mind when you move from a project created in MediaLog to a project created in the Avid editing software.

Keep all tapes in the same project in which they were logged. This means you shouldn't create a new project when you get to the capture stage, just copy over the project you started with in MediaLog. If you have logged on software that doesn't make an Avid bin, you will have an ALE file that you import to the project. It automatically becomes part of your project, but if you import it twice, the system thinks these clips are all on a different tape with the same name. This will have consequences for your EDL, so don't do it.

There is one more challenge with this method. When your tapes were dubbed, four 30-minute-long field tapes were possibly put onto a 2-hour VHS. When the field tape changes on the VHS dub, treat this as if you had just changed tapes and create a new tape name. When the tape changes, pay attention to the kind of timecode on this new tape. The tape name you created when you logged the first field tape on the VHS will accurately track back to the original field tape.

If you do not change the tape name and then log another field tape, you will have problems capturing. The computer tries to find Tape 001 at timecode 5:15:30:00 when the tape may have changed several times on the logged VHS dub. The real Tape 001 ended at 1:29:15:00. When capturing, Tape 001 fast-forwards to the end of the reel as the Avid searches for a timecode that is actually on the next tape or several tapes later (probably Tape 005). You will get an error message that the Avid "failed to find timecode coincidence point." You must then tediously find all the shots that start with a five-hour timecode and modify the source name to be Tape 005.

You may occasionally get the following message when capturing: "The tape you have entered is drop-frame and has been logged as non-drop-frame or it is non-drop-frame and has been logged as drop-frame." Basically, this means that the field tape's timecode does not match the tape name that was logged. If you have really inserted the right tape, then a mistake was made during the logging. When the tape name was assigned, the system was either reading the wrong type of timecode or told to assume the wrong type of timecode because of the Deck Preferences setting. You must modify clips so they are the right type of timecode and assign them all to a new tape name that has the same type of timecode.

Alternatives to MediaLog

You can use any word processing, spreadsheet, or database program to log your project as long as you format the data correctly. There is a standard way to set up the headings and then tab-delimit the information according to an Avid format set forth in the Avid user documentation. You can then export the file as a text-only document and import to Avid editing software as a shot log.

If you export from Avid editing software as a text file that is tab-delimited, you can import into a database program. This enters the column information into the correct headings in your database template. You can then search on 15 in the Scene field of the database template. This takes some setting up, but can be very powerful.

MediaLog ships with every Avid editing system and is by far the simplest to use since it is almost identical to the capture interface of the Avid editing software. You could also use Xpress DV for logging since it is easy to use on a laptop and has bin compatibility with the other Avid editing systems. Some people like other programs because with MediaLog you can't start entering information about a master clip until you mark an outpoint. You can fast-forward to the end of a clip and mark it and then enter all the heading information. But if you want to view the tape in realtime and enter the heading information at the same time, then you might want to consider a database or a spreadsheet program. There are also programs on the market, like Executive Producer by Imagine Products, which create complex databases using the Avid text format with screen grabs and multiple ways to format the information for presentation purposes.

Organizing Bins

An editing system so heavily dependent on the computer means you must spend more time getting organized at the beginning of the job. It may appear like you are off to a slow start, but again, when you get to the fast part—the editing—you will look like a mind reader. The most successful method for organizing bins is to start with a bin for each field tape and then clone the clips (Alt/Opt drag them) and move them to content-based bins afterward. This allows you to print a copy of the contents of each tape as it is logged and keep a central notebook of all the tapes. You can go back to the tape months later and know exactly what is on that tape when you archive it. If the worst should happen and you cannot continue the job on the Avid system, then you will have something tangible to take with you to another edit suite.

Of course all of these strategies assume that you have time to prepare! There will be times when the client just walks into the production house with a box of tapes and literally drops them on you. Now you must log, capture, and view all at the same time. Some people, not wanting to miss anything, will start to capture by figuring out how much space is available and how much material there is and capture every tape from beginning to end. Capturing everything is a good strategy if you are not going to be the editor because it leaves nothing out.

There are times when you must make creative decisions on the fly before you have seen all the material, and here you can operate only on your gut instinct and experience. Scan through the tapes as fast as you can and log what appears to be the best and second-best take of every setup. Go back and batch capture after you have scanned through everything. The benefit to this method is that by the time you get around to actually capturing, you will have seen everything and can weed out the redundant and superfluous. Be sure to grab some shots that you think are cool and have no idea what to do with—they may be the most useful for tying sections together with graphics.

By far the worst outcome of this scenario is when the client tells you exactly what they want, exactly where it is, and that is all you capture. Don't be fooled! You will be looking for shots on tape all day, and you will lose almost all of the power of your random access tool. It is imperative that you see everything. They brought the job to you because they trust your judgment, but if you can't see everything then you can't make the creative leap they expect from you under pressure.

Batch Capture

When you finally get to the capture step, you need to have a standard procedure and follow it closely. You are responsible for technically reinterpreting this footage for everyone else who is going to work on this project. Make sure you can justify every level you adjust. The last thing you want to hear while you are editing furiously is "Does it really look like that on the tape?" There is a button on the Media Composer and Symphony just for those times, called Find Frame, which asks for the original tape to be mounted and then fast-forwards to the exact frame (see Figure 2.5). If you have a roomful of nervous people or a

Figure 2.5   Find Frame Button

client who is inexperienced with compressed images, you may want to map this button and keep your tapes close by. If your clients are unsure, then you should take the extra effort to reassure them, especially at the beginning of the project. As they learn to trust you and your judgment, they should request this less (unless there really is something wrong with the originals).

Also, keep in mind that the Avid allows you to save those video levels on a tape-by-tape basis. Those video levels may be the ones used when it comes time to finish. Chapter 10 discusses the setting of levels.

You can control the batch capture process through the use of sifting and sorting. Create a custom bin view for capturing. Include Tape name, Start time, Tracks, Video, Duration, and Offline at a minimum. Put the Tape name column to the left of the start time in the bin heading. You can Sort (Ctl/Cmd-E) by clicking on the Tape column to put the tapes in alphabetical order. If you shift-click to select both the Tape name and the Start time you can sort each tape based on the start time. In other words, all of Tape 001 will be in ascending order followed by all of Tape 002. Alternatively, you can sift based on tape name, select all the clips that appear after the sift (Ctl/Cmd-A), and batch capture until that tape is completed. That way if there are any shots with problems, you will see them before you move on to the next tape. By mapping this series of commands (sift and batch capture) to keystrokes using the Menu to Button function of the Command palette (see Chapter 3 for details) you can increase the speed of capturing a multi-tape project.

A digital project is never really finished. The mess you leave today will come back to you in a month or two as the clients come back for version 2! Or the next editor will curse you loudly as he or she tries to figure out why you put things in random bins and left all those unnamed clips and sequences. Mastering the first part of this workflow will impact the rest of the job. It will make the difference between always rushing to catch up as you search frantically or staying one step ahead, concentrating on creativity when it is most required, during the editing.

EDITING FLOW

This is the part you should give thanks for every day, freeing you from the restrictions and limitations you have been struggling against since the first time someone showed you how to load a tape or pull apart a splice. Editing is also the most intimidating step because, without the old restrictions, many of the excuses for not being perfect have long disappeared. Some editors fight back by quickly claiming something just can't be done, but be careful, the next editor who works with this director may know how to do it! This is the double edge of the new tool: You are always struggling between what you know how to do fast and confidently and that nagging feeling in the back of your head that if you just had half an hour you could figure out a better way.

Part of this anxiety comes because we are still trying to use the new tools in the old way. Sometimes we focus on the one best way to do something and bring all of our old skills to bear on the new technology. These skills were honed and perfected because of the demands and limitations of the old tools. This is one of the biggest reasons why experienced editors don't want to spend the time to learn all the tips and tricks. The new tools seem too technical, and they just want to edit.

Think about how to take what is truly valuable and brilliant about the way you work now and translate it to the nonlinear way of working. In the process, you should take a long, hard look at those skills you developed only because they were necessary, and discard them if they are no longer so. I used to be proud of how fast I could thread a one-inch machine. The one good part about threading a one-inch machine is that it gets you out of the chair, so you can stretch and think for a minute. That is the real importance of the skill and the valuable part to keep!

A lot of personal choice is involved in discovering the "right" ways to do particular functions on the Avid editing software. The more you work on the system, the quicker you can identify when a technique is not as good as another in a specific situation. Remember the technique or feature that is easiest and most intuitive for you right now. Get one way down and feel confident about that. Later, work on finding the best way, but don't let your fear of not knowing the absolute best way paralyze you and keep you from using the way you know. I used to pretend to feel ill, then grab the manual and read it in the bathroom.

The more complicated techniques are valuable since, at some time in the future, you can count on getting a job that will require them. At the beginning, focus on the content of what you are working on and don't let the machine get in the way. The faster you can reach this stage of transparency with the machine, the better you will feel about the new tools. Only then should you think about memorizing the more advanced techniques.

Refine, Refine, Refine

Do what you do best—edit—and let the machine do what it does best—give you access to your images and allow you to see what is in your head faster. For some editors, if not most, the vision they have comes about through refinement. They have a rough idea of where the scene is to go or the best flow for the material, and they start off that way only to find something better, or something missing. Unless you are working with an incredibly tightly scripted show and very little footage, there is a lot of exploration and discovery as you get more familiar with the material. Just like the speed of changing a one-inch reel becomes less important, the requirement to get a cut absolutely perfect on the first try also becomes less important. The digital nonlinear technology is perfect for this type of approach.

The first pass on a sequence is truly a rough cut. Although we have used this term in the past, it usually meant that you tried your best to get everything perfect and afterward there were some changes. Now it means more precisely that you have roughly assembled the parts and are planning a series of creative refinements.

I am tempted to not show my first pass to anyone. In particular, I don't want to show it to producers who are not familiar with the possibilities of nonlinear. They might think I was the worst editor they had ever worked with and think about firing me on the spot! I watched a young producer squirm as the client's VP of communications saw the first rough cut. It had no effects or graphics and had burned-in timecode. The producer explained very carefully about what they were about to show, but the VP still blew his top. It took everyone in the room to reassure the VP that this was all perfectly normal.

Your first nonlinear assembly should be even rougher than this. Many times just taking all the good takes and lining them up one after the other should be your first step. I have worked with clients who were always vaguely unsure if they were using the right take, even after going through an exhaustive review. This is a legitimate concern, especially since every shot depends on all the shots around it. Changing shot 7 may make you reconsider shot 15. In the past, there was a lot of tape or film shuttling, and this was a good way to remind yourself (and everyone watching) what you had to work with. It was also a good excuse to occasionally just look at all the takes one more time. Again, this is one of the valuable old skills that should be brought along to the new tools. Do not discard the time spent watching and thinking just because you are not forced to view the material over and over.

Take the circled takes and string them together. Take all the best cutaways and make a true B roll so that you can use this intermediate sequence as a source. Use the match frame command, not to extend your shot as so many beginners do, but to remind yourself what came after your original cut point. Use the time gained by not having to dub or rip apart splices to think, watch, and explore. You may not end up cutting faster, but you will cut better and be happier with the end result. Refine, refine, and refine. Start with big chunks and whittle them down. During the first several passes, work for content, then work on flow and timing (if you can separate them).

Use a strategy that incorporates each step only when it is necessary. You should not be worrying about the graphics when you are still trying to work out the sound bites. Some people call this stage the radio cut because it is based on assembling all the sound bites together to tell the story based on spoken word. This is an excellent time to get a sense of the timing and drama of a sequence without the distractions of images. Finally, go for the effects and the smoothing of the audio mix. This particular order is totally up to you. If you cut with a full sound mix as your style, then put that as the second pass, but don't try to do everything at once. You will find yourself watching the scene or segment over and over. In the past, you may not have taken the time to do that until you were almost done. Look at the big picture, pull yourself out of the details until the final stages where they belong. Watch yesterday's work in the morning, take notes, and then move on to something else until the afternoon. Then look at it again. This lets you see things with fresh eyes.

There is room for reconsideration and the type of visual juxtaposition that you might never have thought of unless it was right there in front of you. The ability to hold a scene in your head is a very valuable skill, somewhat like memorizing the opening 50 moves of a chess game. But that is a mental skill, not a visual one. The instruction I give beginners as I see them staring at the screen is: "Don't guess!" Don't create scenarios in your head when you can create them with your eyes. The speed with which you can put two disparate and unexpected images together is another tool for shaking up the other side of your brain. Work beyond the everyday logic and justifications you would have used before even attempting a sequence. Seeing new things with your eyes and not your head will inspire you.

WORKFLOW OUTSIDE THE SUITE

A really good producer can foresee what parts of a project should be left to an editor at an NLE workstation and what parts other experts should take care of. There are two basic reasons to send parts of the project out of the suite—time and expertise.

WORKING WITH GRAPHICS

If the graphics can be done in tandem with the editing, when they are needed at the end of the finishing stage, they can just be inserted. Contrast this with the producer who gets to some midpoint and says, "Let's come up with 'a look' now." All editing stops and you become a graphic designer, a 3D animator, and an expert on four or five third-party programs. Some editors like this new challenge and the creative control, and more power to them. The editor is probably the best person to understand how all of the effects and graphics should be incorporated into the show and the right formats and work parts necessary to make it happen. But editors must also make the correct estimate as to how long the third-party design and render time are going to take. Don't let a producer with a short attention span pull you out of your best workflow. A wrong guess or the failure to build the time into the overall schedule will grind everything to a halt while everyone in the room watches a render-intensive process and decides whether to come back tomorrow.

Sending all the graphics out takes care of the time and expertise problems, but replaces them with new ones. What if the graphics are not right? What if they are the wrong color or the animations are not long enough to cover the new voice-over? The time involved to send them back to the (unsupervised) graphic artist will surely halt the post on your project. A smart producer needs to be able to monitor the bits and pieces being done out of sight just as much as keeping an eye on the progress of the edit. This planning process must also schedule the time to add the graphics at a stage when they will mesh nicely with the editor's final finishing schedule.

All of these concerns are not new. The problem is that in the past, sending the noneditorial bits out to another artist was the only way to get a highly professional finished product. With the onset of the "everything in a box" concept, producers seem to be ignoring their time-honored knowledge, and all those who are getting in for the first time have no experience with anything except this new desktop method.

SOUND MIXING AND COLOR CORRECTION

Many times people ask about color correcting and sound mixing. Yes, the "box" can do it. Who is running the box? Do you play musical chairs as the sound mixer sits down in your chair and fiddles with the sweetening for a while before they let you back in the room? Not likely. Either you are expected to be a sound engineer or a colorist or the process needs to be segmented into parts.

There are some basic challenges with the workflow of sound mixing. You are dealing with a chicken-and-egg scenario. Which comes first: the sound or the pictures? The answer is "Yes." Thankfully, there is Avid's nonproprietary exchange format OMFI (Open Media Format Interchange) and the new emerging industry standard AAF (Advanced Authoring Format). Both OMFI and AAF allow you to send compositions and captured sound media to digital audio workstations like Avid ProTools. This means that there can be parallel processing of sound while the cutting continues. I have held a telephone up to a speaker so someone in New York City could record and use it for a scratch track, but this wasn't something anyone actually planned for. The producer decided that a New York City talent was the only thing that would do and we waited, not so patiently, for the next shuttle to Boston to bring us the master mix. It did, in some small intangible way, help get the governor elected.

The give and take of the sound track cannot be underestimated. That extra few piano notes before a transition might change the length of the dissolve or the amount of time in black. You are shortchanging yourself if you just send the project out for a mix and never see it again until after it is finished and dubbed. There should be time built into the schedule to tweak the cut with the new music in a perfect world scenario. If you can get your timings and transitions as close as possible using scratch tracks and click tracks, then you are way ahead of the game. But wouldn't it be great if you could use the time that you are rendering or waiting for final client approval to coincide with the sound being mixed? What a perfect excuse to wait until tomorrow and get a chance for another "final" pass at the finished product with the finished sound track in place.

The alternative to sending the sound out is, of course, do it yourself. Lately, there have been some great additions to the sound mixing capabilities of the Avid editing systems. Realtime level adjustment with keyframes, equalization, waveforms, and audio punch-in are all features you have come to expect (see Figure 2.6). Using the faders of an external MIDI control surface like the JL Cooper FaderMaster Pro or Yamaha 01V you have realtime rubberbanding capabilities that editors in linear suites could only dream of. Avid editing software remembers those levels and allows you to refine them further. You end up with lots of keyframes that can then be filtered down to the minimal amount you need. There are also lots of AudioSuite plug-ins for all the special effects you couldn't do before, like reverb and pitch shifting.

Many people feel that sound mixing will never be complete without subframe editing, the ability to make changes at the sampling level, which gives much more control than what is available at 30, 25, or 24 fps. Someday in the not too distant future, Avid editing systems will do this (DS Nitris does it today). Fortunately, Avid captured sound media is either in the native SDII format (Sound Designer), AIFF, or WAV, which is compatible with other sound mixing programs like Avid ProTools or AudioVision. You can

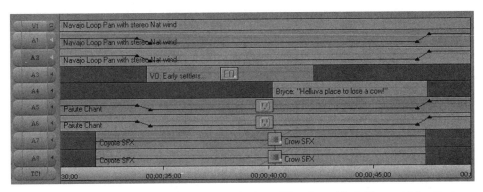

Figure 2.6   Rubberbanding, EQ, and AudioSuite Plug-ins with 8-Channel Monitoring

export sequence information as OMFI or AAF, open existing audio media, and start mixing. This book will discuss audio importing and exporting issues in Chapter 6.

Color Correction

Color correction has been, up until now, a linear process—to adjust shot two, you need to see shots one and three. In a nonlinear project, shots are captured in an order that reduces tape shuttling and tape changes, but does nothing to allow the editor to see two juxtaposed shots one after the other until later. There are built-in color correction features that allow a wide range of adjustment in the whole range of Avid editing systems. Those tools have been continuously improved over the latest several releases and include split screen, safe white and black clips, and gamma adjustments. The color correction effect features red, green, and blue adjustments—more adjustment than you have with just a straight time base corrector (TBC) (see Figures 2.7, 2.8, and 2.9). In Symphony, there has been a breakthrough in color grading functionality that has migrated to all the other Avid products. I will discuss color correction in more detail in Chapter 12. If you are primarily an offline editor, you are probably not also a full-time colorist.

Figure 2.7   Color Effect Interface

However, more editors want total control of the look of their project and use the color to tell a story as well as to salvage a scene. If this is the case, then seriously consider getting proper waveform and vectorscopes, and setting the lighting of the room to be properly neutral.

There are still two basic questions about your workflow: Who is running the box and when? Let's look at a few different scenarios.

Scenario One

You capture everything for offline, and now go back to capture at high resolution. You set everything to color bars, but there are a few shots that just do not match. You take the time and your best shot at making these simple changes using the color correction mode or the automatic functions for basic adjustments. The client signs off and you go home early.

Scenario Two

After capturing at high resolution, you realize that significant sections require color correction, either because of mistakes in the shooting, a difference between cameras, or a blue/green color cast just sets the proper mood. Now you have a more serious choice. Is it just the juxtaposition of certain shots? See Scenario One. More likely, entire tapes need to be tweaked. In this case you could create a color effect template and apply it every time that tape is used. You could recapture using an external color corrector. Some inexpensive digital color correctors can be bought or rented just for this purpose. You need to capture each shot and evaluate it or, after some experimentation, you can find a setting for each tape that is a best light, or tape-based setting. Then you would match shot by shot when you see the material in context. This calls for a specialized skill that is developed only over time. You might consider finishing on a Symphony or DS Nitris system with color grading capabilities for this level of work.

Scenario Three

You know you are going to be using lots of different cameras under many different locations and lighting setups, and you plan for tape-to-tape color correction at a professional facility after the final show is output. If you are planning on this method, you should also plan to master on a digital tape format so there will be no further loss of quality during the tape-to-tape transfer. Again, this is where a Symphony or DS Nitris system would allow you to keep this stage of the work in-house.

You may have a colorist on staff, hire a freelance colorist, or have an online editor who has some experience in matching scenes. Have them learn the Avid system if you need that next level of expertise. By avoiding the tape-to-tape stage and keeping it all nonlinear, you can cut costs, finish faster, and keep the

Figure 2.8 Spot Color Effect on Symphony Combines Paint and Color Correction

final stage all in one room. You can sell this service to your clients for a rate lower than a standard tape-to-tape color correction session or keep all of your rates the same and pocket the difference.

A Symphony tied to a media-sharing solution like Avid Unity will allow the colorist and the editor to work on the exact same frame of video at the exact same time. You now have the flexibility to work in parallel and merge the color grading information with the final version of the sequence at the very last stage. This parallel workflow with shared media solves the most difficult color problems with the talent of an expert and does not affect the editing schedule.

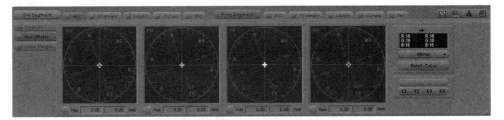

Figure 2.9   Hue Offset Interface on Symphony's Color Correction Mode

Scenario Four

Let's face it: Color correction and color grading are addictive. Once you get a taste you will want to tweak every shot. Be prepared for this because once your producer or director sees how you can make their marginal footage pop off the screen you will have to do it for every project. Create favorite templates for basic improvements and corrections and learn to use every automatic function in the color correction interface. Map everything possible to keystrokes and work to improve your eye toward color. This is wonderful if you are charging hourly and wonderful if you are using it to attract customers to your facility because of your "save it in post" talents. Stylize the project and give it some attitude. Attract customers with your skills, like color correction, and you will worry less about dropping your rates to beat the competition.

RENDERING

What would a digital nonlinear project be without rendering? A lot faster. This is why Avid designs have been so focused on multiple streams of realtime effects. Part of the strategy is knowing what to render and when. Use the draft mode to get more streams in realtime to refine the timing and shot content. Then worry about high quality and finishing details that may cause you to render. With the invention of ExpertRender™, much of the rendering guesswork has been eliminated. By choosing an ExpertRender in to out, you have the benefit of the system itself analyzing the sequence for you (see Fig. 2.10). Most times ExpertRender will show you the least amount of effects to render and allow you to take better advantage of the large amount of realtime capabilities. The times that you may disagree with the Expert, you can easily override the choices and perhaps find a better way to render the least.

Are there times when you don't need to see all the layers? Take advantage of them and move your video track monitor to view just the minimum. Sure, you need to see the text on V3, but not the color gradation on V2, so render the dissolves on V1 and leave the other tracks unrendered. The dissolves play fine, the titles come up in all the right places, and they are spelled correctly. When possible, you can Control or Command-click on a video track

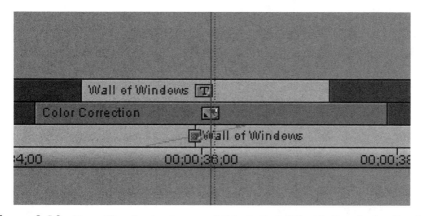

Figure 2.10 ExpertRender Analyzes and Selects Only What Is Required to Render

monitor and turn on video solo to view just one effect, which will now play back in realtime. Of course, you need to explain what you are doing and use your best judgment. Determine how much your clients can see (or not see) before they start to get cranky and you have to stop and render everything to show them what it will really look like. Think about the kinds of changes that force you to rerender and whether you can make a series of trims and keep everything intact that is already rendered. These trimming considerations are skills that are more advanced. They require you to ensure all the affected tracks are active when you make changes when you have lots of rendered effects. This is why you want to get as far into the project as possible before you start to add the effects. You have enough to keep in mind when you are still refining the opening montage without worrying about the rendered layers at the end of the show.

Just how much do you really need to render? Most people render too much, way too much, and it is all because of misinformation or ignorance. Think about the minimum necessary to render and the minimum to monitor. The less you see, the less you wait. And be clear what will play back in realtime and what will not. This book will discuss rendering in more detail in Chapter 7.

FINISHING

Throughout this process, there needs to be one thing in the back of your mind: How are you going to finish this thing? Are you going to a linear edit suite, the negative cutter, an Avid finishing system, or does it stay "in the box"? Good form means that no matter what the decision for finishing will ultimately be—and the producer will at some point be forced to tell you—it shouldn't make too much difference. This means you need to name the tapes for an EDL, but make the names obvious so that you or your assistant can find them again easily for recapturing.

The amount of time you spend on effects should be more concentrated on the show that you know will be recaptured on the Avid for finishing. Otherwise, you are just making a visual guide for re-creation by the online editor. This is a decision about the amount of refinement and a focus of time and effort. It is not unusual to spend an inordinate amount of time on the final five percent of a project if you are finishing. This time would normally have been turned over to an online editor and that would have been their sole job. It is this last stage that most beginners completely underestimate.

There are several ways to save time and money at the final output stage. If you can keep the project in-house, then you can reap extra profits. You can charge for the extra hours involved in finishing instead of figuring in the $400 per hour cost of turning the project over to someone else. Investing in the fastest computers and fastest drives can pay off for keeping everything in the Avid, but truly saving money in the finishing process requires a sign-off from your client that there will be no further changes (get it in writing!). If the project must go to a linear online then the client should realize that you can even save them hours there, too, by manipulating the EDL (see Chapter 8). A sign-off is the key to the true auto-assemble, but the benefit to finishing with a NLE is that you really can keep changing your mind without losing it.

KEEP THE AUDIO

There is no compression of your audio when you capture, no matter what your video looks like. You can use this audio in the finishing stages or the linear edit suite—you just need to get it there. Copying those audio files to a removable drive and carrying it to a digital audio workstation is becoming more popular as it gets simpler. You may also want to hand-carry or copy over a network your audio media files from your offline system to the Avid finishing system. The Avid Unity MediaNetwork or LANShare systems make this process as easy as getting access to the Unity workspace since everyone can share media. An older method is to record the Avid 8-channel audio to a timecode-controlled DAT or DA-88/98, put it to digital videotape, or use audio tracks three and four of a Betacam SP. You can lay the audio to the new master tape in the linear suite and then make your EDL video-only. You can keep the mix you created in offline and just keep an eye on video levels and effects in the linear suite. Chapter 8 describes this in more detail.

The last important step when you are going to finish on the Avid editing software is the tape itself. You need to black a tape and, unless you have two decks, you need to black the tapes for protection masters as well. You don't need to black the entire tape and can use the assemble editing function if you are rushed for time. Chapter 14 discusses these techniques.

PLANNING FOR CHANGES

When everything is finished, the product is shipped, and the check is in the mail, you must concern yourself with one final nightmare scenario. What if they come back? Well, if they are bringing new work this is a good thing, but if they are coming for changes you need to be ready. Can you charge for the changes? Whose fault is the typo? (This is a good reason to get them to type all the text and paste it into the Title Tool!) But the real trick is being able to make the client's changes quickly and painlessly, and they will certainly keep that in mind when they are projecting costs for their next project. How are you going to archive this project? Don't be tricked into thinking that an EDL will do the job. You must back up the project file and all the bins. Even if the software changes, you should usually be able to bring any job forward even if it is a year or so later. Large projects require large backups, and it may be time for you to invest in removable storage like Iomega, a recordable CD-ROM or a DVD just for the bins and graphics.

Consider also whether you need to back up the captured media. If you are finishing on the Avid editing software, you should build into the cost of the project the amount of time to back up overnight and the cost of an archive master. With all the new choices like Archive to Tape, DVD, and Digital Linear Tape (DLT), you are still dealing with an incredible amount of material. As the cost of slower drives like firewire (older systems will not allow consolidating to firewire drives) or IDE drives gets closer to the price of tape (I expect this to happen sometime in 2004), your best bet for archiving may be another drive! Having a few hundred gigabytes of inexpensive firewire drives for archiving media quickly is becoming an excellent choice.

If you have spent the time to color correct and render and everything is just right, then you should spend a little extra time and save it. If you have a few hours' warning, you can have the entire project ready to pick right up again by restoring it over a network to shared storage in the background. The potential downside is that when the technology changes, it will change video quality faster than anything else. Eventually, video-capture boards will change and your new video most probably will be incompatible with the old video if you have upgraded. With luck, there will be a conversion utility, but the quality will never be as good as if you recapture. Not everything ages as well as wine and cheese. So plan for an adequate amount of time to review and make changes because there will be lots of each in a nonlinear project. Have a plan to back up and restore media and twiddle your thumbs occasionally. Everything takes longer than you think it will.

CONCLUSION

Workflow encompasses much more than just input, edit, and output. It is full of variations and possibilities and is guaranteed to be unpredictable. Plan for it and be ready for the places where the changes could cost you money and time. If you understand the entire process, you will be better prepared to avoid the traps and exploit the advantages of the changing technology.

3

Intermediate Techniques

It may come as a surprise that most beginners of Avid editing make the same basic mistakes. I don't mean mistakes caused by the software being too difficult, but mistakes from working hard to grasp some fundamental ideas. They are sometimes crucial mistakes, like not knowing exactly what to back up and then trying to restore a project with no bins. Sometimes it is a subtler mistake, like not using the power of a new tool because "That's not the way I work." You may be missing a huge opportunity to improve your speed and understanding.

I have heard people say that the Avid is difficult to learn and that the interface has a steep learning curve. This is only partially correct. You can be mousing around the screen in only a few hours and really editing by the end of your first day. But as with any professional tool, you want it to go faster and do more and the Avid interface rewards this yet very few people use everything the software has to offer. If you have a particular task to perform, there are the tools designed to facilitate that task in a straightforward way. If you need something a little different, there is a lot of room for variations. The variations take the most time to learn, but are the most rewarding.

The most basic mistakes are made right at the beginning, when editors are still trying to learn how to navigate through their material. There is a lot of translation going on between where they want to go, how they used to do it, and the two or three techniques they know how to use. They end up settling for the dog paddle before they have mastered the breaststroke, the crawl, the backstroke, and the sidestroke. They will always poke along unless they unlearn the method that wastes energy and, let's face it, gets you there without much style.

USING THE KEYBOARD

You may have guessed by now that I am referring to the overuse and abuse of the mouse or trackball. This is where you should start to improve your technique. When you first learn, use the most obvious way, the mouse. This helps you get over the beginner's problem of trying to remember what you want to do next and where it is on the screen. After this beginner's stage, you instantly forget how magical it all is and want to go as fast as possible. Then you must cast down the mouse! Use your keyboard!

Force yourself to use the edit keys and keyboard equivalents as soon as possible. If you haven't put the colored keycaps or stickers on your keyboard yet, you are missing a whole world of speed. Look at the Ctl/Cmd key equivalent for the functions you use the most and think up funny little ways to make them stick in your head. Ctl/Cmd-Z to undo and Ctl/Cmd-S to save should be comfortable before your first day is over. Then start to use Ctl/Cmd-W to close windows and Ctl/Cmd-A to select all. Use F4 to start capturing once you are in the Capture mode and, as an ongoing project, memorize the Tools menu. You should mark in and out mainly from the keyboard so you can keep your material rolling and mark on the fly. Trimming can be done in several ways from the keyboard. You don't want to give yourself a repetitive stress injury, though, so always try to make the environment as friendly as possible for your wrists. Get the keyboard at the right height, get a wrist pad if you need it, and give your wrists the rest time and exercise they need to keep functioning. Use the extra time gained with these keyboard techniques to watch the sequence back one more time and think.

Custom Keyboard

There are many ways to personalize the keyboard. I don't recommend any special keyboard layouts since I believe they all should be created organically from observing the functions and keys you use the most. You can take any button from the Command Palette and put it on any key (button to button). I recommend learning the colored keycaps and modifying only the function keys or the shifted functions of keys that make alphabetic sense to you. Open the Keyboard setting and hold down the Shift key. You have a whole new keyboard to create! You can put the Render button on Shift-R or the Subclip button on Shift-S. (See Fig. 3.1.)

Some functions on the pulldown menus do not have keyboard equivalents or buttons. You will need to map them in order to use them as a single keystroke or in conjunction with a third-party macro creation program. A good example of this would be the Fill Sorted function in the Bin menu. By mapping this to a key you can tidy up a cluttered bin instantly. (See Fig. 3.2.)

Here's how to map a pulldown menu to a key:

1. Open the Keyboard Setting and the Command Palette at the same time.
2. Click on the Menu to Button reassignment button on the lower right of the Command Palette.

Figure 3.1 A Customized Keyboard

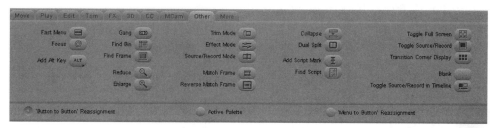

Figure 3.2 The Command Palette

3. Your cursor will change to an icon of a mini-pulldown menu. On the Keyboard setting window, click the key you want to map. Hold down the Shift key if you want it to be a shifted function.

4. Choose the menu you would like to map from the pulldown menu choices. The initials representing that function will appear on the key.

5. Save your settings when you are done. Highlight the Project Window and press Ctl/Cmd- S. You may want to locate your user settings (which are in the Avid program folder, in the Avid Users folder), and save a backup copy of the settings on removable media.

In older versions of the software, pulldown menus must be mapped through a special combination of keystrokes:

1. While looking at the open Keyboard setting, hold the Option and Command keys down at the same time; you will get a cursor that looks like a small pulldown menu.

2. Click the function key in the Keyboard setting where you want the function to go.

3. Choose the pulldown menu you want to map, and the initials representing that function will appear on the key.

Using the keyboard while editing is touch typing for concepts. Once you feel comfortable with where everything is on the screen, push yourself to resist the mouse and keep your hands on the keyboard! You may find that you need to create several keyboard settings and use them for different types of projects.

If you have a custom keyboard, be aware that it works best for the version of the software you were using when you created it. If you go to an earlier version of the Avid software, your keyboard, like all User Settings, may have features that do not exist in the earlier version. Usually you can go forward to the latest software version with Keyboard Settings, but this has been known to occasionally create odd, unrepeatable problems. It is best to take a screen shot of your Keyboard Setting (use a shareware screen capture program to save the Keyboard Setting window as a graphic file), print it out, and make the custom keyboard again.

J-K-L

Use keyboard control for shuttling. Some people love the older MUI or the AvidDroid because it feels like what they are used to, but those devices are no longer supported. There is a whole range of new devices made by companies like Contour Design (www.contourdesign.com) and JL Cooper (www.jlcooper.com) as well as custom keyboards that have shuttle knobs built in. Those devices are better than using the mouse to shuttle through tapes or the timeline to view material quickly. If you don't have access to that extra hardware, then you will need to rely on the J-K-L keys for shuttling tapes and efficiently moving through captured material. J-K-L is also great for fine-tuning during the refinement stages.

After you get adept at using the Jog buttons for minor adjustments, stop using them! Press the L key and move forward at 30 fps, then press it again and speed up to 60 fps, and then again, up to 90, 150, and 240 fps. (PAL numbers will be multiples of 25 and film is 24.) Eventually the sound will cut out (thankfully) above three times normal speed. Press K or the space bar to pause. Now use the J key to go backward with all these different speeds.

Soon you will find yourself cooking through the material at double or triple speed and following the script. Surprisingly, you can understand what people are saying and can work consistently at the higher speed to fly faster through the material without the distracting rollbars on the screen that you would see with tape.

This is a linear way of moving through the material. J-K-L is an example of how the computer has improved on an old technique by just making it easier and faster. Ideally, you type your timecode logs and go to the correct place instantly. Take advantage of the random access powers. You search back and forth once you get there; however, if you really need to listen and watch the material, there is no replacement for just shuttling through, especially if you were not the person who logged the shots.

The true beauty of J-K-L comes when you want to make that final editing decision. Holding down the K and the L keys at the same time will "scrub" (move slowly), with audio at 8 fps (5 fps PAL). You also get the familiar analog sound effect where the pitch drops and it sounds like Satan has possessed your talking head. Most people like this sound, especially if they are used to editing audio in an analog environment.

J-K-L for Trimming

Using the keyboard, you can move incredibly fast or slow without taking your fingers off the keyboard. J-K-L for trimming is what sets Avid apart from any other editing system and has been a hallmark of how well they understand editing. It is the best way to get a sense of timing of the transition with direct control and is the best way to make sure you have eyeline continuity in dramatic scenes.

Just wait until you get to the left hand! Let's explore J-K-L and its most powerful (and potentially dangerous) use — trimming. The first time you try this there is a little surprise — while watching the show play back, you are trimming! There are two important benefits to this. Don't scrub and then mark the points for your edit. Make the edit first and then scrub the trim. You will be able to make the fine decisions with the shot already in place. Second, if you mistakenly make a clip way too short, say by a paragraph, you can lengthen it and listen to the audio at 90 fps at the same time. This is an excellent way to hop through the sequence, one transition at a time, and tighten everything just a bit before moving on to the next level of refinement — always keeping your hands on the keyboard.

Scrubbing and Caps Lock

You can achieve a choppy, digital scrub sound by holding down the Shift key or pressing Caps Lock and then using the Step or Trim keys. If you like this better than J-K-L, it is precise and still has its place. This technique should be used at the last stage of determining a trim if you find you cannot get the point you want with the slow scrub. It is good to know first because you can always fall back on it, but with J-K-L you may be able to get the breathing, pause, or other timing just right more quickly for normal soundbites.

There are some other restrictions with using the Caps Lock key. First and most important, the Caps Lock key takes up a lot of RAM; a lot more than you would think. Using absolutely all of the RAM in your system may cause unusual errors or keep you from seeing certain realtime effects when they are definitely there, like many dissolves in a row at AVR 75. Now that systems are configured with a gigabyte of RAM, this isn't as much of a problem. It is something to consider if you have an older ABVB system and less RAM. Also, RAM is much cheaper than it used to be. If you can, put this book down, go to the phone, and order another 128 megabytes or so. This is the cheapest upgrade possible.

The second problem is that leaving the Caps Lock key down is incredibly annoying. OK, you've escaped from having to hear high-pitched voices screaming backward when you abandon tape, but now you have blip, blip, blip all day every time you change position in the timeline. Stop that! Turn it off!

There, did I sound enough like your mother to get your attention? Hold down the Shift key when you need it; when you don't, it automatically turns itself off when you remove your finger. And start using J-K-L.

AUDIO MONITORING

Many beginners don't see the connection between what they are viewing and why they can't hear the sound anymore. Being able to turn audio track monitors on and off selectively means you can concentrate on just the sound effects on track 5 or make five versions in separate languages and monitor one language at a time. But it also means you always need to keep an eye on which tracks are being monitored because they could be different from the tracks you are trying to use.

You can monitor up to eight audio tracks at once so look closely at the monitor speaker icon. On older systems one of the icons will be hollow, represented by an outline instead of solid black. On current systems, some icons will be yellow (which is easier to see than a hollow icon with the high-resolution monitors). (See Fig. 3.3.)

Using some of the increased bandwidth available on the latest computers, Avid has been able to increase the number of audio channels that can be monitored at "off speeds" (this includes backward and slow speed scrubbing). You can monitor all eight channels 1X backward, slow scrubbing, and 2X forward speed. When you go 2X backward or 3X forward you will hear the two channels with the yellow speaker icons (all audio still cuts out above 3X normal speed). The yellow audio monitors in the timeline distinguish those tracks from the normal black audio monitor icons. Any track can be made to be an off-speed monitor.

To change a track from black to gold you can Alt/Option-click on the already active black icon. You can also turn monitoring on or off from the keyboard if the timeline is the active window. The audio track keys on the default keyboard are 9, 0, -, = for audio tracks A1, A2, A3, and A4. Shift 9, 0, -, = are used for A5, A6, A7, and A8 without any customizing. Hold down the Alt/Option keys when pressing the audio track keys and the track monitor will toggle off or on.

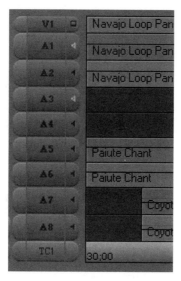

Figure 3.3 Audio Monitoring

Occasionally, having all eight audio channels playing at once is distracting when you just want to find the blip. Even though you may have eight-channel potential for monitoring, sometimes it is faster to solo a track for a critical trim. You can solo a track (or multiple tracks like a sound effect and a verbal cue) by Ctl/Cmd-clicking on the desired audio monitor icons. This turns the entire monitor icon area green.

To summarize:

- The audio track key will enable/disable the track for editing.
- Alt/Option with the audio track key will enable/disable audio monitoring.
- Alt/Option-clicking changes the off-speed monitor icon to yellow or black.
- Ctl/Cmd-clicking on the audio monitor solos the track. Multiple solos are possible.

It is a significant workflow improvement to have all these tracks monitored when you are using J-K-L keys to fly through lots of material or a sequence where you have "checkerboarded" the dialog. Checkerboarding is a great dialog technique of putting each of two actors on separate audio tracks. This makes it easier to slip overlapping dialog without chopping off the previous spoken line. You can now listen to both tracks of dialog when you move at 3X, −3X, or slow speed scrubbing. With one line on V1 and the next on V2, you can monitor something that sounds like the finished audio when you are trimming and continuing to tweak the sequence. No more lip reading at 2X!

THE OTHER HAND

Now let's cover the other hand. The E and the R keys should be the main resting place to mark in and out on the fly (or I and O if you are reversing all this). Using the keys to mark in and out allows the video to keep rolling while you continue to search and change your mind. Combine this with Make Subclip mapped to a function key and you can fly through material and organize at the same time.

- Press Enter or Return to load a selected clip from the bin.
- Press E and R (or I and O) for marking.
- Map a function key for subclip function from the Command Palette (or map Shift-S, T, or P).
- Type the name of the subclip in the bin.
- Press Ctl/Cmd-4 to take you from the bin back to the source window to make the next subclip.
- Press the Esc key to toggle between the source and record windows if you want to keep going through the sequence.

With this technique you can subclip and name an entire bin without taking your hands away from the keys!

Alternatively, use your thumb for the V or B keys to actually splice or overwrite the material into a sequence. If you are working from long source clips with many different shots in one master clip, like a film transfer, you literally can type your show together. Consider creating a custom keyboard just for organizing and subclipping material before you begin to edit. Use the mouse mainly for accessing the graphic and nonlinear functions of the timeline.

MULTIPLE METHODS TO SOLVE ONE PROBLEM

There are other ways to do the fine adjustments and still get the analog scrub sound, but with the mouse. It's called the mouse jog, and with Media Composer and Symphony it's a finer scrub than J-K-L. Press the N key to enter the mode and the space bar to exit. Holding down the left mouse button in this mode gives you a faster shuttle than J-K-L. Lift off the left mouse button and you are back to the very fine scrub of mouse jog.

Here is another fundamental tenet of using the Avid systems: You can use multiple methods, one after the other, to fix a problem. Editors who consider themselves novices may have acquired one or two methods that they use under all circumstances. The more experienced editor uses a method of trim to get close as fast as possible, depending on the particular timeline view and level of sequence complexity. The editor then reanalyzes the problem and easily switches to another method to put the final polish on the fine points. The "right" way is the way that accomplishes what you want in that particular situation in the fastest way possible.

NAVIGATING NONLINEARLY

Another beginner's challenge is to not think linearly when jumping large amounts of time. Editors make a compromise and move at the speed of human comprehension in order to grasp a particular point in the material or listen to the performance. But what if you just want to get there as fast as possible? I see beginners actually dragging the timeline's blue bar through their material or sequence to get to the end! It's random access; get random. Jump to the end with the End key and jump to the beginning with the Home key.

FAST FORWARD AND REWIND

The Fast Forward and Rewind buttons are useful if you just want to jump to the next or the previous edit, but these buttons are usually left to the User setting default of being "track sensitive." This means that the default for the Fast Forward and Rewind buttons jumps to the next edit that uses all the tracks that are highlighted in the timeline. If video track 1 is highlighted, then you jump, in a sequential way, from cut to cut on video track 1 only. Turn on all the tracks (Ctl/Cmd-A with the timeline highlighted). Now you jump to every edit where all the tracks in the sequence have a cut in the same place. No straight cuts on all

Figure 3.4 Fast-Forward and Rewind User Settings

your tracks in the same place? With all your tracks highlighted, Fast Forward jumps all the way to the end of the sequence! That can be confusing since there is no easy way to get back to where you were in the timeline. No undo for jumping to the wrong place!

If you find yourself jumping to the end of the sequence a lot by accident, you can change the Composer User Setting on Media Composer and Symphony so that it jumps to every edit regardless of which tracks are highlighted (see Figure 3.4). You can change the settings to jump to every locator, too. You can reverse this default Composer setting instantly by holding down the Alt/Option key in combination with the Fast Forward or Rewind keys. This may be faster than going to the setting, especially if you need this option only occasionally.

USING THE TIMELINE

Modifying Fast Forward and Rewind still misses the larger point, which is that you don't need to step through lots of edits just to get through the timeline. If you want to get near the end, just click there! But more important, you need to be able to see where you are going. Learn to change the scale or view of your timeline quickly and easily.

There is a marvelous and intuitive drag bar to resize the timeline. Drag the slider to the left and the timeline compresses; drag it to the right and it expands. You have fantastic fine control and can change the view by large amounts quickly with the same function. On Media Composer and Symphony there are more keyboard controls so don't neglect these:

- The Focus button (H). The Focus button (H) is especially useful for troubleshooting small problems like flashframes because it is a one-step zoom to a preset amount to analyze a small section. Since it is a toggle, pressing it again takes you back to where you were.
- The keyboard equivalents to the drag bar (Ctl/Cmd-[, Ctl/Cmd-]). Xpress and Xpress DV use the Down and Up cursor keys to enlarge and reduce the timeline. This is so much more intuitive than the [and] keys that you will want to map the More Detail and Less Detail functionality to the cursor keys on Media Composer and Symphony.
- Ctl/Cmd-/ to show the entire sequence. This is always a quick reference if you get lost while zoomed in too far.
- Ctl/Cmd-J for "Jump Back."
- Ctl/Cmd-M for Zoom In or "More." Ctl/Cmd-M allows you to drag a unique shaped cursor around a specific area in the timeline where you want to zoom in. This technique is very useful for pinpointing a segment that needs more refinement.

There are so many easy ways to change the scale of the view in the timeline because it is vital to using the power of the system. The timeline is not just a pretty picture. It is an important tool for navigation and should always be sized to fit your needs for that moment. Swoop in to do some fine trimming, step back a little and look at the whole section, then fly off somewhere else to fix the next problem. You should be considering the scale of the sequence view at every stage of work.

JUMPING PRECISELY

Even though you may see precisely where you are going, you may not always get there the fastest by just clicking the mouse. Don't dog paddle through the sequence; you need to combine the power of the random access navigation of the timeline with the precision of the Fast Forward and Rewind keys. How can I jump huge distances in a single bound and still end up on the first frame of the cut? There are a series of modifier keys without which life as you know it could not exist. The most important is the Ctl/Cmd key. When you hold down the Ctl/Cmd key and click in the timeline, the blue position bar always snaps to the first frame of any edit on any track. It also snaps to marked in or out points.

If you hold the Ctl/Cmd key down while dragging your cursor through the timeline, it snaps to the head of every video and every audio edit. If you click anywhere in the timeline with the Ctl/Cmd key held down, you are guaranteed to land at the head of the frame or a marked in or out point. This can eliminate missing a frame here or there and creating flash frames. You must still combine this with the timeline zooming techniques, especially with a very complicated sequence. You may be surprised that you have snapped to the audio edit on track 7, which is three frames off from the video edit on track 1 you really wanted.

Important Modifiers When Dragging Segments

A description of the most important modifier keys and what they do in the timeline follows.

Ctl/Cmd
This key guarantees snapping to the head of a frame when dragging. This works for all models.

Alt-Ctl or Option-Cmd
This key combination allows you to snap to the tail or last frame of an edit as you drag. This works for all models.

Alt or Option

This key gives you fine dragging control even though you may be zoomed way out in the timeline view. This works for all models.

Shift-Ctl or Ctrl

Note that this is the only use of the Ctrl modifier key for dragging segments on the Macintosh versions. This key combination will drag a segment vertically without slipping out of sync and without snapping to any marks or existing segments around it. This technique is perfect for multilayered effects. Although the Ctrl or Command keys alone will allow you to do something close to this, it is not as exact and may cause you to snap to an edit hidden on audio track 9.

Lasso with Modifiers

This trick is one of the most important for using the timeline precisely and as a true nonlinear graphic tool. Holding down the Alt key on Windows or the Control key on Macintosh allows you to lasso any transition on any track and go into the Trim mode at a specific place. In combination with the Shift key, you can make multiple selections easily and, in a graphic way, extend the use of the timeline to get exactly what you want.

Changing the Track Name

A hidden timeline feature that is quite useful when working with lots of layers is the ability to name the track you are working on. You can right-click or Shift-Ctl-click on the track number in the timeline. Choose rename track (this is not available with Xpress or Xpress Pro.).

THE IMPORTANCE OF THE LAST FRAME

Avid editing systems work from the film model that insists that every frame exists and is important to duration calculations. In contrast, the established linear video model allows you to have an outpoint and an inpoint on the same frame of timecode. How can two shots use the same frame on the master tape? Don't ask—in linear video editing they just do, and video people occasionally are confused when the Avid editing system counts every frame separately. Mark an in and out point on the same frame. What is the duration? One frame. If you are looking at a frame and mark it as an in point, then say "Go 15 frames from here" by typing "+:15," when you mark out, you will have a duration of 16 frames. You have said, "Take this first frame and 15 more," which makes perfect sense to a film editor and seems suspiciously like a bug to a video editor.

There are two ways to deal with this fundamental difference. If you fast-forward to the beginning of one edit (or Ctl/Cmd-click in the timeline) to mark in and then fast-forward to the next video edit to mark out, how many frames do you have? You have one too many. You have included the first frame of the next edit. If you do an Alt-Ctl or Option-Command-click, then you are at the last frame of the shot you want (or just back one frame). Your in points and out points are exactly equivalent to the duration of the shot.

There is an easier way to do this, of course, and that is using the Mark Clip button, which is also track sensitive. You can make the Mark Clip button apply to the closest transitions on either side of the blue position bar on any track by holding down the Alt/Option button at the same time. This technique avoids the entire problem of grabbing an extra frame by just taking the first frame and the last frame of the edit you are working with. Leave the next edit alone!

TRIMMING

Beginners bring linear thinking to the trimming process as well. This is potentially the biggest mistake you can make, short of deleting all your media. The first place I see this is when beginners misuse the Match Frame button. Think of this really as the "Fetch button" because the Match Frame name is too close to the function that linear tape editors have been using since the beginning of computer-controlled timecode editing.

The traditional tape method is to get the edit controller to find the same frame on the source material as where you are parked on the master tape. The source tape cues up, you adjust the video levels to match what is already on the master tape, and then you lay in a little more of the shot, usually a dissolve or another effect. You can do this in the Avid as well. This is logical and simple, but it completely misses the point. Every master clip that you add into the sequence is linked to the rest of the captured material. You don't need to go get it because it is already there. Think of the extra captured material as always being attached to every edit in the timeline all the time. Each shot in a sequence is a window onto the original source material. The window can be moved, enlarged, contracted, or eliminated in the sequence, but the original source material is still there.

The best reason to use the Match Frame button is to call up a shot to look at the whole thing. It should be used for reviewing material, not as an integral part of the trim process. Used the wrong way, Match Frame is another dog paddle.

Trim vs Extend

Going into the Trim mode is the more sophisticated, but potentially more confusing, way to extend a shot. It allows you to do more in a complex situation. I personally think that the Trim mode is the heart of the Avid editing systems' functionality. Master trim and you have mastered the system and changed the way you think about editing forever. Trim is creative, not just corrective.

The Extend button in MC and Symphony is the simpler, more straightforward way of extending a shot. Extend can also be faster than trimming. Mark where you want the edit to extend to, turn on the tracks you want to extend, and then press the Extend button. It will not knock you out of sync because it is a center trim (this is a trim with two rollers; one on each side of the transition). It trims both sides of the transition simultaneously. I find it most useful for mechanical trims that go to a direct and easy-to-mark point. It is best used, for instance, to extend a B-roll shot to the end of a sound bite. Any finessing and I go to the Trim mode.

Mastering Trim

The main reason that trimming is so much better than just extracting the shot and splicing it back in is that you have the immediate feedback of seeing the shot in context. Fixing a shot while it is in place gives you the instant sensory feedback that is so important when using a nonlinear editing system. When expanding your use of the Trim mode, stay in sync as much as possible. Now obviously, there are times when you want to go out of sync, for cheating action or artistic purposes—I'm not talking about that. I'm referring to the skill of understanding the relationship between what tracks are highlighted and what kind of a trim you are doing. Some people get so flustered the first few times they try trimming with sync sound that they abandon it altogether and invent elaborate workarounds that are easier for them to understand. Lots of energy, not much style. This is one of those skills that film editors (those who have actually touched celluloid) have over video editors. It is pretty hard to knock yourself out of sync with a tape-based project, so thinking in terms of maintaining sync is quite foreign. But film editors must learn that whenever they add something to the picture—a trim or a reaction shot—they must add a corresponding number of frames to the sound track.

With the extra power of the nonlinear world (and film was the first nonlinear editing system!), there is the responsibility of keeping track of sync. The Avid editing systems do a pretty good job of telling you if the video and audio you captured together or auto-synched together (matching sound and vision from separate sources after digitizing) have lost their exact relationship. They are the white numbers called sync breaks (see Fig. 3.5). I think of them as a silent white alarm that, when I see them ripple across my timeline, tells me I most probably have made a mistake. The only time I want to see sync breaks is when I have cheated action or I am dropping in room tone.

Staying in Sync

The best way I have discovered to think about trimming is to imagine moving earlier in time or later in time to see a different part of the shot. Coincidentally, as you move earlier you may be making a shot longer or shorter. Any trim that

Figure 3.5 Timeline Sync Breaks

adds or subtracts frames—any trim that is on one side or the other of the transition—changes the length of that track and must have a corresponding change on all of the other tracks in the sequence. Not all trims change the actual length of a sequence, but the ones that do—the trims on one side of the transition or the other—knock you out of sync if you don't pay attention. This means you must look to the tracks that are highlighted when you decide to add a little video. Don't make the beginner's mistake of thinking that just because you are adding a few more frames to lengthen an action, it is a video-only trim. All the sound tracks must be trimmed if you make the sequence longer or shorter in any way.

Sync Locks

The easiest way out of this dilemma is to turn on the sync locks. The sync locks allow the system to resolve certain situations where you tell it to do two different things: Make video longer and don't affect the sound tracks. The system adds the equivalent of blank mag—silence—to the sound track. This may be safer than trimming and accidentally adding the director shouting "Cut!" but it will also leave a hole that must be filled in later. Blank spaces in the sound track are really not allowed! You will find yourself having to return and add room tone or presence so the sound does not drop out completely.

There will be a time when you tell the system conflicting things. You tell it to make the video shorter, don't change the audio tracks, and stay in sync. This is beyond the laws of physics. In this case, the system cannot make the decision for you where to cut sound in order to stay in sync, so it will give you an error beep and do nothing.

Sync locks work best if the majority of your work is straight assembly with little complex trimming. It is very effective, however, when you are sync locking a sound effect track to a video track. The crash and the flying brick need to stay together. Also sync-locking multiple video tracks together to keep them from being trimmed separately may keep you from unrendering an effect.

Sometimes you will be cutting video to a premade sound track. The video and audio parts of the sequence do not give you sync breaks when you change their relationship. Here, you must be even more conscious of maintaining sync. Don't fall into the trap of thinking that you can knock yourself out of sync now and later; when you get a chance, go back and fix it. Believe me, by the time you get the chance to go back, you will have created a situation that takes much longer to fix than if you did it right in the first place.

Top and Tail

Top and Tail are fast, straightforward ways to cut and trim interviews and other nonscripted material. They are two buttons inherited from the News-Cutter and work best with simple sequences or during the early "radio cut" stage of a project. If you are not sure how much of a sound bite you will need there is the tendency to cut in more than you really have time for. Play the interview back, and when the speaker finishes an important point stop the playback and press the Tail button. This activates a macro function that will mark in point at the present blue position bar, go to the end of the master clip in the sequence, and extract. This eliminates the rest of the sound bite. You can do the same thing by listening to the beginning of the statement and realizing you can lose all the "ums" and "ahs." Stop the playback before the first crucial word of the interview and use the Top button. This will mark an out at the present location, go back to the beginning of the master clip, and extract.

Top and Tail are both excellent ways to quickly pull out that little bit at the head of an interview to make it tighter or to end a rambling statement, all with a single button. This allows the user to work best with the concept of refine, refine, refine. Throw in too much material; evaluate what makes sense once the material is in context and make adjustments as fast as possible. Top and Tail are also track sensitive so try it with items not on the same tracks like items on V1 and A1 mixed with items on V1 and A2. These tools are optimized for quick content decisions and are not designed to work well with more complex trimming techniques like split edits.

Trimming in Two Directions

Trimming in two directions, an asymmetrical trim, is a subtle and very powerful technique. You may have a situation where you must trim a video track longer, then add extra material to a sound effect or music that would keep you in sync but ruin the edit. Here, keep in mind that just because you are adding video on V1, you can still trim the black before the clip on A2. Both tracks get longer, but one adds more captured video, fixing a visual problem, and the other adds more silence, keeping the two tracks in sync. (See Fig. 3.6.)

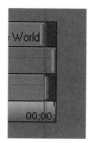

Figure 3.6 Trimming in Two Directions in the Timeline

Slipping and Sliding

Beyond the basic situations, most of the difficult sync problems can be fixed by two methods. The first method is to ignore the fact that trimming a shot must make the sequence longer. Trim in the center of the transition, basically not affecting the sync, then use the Slip function. Many people have a difficult time grasping the Slip function because it is so tied to the nonlinear, random access concept. (See Figs. 3.7, 3.8, and 3.9.)

You can enter the Slip mode by multiple methods. With Media Composer and Symphony you can double-click on a clip once you are already in the trim mode (as long as the timeline view allows you to see a black arrow cursor). In all models you can also get to Slip mode by lassoing the entire clip from right to left. You may need to hold down the Alt/Option keys to select the exact clip in

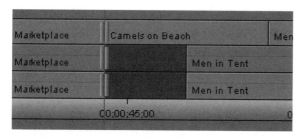

Figure 3.7 Selecting the Transition to Lengthen

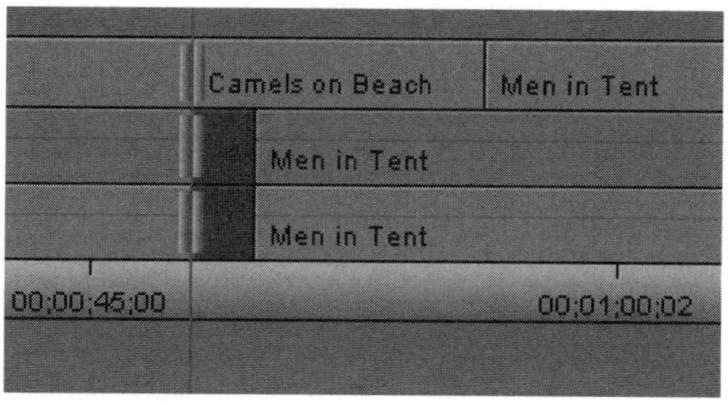

Figure 3.8 Trimming the Transition 10 Frames Forward

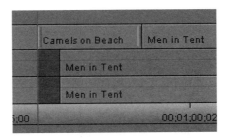

Figure 3.9 Slipping the Shot Back 10 Frames to Get the Same First Frame

a complex timeline. In Xpress DV, right-clicking or Shift + Ctrl + clicking on a segment in normal trim mode offers Slip Trim and Slide Trim choices in the context-sensitive menu. I generally get there by double-clicking in Trim mode because I use it as a second step in a difficult trim situation.

Think of slipping as a shot on a treadmill. The shot slips forward or backward, showing an earlier or later part, but the place in the timeline never changes. A slip will change the content of a shot by revealing new material, but leave the duration of the shot and location in the timeline the same. Because you usually have more video linked to any shot used in the sequence, you can slip that entire shot back and forth.

So if you trim the beginning of the shot 10 frames as part of a center trim, you can slip the shot back into position so that it still starts with the same frame. If the first 10 frames of shot B are important, then slip them back into place. Your center trim moves the frames viewed in the A and B shots of a transition to be 10 frames later. Although shot A gets longer and shot B gets shorter, the length of the sequence is not affected and the sync is not disturbed. Shot B has gotten shorter, but after you slip, it still has the same starting frames. I have worked with producers who have edited their programs on Avid systems for years who had never seen Slip mode! Although it seems complex at first, it is truly a powerful tool used in the right place.

The corollary to Slip mode is Slide mode. You can enter Slide mode by Ctl/Opt dragging from right to left, or Ctl/Opt double-clicking in Trim mode (this last method is not available with Xpress).

Slide mode moves the shot neatly through the sequence by trimming the shots on both sides of the selection. It affects the location of the shot in the timeline, but not the content or the duration. It is a good alternative to dragging a shot with the segment mode arrows if you are making a smaller change, but not nearly as useful as Slip mode.

Alt/Option-Add Edit

The other main technique to deal with difficult trim problems and maintain sync, especially with many tracks, is the Alt/Option-Add edit. It is very easy to just lasso all the transitions around the area you want to trim or, with the timeline

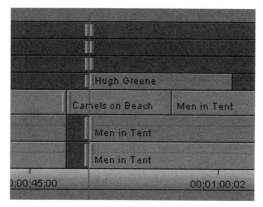

Figure 3.10   Trimming Multiple Tracks with Alt/Option-Add Edit

window active, press Ctl/Cmd-A (select all), turn on all the tracks, and then go into Trim mode. This can be unpredictable if you have zoomed in on a particular transition to work on it closely and the next edit on V2 is over 10 minutes away. You don't really know how you are going to affect V2 unless you zoom back out to see. You might accidentally add more video to a shot on V2 that you think is already perfect or unrender an effect.

The best method for trimming multiple tracks, but with empty spaces on some tracks, is to go into Trim mode and use Alt-Add edit or Option-Add edit. This modifier key and button combination automatically adds an Add edit in the blank tracks and gives you a place to trim where it would normally be just black filler (see Figure 3.10). If you do this in Trim mode, all the Add edits are selected automatically for trimming. This method is so useful you should map the Alt or Option key to the Add edit button.

The Alt/Option-Add edit gives you the handle to stay in sync when trimming multiple tracks. You are left with Add edits in the filler — fake transition points — but you can get rid of them by choosing Remove Match Frame Edits from the Clip pulldown menu. You can isolate which ones to delete by putting mark in and out points around the Add edits before choosing Remove Match Frame Edits. When you have selected the Add edits in Trim mode, you can also press the Delete key. But my advice is just to leave them.

Add edits provide a continuing ability to make changes at that point in the sequence and are a good indicator so that you can pinpoint exactly where you knocked yourself out of sync. They won't show up in your EDL unless you put a dissolve on them — unlikely for Add edits in black.

So the most important aspect of trimming is to be aware of when you are going to change the length of the sequence by trimming on one side or the other of a transition and which tracks will be affected. When you grasp these points and overcome the fear of going out of sync, you will have a much more powerful tool and feel much more comfortable with the workflow concept of refine, refine, refine.

Customizing Your Interface Environment

There has been so much progress in the user's ability to change and optimize the user interface that now we need to discuss the most effective trick to cut through the options. Most interface changes can be saved and recalled quickly through the judicious use of Workspaces, Toolsets, and Custom Views. The trick to finding and using your perfect setup is how quickly you can keep changing it to exactly what you need at exactly the right moment.

General Modality

A general philosophy for modal editing systems is to have only the functions you need in front of you when you need them. The idea of modes is very powerful and should not easily be dismissed by marketing hype. If you have all the functions available all the time, what is the possibility that you will need more than a very small percentage of them? The rest are wasted and clutter the valuable screen real estate. If pressed accidentally, these unneeded functions may cause more harm than good, sending you into a function or display change that is not desired. Did you mean to trim that shot or just navigate there to look at it? A mouse click in just the wrong place will give the undesired result. A very careful mouse click in just the right place eventually will cause carpal tunnel syndrome. A nonmodal interface also may obscure the more needed functions at just the most critical time when you must have them close at hand. If you can't find a function, it doesn't exist.

A modal system gives you a series of streamlined, focused interfaces for the most used functions. To go to or from a mode should be seamless; this is where the real challenge comes. If you can't get to a mode easily, then you may feel that you need to have the important parts of that mode available all the time. The Avid editing interface is based on the Source/Record mode being a type of home page. By pressing the Escape key you can get to it instantly from any other mode. You can get to the other modes, trim, effects, and color correction through dedicated buttons that can be mapped to the keyboard. These mapped keys are critical to using the modal interface to its most powerful advantage.

Custom Views

There are custom views for the timeline as well as for the entire user interface. You can change colors, track size, track position and information displayed for the timeline to display just the information you need; for instance, for audio mixing or effects creation. Create the view in the timeline, then click on the default name of the view and choose Save As. You can use the general user interface to eliminate or enlarge buttons and use color as a key to the functions you are using and sign as to what custom setting you are using. Go to Interface in the Project window, then the Appearance tab. If you really mess up the view (and potentially can't see anything to change it back) go to that view and right-click/Ctl-Shift-click and choose Restore to Default. When you are done customizing a setting, name it something useful so that you can tie it together with a workspace or toolset. If you have other User Settings that are meant to be used at the same time, name the interface the same thing.

Workspaces and Toolsets

One way to conquer the complexity of modes and settings is to create a series of snapshots of your favorite configurations. These are Workspaces and Toolsets.

Workspaces can be created from scratch, whereas Toolsets start with some preset modes like Source/Record, Color Correction, and Effects. You can modify both types to reflect your personal choices for buttons layout, screen colors, text, and button size. Think of these as user interface setup macros since they can even contain Project and User Settings. If you find yourself switching between any two User Settings on a regular basis, just program them into the Workspace. If you need to have a timeline change the size, color, and information displayed when you start to do audio mixing, link the Workspace or Toolset to the particular timeline view.

To link a Toolset or a Workspace to a setting you must first create the setting and then give it a name. For Workspaces the name of the linked setting must be the same as the Workspace. For the Toolset you can link a preset Toolset to any User Setting name. If you have multiple user or project settings that you want to change at the same time, you must give them all the same name. You can name Workspaces and other settings by clicking the empty space to the right of the default setting name in the Project window. You can create multiple versions of any setting by clicking once to make it active and then using Ctl/Cmd-D to duplicate it (the duplicate command is also under the Edit menu and is a right-click/Shift-Ctl-click choice). Then change the setting and rename it. To see your new setting make sure that the Project window is displaying All Settings, not just Active Settings. Otherwise you will continue to duplicate settings and never see them!

The following method links User Settings to Workspaces:

1. Create a timeline view that is designed especially for audio. Turn on important audio graphic information, make the video tracks smaller, and move the timecode track in between the video and audio tracks (Ctl/Cmd drag the timecode track vertically—not available with Xpress).
2. Save the timeline view and name it Audio.

3. Open important audio tools that you like to use, like the Audio Mix and Automation Gain Windows. Position them where they are best integrated with the rest of the interface.
4. Click a Workspace in the User Settings window and duplicate it (Ctl/Cmd-D).
5. Double-click to open the Workspace.
6. Choose Activate Settings Linked by Name and Manually Update This Workspace.
7. Click on Save Workspace Now.
8. Click the empty space next in the User Setting Window on the Workspace setting to name the Workspace. Name it "Audio". If this is your only Workspace then it is Workspace 1. Workspaces are numbered based on alphabetical order in the User Settings window. They can change numbers dynamically when you add new workspaces so be careful when you name them. You may want to name the workspaces with numbers like "1 Color Correction Workspace" so that when you create an Audio Workspace the order won't change.
9. Open the Keyboard Setting and the Command Palette at the same time.
10. Go to the More tab and grab the button for W1. This Audio workspace is Workspace 1. Map it somewhere you can remember easily like Shift-1.

When you press this button you will call up all the audio windows and the timeline will change. Make a Workspace or a Toolset in a similar manner for all your important functions. The Toolset actually is easier since you can choose Link Current to... from the Toolset menu and get several options for linking to different User Settings, although you have fewer Toolsets to work with. (See Fig. 3.11.)

Figure 3.11 Linking Toolsets

ORGANIZING

One thing that working with a computer forces you to do is organize. If you don't have a plan from the beginning that is easy for you, you just won't do it and will find yourself relying on the frame view to find shots. Although this may make it easier for clients to put their fingers on your monitor and shout "That one!" it is slow. The bins that come from the telecine need to be named for scene and take, but after that, everything follows the standard script notation. In a traditional film edit room, organizing has more to do with tracking the physical film and keeping the right scenes ready for cutting. In a nonlinear edit, a film project must keep everything from a scene together. Documentaries or other formats where the form takes shape during the edit must live or die by good organizing of shots in a computer-based edit. I once edited a show with 200 sound bites and no script. To use the tools that are available for finding shots, you need to enter the information in a way that allows you to search for it easily.

There are two things to keep in mind when creating bins: tape name and bin size. Generally, the tape name is most important when you are starting to organize, because this is initially what you are handed from the field. Tape names should have some criteria for allowing you to trace back shots. For instance, with day and location coded into the number used for the tape name, you more easily can follow a series of shots to a common starting point. Also, as mentioned before, you can start with a tape-named bin for digitizing and then organize based on content.

If your bins are too large, however, you are defeating a lot of the benefit of the organizing. It takes longer to open and close large bins, and once they are open, it takes longer to find exactly what you need. Most likely, you will use the Find Bin button frequently, so you want it to be opening small bins to speed up the retrieval process. To use the Find Bin function, you must have that bin listed in the Project window (or in a folder in the Project window); it needs to have been opened once in the project. Also, you can go directly from the Sequence window to the source bin by using Alt-Find Bin or Option-Find Bin. Again, Match Frame is also a useful tool for calling up a shot if you are not interested in the bin. It calls up the source clip regardless of whether the bin even still exists! All Match Frame needs is media on the drives.

THE MEDIA TOOL FOR EDITING

The Media Tool is a good way to find shots, especially if the bin has been deleted, by sorting or sifting in the Media Tool window and then dragging the shots to a new bin. If the media is online, you can get it through the Media Tool.

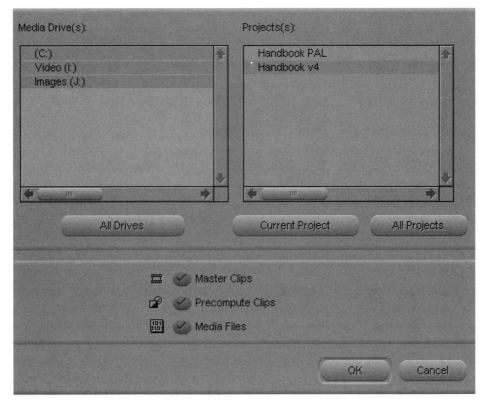

Figure 3.12   Media Tool Choices for Finding Shots

Most people use it for media management, but it is an easy way to find hidden shots quickly by searching across projects.

On a very large job, or a system with lots of media, opening the Media Tool can be a time-consuming process causing some people to avoid it completely. That is why you can choose exactly what drives or what projects you will search through before you open the Media Tool. The smaller amount of media that is searched through, the faster the Media Tool will open. Opening the Media Tool causes the system to read the Media databases from every active MediaFiles folder and load that information into RAM. That can take drive access time and potentially use up much of the available RAM (if you have hundreds of thousands of objects or several terabytes of storage). Of course, if you are really desperate to find a shot, you can just press All Drives and All Projects, and load the entire Media database. Once loaded completely, all searches in the Media Tool will be instantaneous for the rest of the session or until you shut down the computer. (The exception here is if you have Avid Unity—you always will have to reload the media database since the assumption is that someone else has been using the same storage. The best answer with Avid Unity is MediaManager, which will be discussed later.)

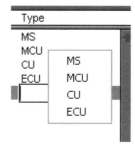

Figure 3.13   Reusing Standardized Data with Alt/Option-Click

Early in the process of organizing a new project, if the Media Tool is not too large yet, I leave it open to avoid having to keep opening a bunch of separate bins when looking for a shot. Think of the Media Tool as the largest bin of media you can have at one time, and when it gets really large, open it only when you must.

To search more effectively, make use of the customized columns in a bin and try to standardize the way you enter particular information. If you create a custom column called Shot Type and enter both Wshot and WS, you will have a harder time searching. Once you have entered a particular notation in a custom column, whenever you go back to that column and Alt-click or Option-click, you will get a list of everything you have entered (see Figure 3.13). You can choose a common entry from this temporary menu. This lessens the chance of spelling errors and speeds data entry. Workflow is key to finishing mechanical tasks quickly.

So far we have discussed stylistic techniques; the benefit gained is primarily in speed, efficiency, and making you look good. But a few common techniques, if not followed regularly, can create real technical problems. Chapter 13 describes some of these problems, but let's cover a few that can be prevented easily by just following directions.

BACKING UP

If you are a beginner to the computer, you may not realize the seriousness of backing up, but first imagine the cost of losing a day of work. Then imagine losing the entire project. Until you lose your first project, you might not back up on a regular basis. The fact that for about a dollar and five extra minutes you easily could have saved all the project information is a very compelling argument. There is nothing more sickening than returning to a project after a short break and not being able to find a sequence. Back up everything important, even twice a day, using rewritable CD-ROMs. Always have your work somewhere else when something goes wrong. Auto-save can fail if the drive is too full and random crashes can destroy boot drives with projects and sequences. I back up constantly, but not to the same disk! Use a separate removable disk for every day of the week, and you will have seven chances to find a usable version.

What should you back up? It is usually difficult to back up all of your captured media in the middle of a project. But with a DLT or other data type format and backup software that allows you to back up invisibly while you edit, backing up media isn't out of the question. There is also the option for creating an archive to videotape on Media Composer and Symphony, which we will deal with in Chapter 4. Backing up captured media may not be top priority in your project unless you are working with lots of nontimecoded material. Batch digitizing can be faster than any other form of restore—if you have the source tapes! What you really need to recreate your job is the project folder on the internal hard drive. Take all of it, not just the bins or the project icon. Many short-term projects should have a project folder that can fit onto a high-density floppy. If you are working on "History of the World" or need to back up lots of graphics and music tracks from CDs, however, you need to back-up to CD-ROM or DVD. CD-ROMs hold 650 megabytes of data and a DVD will hold 4.75 gigabytes. The cost of DVD recorders has dropped quickly, but the price of the blank media is still not as cheap as CD-ROMs. You only need DVD-RAM to back up data, but the time will come when you want to compress sequences for playback from DVD, so shop around for a full featured DVD burner.

If the backup process is unpleasant, you won't do it, so check the file size before you begin. You can find the size of a folder by clicking it and then using Command-I to Get Info on the Mac or by right-clicking on the folder to get Properties on Windows. Compression programs allow you to segment any file into smaller pieces and save it across multiple floppy disks. Some people are nervous about compacting files across multiple floppy disks and for a good reason: If one of the floppies fails or grows legs, then the entire archive—your project—is unreadable. This method is tedious and slow if there is a lot to back up (more than two or three floppies worth, or around 3 MB). Although the new USB memory sticks or flash memory may be appealing for smaller backups, you still will need multiple copies to be completely safe, and that can start to add up compared to CD-ROM or floppies.

Alternatively, you can send your project over a fast network to a system that contains a drive suitable for backup. On a particularly big job several years ago, I backed up to a Zip drive on an Ethernet server on the hour. Although we lost power almost every afternoon while editing on location, we never lost any data.

Backing up to another hard drive on your system or another computer at your facility may not be good enough. I have had both the Boston Fire Department and Mother Nature ruin two different suites where I was working. (To be fair, the fire department was trying to save a historic building.) A particularly successful film assistant I know in Los Angeles makes two backups of the project as the film gets close to picture lock. He takes one, his assistant takes one, and they take separate routes home. It is Los Angeles after all. Avid technical support has a category in their database to record reasons for equipment failure. Earthquake is one of them. And having a floppy in your pocket also means you are more likely to get paid as a freelancer.

NONTIMECODED MATERIAL

In the rush to complete a project, people throw anything and everything into their project just to get it done. Scratch audio is recorded straight to the timeline, VHS material is cued up by hand, and CDs are played directly into the Avid without any thought about recreating the job or, worse, starting over again if disaster were to strike. If the time is available, then seriously consider dubbing all nontimecoded material to a timecoded, high-quality format source. The potential slight loss of quality involved in dubbing will be the difference between quickly recreating what you have done or matching things by eye and ear. If you cannot timecode all your sources, then seriously consider copying to the DLT or a large inexpensive firewire drive for the evening after you have finished digitizing all your media. Remember, a digital nonlinear project is never done, you just run out of time, money, or both.

CONCLUSION

Beginners don't grasp these techniques from the entry-level course. The techniques are usually a combination of changing some of your work methods and taking advantage of some unfamiliar functions. There are many other keys, modifiers, and tips and techniques in the Avid editing systems, but these are the main areas to concentrate on. Use the keyboard more and spend more creative time using trim. Start to think nonlinearly about navigation and the structure of the sequence. You will find your speed incrementing in leaps and bounds.

4

Administration of the Avid

Some people are lucky enough to have their Avid system administered by someone else—hand them this chapter and go back to cutting. However, the day-to-day reality is that most people are responsible, at some level, for their own system. After all, if you have created a difficult situation for yourself, it is you that cannot go home until it is resolved. A smooth running session makes you look good, period.

This chapter explores the peripheral issues of owning and maintaining a professional editing system. This includes environment, media management, and networks. For the Avid administrator or postproduction supervisor, getting media in and out of the system is as important as anything you do with it while editing.

ROOM DESIGN

It is easier to administer a system that is set up correctly and put into a room that is well designed for the Avid. Fewer constraints are placed on the design of a digital, nonlinear suite than a traditional, tape-based suite, but this doesn't mean you can just plunk the equipment down in a pile! There is flexibility to move all the equipment into another room and keep the bare minimum in the suite or move the noisiest parts, like drives. There are technical implications to making cables longer, as you will see. Suites need to be designed so they can be serviced and simple maintenance can be done without disruption. If you plan to use the suite as an online finishing facility, then you should treat it like an online suite. Use an external waveform and vectorscope, high-quality speakers, a high-quality third monitor, and seriously consider a patch bay. There is nothing worse at 2 A.M. than carrying VHS decks around so that you can do two dubs. If connections are made through a professionally installed patch bay, there is a better chance the dubs will come out well!

Probably the biggest consideration after editor ergonomics is the noise level. Many people make the mistake of putting all their equipment in one room because the SCSI (small computer systems interface) cables that ship with the system are so short. Longer SCSI cables are easy to purchase at your favorite computer or hobby store, but you should not use them for any reason. These will almost certainly cause weird problems and data errors later.

The slickest installations I have seen have completely relocated the CPU and the drives to another room. They have extended the keyboard, or ADB (Apple Desktop Bus) or USB (Universal Serial Bus) cables, and the monitor cables. If you can also custom install a removable drive for projects, EDLs, and so on, it will allow the editor to perform these loading and saving chores more efficiently. The drives and the CPU are likely to be the noisest elements of your system because of their fans so they are the most important to move to another room. Of course, if you have no central machine room and the editor must load tapes and capture, there should be a connection for a video deck's video, audio, and control cables. If all the hardware is in another room, there should be telephone links between the two rooms (preferably with speakerphones for troubleshooting purposes). The best facilities have gone back to the central machine room design after the initial years of isolated Avid suites. It is important to have a range of decks during the course of an edit, so multiple decks need to be available. Tying up a deck all day when it is not really needed is just as bad. A central machine room also makes it easier to get to the equipment for support or upgrading. There is nothing worse than working under a table with a flashlight in your mouth, the telephone in one hand, and a screwdriver in the other. Unfortunately, this happens way too often because of thoughtless room design.

If moving all the parts with fans is too elaborate, I have seen rooms where they used fiber channel drive towers or JBODs (Just A Bunch of Drives). You can use fiber channel cables to put the drives farther away than just a foot or so (the length of the SCSI cables that come with the system). With Avid Unity MediaNet or LANShare you have much more flexibility and can put a central group of drives as far away as fiber channel allows. The first appearance of the now obsolete differential SCSI drives led to one support call where the freelancer insisted that he didn't need drives because "they just plugged into the wall." Show your freelancers the layout before they start the night shift!

Many who build rooms with extended ADB cables use the Gefen Systems ADB Extender. This device amplifies the ADB signal so that it can be sent as far as several hundred feet. Monitor cables for video and audio can be extended quite a distance using standard audio and video cables, but eventually must be boosted with distribution amplifiers (DAs). The new USB devices can be daisy-chained and connected to a hub to make them more flexible than the old ADB, although some USB hubs and repeaters have been known to cause gray screens when booting. Experiment with any cable extenders before you permanently install a system.

One of the nicest things about many of the Avid suites I have worked in is the addition of windows. There is nothing better to clear your head than just to stick it out a window for a few minutes, and having natural light is a nice change; however, I must fall back on the admonition that if you are doing color correction or shot matching of any kind, you need to have complete control of the lighting. Pick the neutral color temperature of the lighting carefully and by all means avoid fluorescent fixtures. Color temperature of sunlight changes during the day so that the shot captured or color corrected in the morning may not look like those adjusted at noon unless the ambient lighting is indirect and consistent. If you have windows, make sure you can pull the room-darkening shades and keep glare off the monitors.

If you also are planning to do final sound mixing, make sure the room has been deadened. Apply sound-absorbing foam around the room or make sure that the room has enough carpeting and wall hangings. Mixing in an empty office is probably a bad idea. Investing in good speakers and a real amplifier pays off quite quickly. You may also want to consider cheap speakers with an A/B switch to the reference monitors or pumping the Avid audio output through a standard home television monitor in the suite. There is nothing worse than listening to your wonderful sound mix at home and having it sound muddy from overpowered bass.

Personally, I feel no suite is complete without two phones, a trash can, a box of tissues, and a dictionary. You'd be surprised where people set up these temporary suites: attics, basements, bedrooms, storage closets, hotel rooms, boats, bank vaults, Chinese laundries, and ski lodges. Try to minimize any environmental impact like heat, dust, or jarring motion (like editing in the back of a moving truck). Even 5 to 10 degrees of difference in temperature can add or subtract useful years from the life of the equipment.

ELECTRICAL POWER

The final and probably most important piece of equipment you need under any conditions is an uninterruptible power source (UPS). If you are running a system now without a UPS, you should nonchalantly put down this book and run to the phone to order one now. You are living on borrowed time. A UPS regulates your power, giving you more when your electric company browns out or less when you get a spike. If power fails completely, a UPS gives you enough battery time to shut down the system in an orderly fashion and avoid crashing, losing work, and potentially corrupting important media or sequences. A UPS makes your equipment run with fewer problems, and you will be able to charge for more productive hours of use.

The real question is not whether you have a UPS, but how much of one do you need? They are figured in the confusing scale of volt-amps. The math is not so hard if you can find out what each piece of equipment needs for electrical power for either volts or amps. The numbers are usually in the manuals or on the equipment itself. There is information on the Avid Web site under the Customer Support Knowledge Center that lists the power requirements of each piece of equipment. But don't neglect connecting the tape deck. What happens to your camera-original tape if the power goes out when you are rewinding?

Here's how to figure the size of an uninterruptible power source that is sold in volt-amp models:

$$Volts \times amps \times power\ factor = volt\text{-}amps \times power\ factor = watts$$

The power factor for computers is between 0.6 and 0.7, so you can look at the same equation as:

$$Watts \times 1.4 = volt\text{-}amps\ for\ computers$$

Watts for a Media Composer 9000 with 4 LVD SCSI drives, a Sony PVW 2800 Betacam SP deck, and a DLT equal 1111 watts, so

$$1111 \times 1.4 = 1555.4$$

And because you don't really trust manufacturer specs for the UPS and you buy more than you need to accommodate future expansion, you add another 33 percent on top:

$$1555.4 \times 1.33 = 2068.7$$

You purchase the 2200-volt-amp model and sleep well at night.

Keep in mind that before you get a true blackout, you will probably suffer from sags and brownouts. These may cause the UPS to use up some of its battery power to keep you going until the Big One hits. Then, when the power comes back on, you can count on a serious power spike. A spike can cause damage to boards, RAM, and drives, and that damage may not show up until days, weeks, or months later as the parts start to fail prematurely. The fact that all power is going through a series of batteries and power conditioners with a UPS before it gets to your delicate equipment should give you a warm feeling in your stomach. Just make sure when you get it all hooked up that the battery is actually connected, since some UPS manufacturers ship that way.

If there is any question about what to put on the UPS, imagine using that device full tilt when the power goes out. Ever see a one-inch machine lose power while rewinding a finished master tape? Not good. A cassette-based tape deck will almost certainly crease the source tape if it is rewinding when the power goes out.

Even if all the lights and your monitors go out, you can always save and shut down quickly using just the keyboard. In an emergency, remember:

- Ctl or Cmd-9 to activate the Project window. You want to save the whole project, not just the active bin.
- Ctl or Cmd-S to actually save everything.
- Ctl or Cmd-Q to quit the application in an orderly way. Quitting will save everything first, but trying to save the project should be your first step anyway. You may have to hard boot the system to get control back after a power hit. Saving should be an automatic first step in any emergency procedure.
- Enter to confirm that you really do want to quit.

This sequence of keystrokes avoids the chance of corrupting the project from being shut down improperly and can be performed (if you really have to) with the monitors blacked out. A UPS has saved me literally a dozen times. After your first serious power hit, what is the real cost of replacing your system one board at a time?

ERGONOMICS

Human ergonomics has been written about at length in other places, so just a quick word about it here. Don't scrimp on chairs. They make the difference between happy editors and editors in pain. Get chairs that can adjust armrests, back, and height. Many people swear by armrests, and with a keyboard and wrist rest at about the same level, there is less chance of wrist strain. Keyboards can be put on sliding shelves below the workspace. Keep the back of the hand parallel to the forearm to reduce wrist strain. The relationship of chair, keyboard, and monitor cannot be underestimated as important to the creative process. Some editors, like multi-Oscar-winning editor Walter Murch, cut most of his latest films on the Film Composer standing up!

MEDIA MANAGEMENT

Let's discuss the nuts and bolts, the bits and bytes, of what happens when you put media on your system. The Avid editing application is an object-oriented program, which means that many things you do create an object. Capturing media, rendering, importing, and creating sequences and bins all create different kinds of objects. It is the relationship between those objects that allows you to combine things in such interesting ways.

The limitations of desktop computers become apparent when you are dealing with hundreds of thousands of objects. Achieving object counts of this scale is not uncommon, and once the system begins tracking around 150,000 objects on older systems there is a slowdown. New computers usually can handle much more than this. If you are tracking too many objects then features do not respond as fast, and you find yourself clicking and waiting more often. Buying more RAM can alleviate much of this. Because feature films require lots of footage to be online at the same time, you may want to add more RAM to your Film Composer—over a gigabyte if possible. The more media you have, the more objects you have. If you are working on projects that require a lot of effects then you are rendering often and creating even more objects. Cleaning off the unused rendered effects is essential to making your system run at speed. In addition, if you experience slow response time, don't overlook just restarting, especially on an OS 9 Macintosh—it defragments your RAM. This was a problem on older Macintoshes and may not be a factor today, but many things are reset when you restart and you may find it beneficial as a first course of action when a system starts to get sluggish.

The number of objects can creep upward if you are not paying close attention or if there are multiple shifts capturing and they are not communicating. There is an easy way to find out how many objects are on the system right now. Go to the Project window, click the Info button, and click Memory (or choose it in the Fast menu at the bottom of the Project window). There is a display of the number of objects the system sees, but this Object window has not been optimized yet for OS X use and may not give you useful information.

The good news is that there are ways to reduce the number of objects without actually deleting material. Deleting is quick and easy, but if the material has no timecode, you are forced to either recreate it later or edit it back in by eye. A painful, tedious process! Why didn't you dub it to a timecoded tape? Being able to clear off drives when they become too full with large amounts of small objects is a good argument for dubbing as much as possible to timecoded sources, but that takes time, too. As an administrator, you need to decide about policies regarding

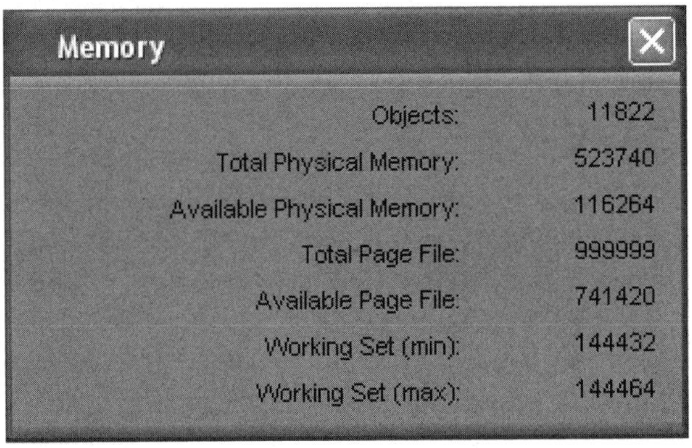

Figure 4.1 Number of Objects in the System

the speed and ease of replacing material and when it is appropriate to clean off drives completely. As you will see later, you may end up archiving certain crucial elements and recapturing the rest.

The editing system sees only the media that is on the media drives in the folder named OMFI MediaFiles, 6.xMediaFiles, or 5.xMediaFiles, depending on the version of software you are using. This folder is created automatically and named by the software when the application is first launched and the drive is used for capturing. The OMFI MediaFiles folder must keep this precise spelling and it must stay on the root level of the file hierarchy. In other words, you cannot put this folder inside another folder and you cannot rename it. If you do, the media inside the folder goes offline and is no longer accessible to you when you try to edit. Also, on any local storage system any folder that is put inside the OMFI MediaFiles folder appears to be offline as well, and this would be considered a mistake; it is not recommended as an organizing technique.

THE MEDIA DATABASE

Media File Databases and Versions 7.x/2.x

Media files are tracked on non-Unity systems using two files in every OMFI MediaFiles folder: msmFMID and msmMMOB. These are forms of a media database, keeping track of all the files in the folder. The MSM type of media file in these later versions is designed to work on large media servers and to handle media in group user situations where multiple editors are sharing media. Unlike the older format MFM, when MSM type media is changed in the OMFI Media Files folder, the editing on all the systems does not stop to update the media database. Imagine being interrupted every time someone moved something around on one of your media drives if they were accessing it from another system! In order for the system to look at what is actually in the OMFI MediaFiles

Figure 4.2 Latest Media Databases in the OMFI MediaFiles Folder

folder, you must choose Update Media Database from the File menu or open the Media Tool. After this, all offline media that is really in the folder shows up as available for use, or relinked. This chapter will discuss relinking in more detail later. In later versions, relinking is more important than ever as a strategic technique.

COMPRESSION, COMPLEXITY, AND STORAGE ESTIMATES

None of this clever manipulation of objects solves the basic problem of running out of space on the media drives. It is only a matter of time before this problem occurs, and you should prepare for it in an organized way. There are several ways to tell when you are going to run out of space. The Hardware Tool under the Tools menu (also under the Info tab of the Project window) gives you a bar graph of how full the drives are relative to each other, the amount of storage empty and used, and percentages full (if you have Tool Tips turned on). This is good for figuring out where to start capturing the next job, but it does not give you the amount of space in terms of amount of footage.

If you need precise numbers, then open the Capture Tool and choose the tracks you think will be needed the most (this may be called the Digitize or Record Tool on your system, but all new Avid systems have moved to use the term Capture). If you are working with material that has sync audio, then turn on all the tracks. But if you are working with mostly MOS (silent) film transfer that will be cut to an existing soundtrack, get the estimate with only the video track turned on. Make sure the compression level or resolution you will be using is set correctly (on Xpress DV this will always be DV25). Video takes a massive amount of space to store, even compressed, compared to audio files. The Capture Tool gives you an estimate of how much time you have on each drive.

You can capture one clip across multiple drives and across partitions of the same drives. In Capture Settings, you can tell the system that when a clip is under a particular time length, say 30 minutes, and you need extra space, then jump to another drive and continue without a break. If the clip is longer than 30 minutes, it tells you that you do not have enough space. Now by Alt-clicking or Option-clicking on the drive icon in the Capture Tool you can toggle the display to tell you how much space is on the drive or how long your specific clip happens to be. There is a 2-gigabyte file size limit on certain older systems so you want to avoid capturing all your media to a single large file. Since you can now specify a group of drives for capturing (use Change Group under the drive selector in the Capture Tool) and because you have a setting for Capture to Multiple Files (in the Capture/Media Files User setting), both the 2-gigabyte file size limit and the drive space of a single drive no longer restrict your capturing.

If you select all the clips in a bin and choose Ctl-I on Windows or Cmd-I on Mac (Get Info), this opens the Console window and gives you a total length of all your clips. This is a powerful way to see whether you have enough space on your drives to recapture everything.

Unlike video or film, compressed images are judged by their complexity. A complex image takes exactly the same amount of space to record on videotape as a simple one! When you capture an analog image to disk or "ingest" an already digital image, the level of image complexity is important. The more information and detail in frame, the more space it takes to store and the more difficult it is to play back. Playback from a disk-based system is a question of throughput or how much information can be read from the hard drive, pushed through the connecting buses into the host memory, and out to the monitors or tape decks. This is why a slower system may not be able to handle high-resolution images—it cannot get the information from the drives fast enough to play all the information in realtime with effects and audio.

Some systems actually drop frames during capture, but Avid does not. Instead, you receive the dreaded underrun or overrun error messages and the system stops. You need to decide how to get that section to capture or play through, and this chapter discusses that later. It is remarkable that the Meridien video board based Avid editing system will never drop frames during playback. This is crucial to trusting Digital Cuts as a visual reference to compare to EDLs and Cut Lists. Not many systems are always frame accurate for output. Although the new host-based systems like Media Composer Adrenaline will drop frames when they cannot play back all the realtime streams, this is clearly displayed in the timeline. There is a timeline setting called Highlight Suggested Render Areas After Playback, and when active there will be colored lines at the bottom of the timeline after playing back a demanding sequence. Areas that are playing without dropping frames, but just barely, will have yellow lines, and areas that absolutely need rendering are lined in red. When you render these areas the system will no longer drop frames. And Avid will never drop frames during capture.

The Avid Meridien video board uses a method of compression called JFIF, which is a variation of motion JPEG. This compression scheme is based on the Rolling Q method for an image that may change in quality while keeping approximately the same amount of compression from shot to shot. This tends to ensure an overall higher data rate rather than an average, lower rate for Fixed Q. Although this high data rate will clearly improve the quality of complex images, it is also effective for improving the quality of simple images. Rolling Q does not try to compress a simple image more than a complex image. However, a simple image may not require that high amount of data and, rather than fill up the frame with noise or interpolated information, those images will use less data than the advertised fixed rate. This is why even with the new JFIF resolutions on the Meridien board you may see different frame sizes with the same amount of compression. The MPEG-2, I-frame compression method used in IMX is similar to Rolling Q, because each frame is compressed based on the qualities of that frame. This means that an average compression will vary but it will always be the best compression rate for each individual frame.

What is the drawback to a fixed data rate? It takes more storage space and requires more sustained throughput. But with the plummeting price of high speed workstations and storage making such financial decisions less important, it seems like a small price to pay. Always taking up the same amount of storage space allows the user to better estimate the total amount of drives required for a project. Expect uncompressed resolution to take up more than a gigabyte of storage for one minute of material (around 675 kilobytes per frame, or kpf), and so expect Avid's 2:1 two field resolution to take up about half that (300 kpf). This means that with a 90-gigabyte drive array you can expect to get about 76 minutes of uncompressed storage and about twice that with 2:1 compression.

There are three categories for captured images: uncompressed, lossless compression, and lossy compression. All compressed images on Avid systems are lossy, where redundant information is thrown away during capturing. As you move closer and closer to uncompressed quality, you pay a higher cost for hardware and disk space. You must carry over every pixel of every frame, no matter how redundant that pixel is. Uncompressed images demand faster computers, wider bandwidth, and much more disk space on faster, striped drives.

Lossless compression is many times touted as better than uncompressed (or noncompressed, as some insist) because it takes less disk space. Lossless compression is associated more with programs like Aladdin's Stuffit or WinZip for compressing documents before posting them on the Web for downloading. The difference when compressing something variable, like a moving video shot, is that as the image gets more complex, the compression is less effective. Potentially, under a wide range of circumstances, a lossless compressed image could be larger than the equivalent uncompressed image (compression information and the less compressible image are added together). If the editing system is designed to take advantage of a low bandwidth as a benefit of smaller file sizes, you may have some playback problems. The system may impose a rollback, where a maximum frame size is imposed by throwing away information (lossy) when the frame size gets too big. If they don't do this, they must prepare to handle even larger frame sizes than the uncompressed system and lose much of the benefit of lossless compression.

Only time will tell which technology eventually becomes the accepted quality, but the ironic situation at present is that although uncompressed is the goal for nonlinear editing, only two of the standard-definition digital tape formats use a component signal without compression. These formats, D1 and D5, are not for those on a low or even medium budget.

The reality of compression is that there are all sorts of tricks to make any method improve. Although uncompressed was a major goal of many productions, as producers saw the quality of 2:1 (actually, *didn't* see the difference with 2:1) they realized that it was good enough for most productions. Keep in mind that everything broadcast today goes through another level of extreme compression before transmission. The main reason to use uncompressed images is for archival purposes since there may be a future compression method that works better if beginning with no compression at all. These uncompressed standard definition images will most likely be upconverted to HD at some point in the near future and any compression may be visible in context at that time.

DELETION OF PRECOMPUTES

Predicting available space on media drives must go hand-in-hand with keeping track of rendered effects, or precomputes. Imported graphics and animation also take up space. Every time you render an effect, it creates a media file on the drive. Even though you may cause that effect to become unrendered or delete it from the sequence, that file still lurks on your media drive. This is actually for a very good reason, for both undos and for all the multiple versions of that sequence, but it means you need to pay attention to how full a drive has become even though no one has captured to it that day.

Deletion of precomputes is one of the most important things an Avid administrator can oversee. One of the most common calls to Customer Support is when, during a session, a system grinds to a halt because it is too full of thousands of tiny rendered effects. Are all those effects necessary? Probably not, and now the editor or the assistant must be walked through the process of deciding what can go. One of your most important responsibilities in making sure that sessions start and end smoothly is to keep an eye on how many precomputes are on the system and how many are really important. The Avid editing systems do not keep track of how many sequences are created during a project. There could be multiple CPUs accessing the same media or archived sequences that are modified on another system and brought back. Since the system is so flexible, there is no way the system could definitively know the number of sequences created. What if the system were to make very important decisions for you, like deleting "unneeded" rendered effects? What would happen if you called up a sequence you had spent hours rendering to find that the software had neatly deleted that media automatically, thinking you were done with it? Just because you deleted the effects sequence from sequence version 15 doesn't mean you don't want effects on sequence versions 1 through 14! There are too many variables, and this decision is too important to leave up to an automated function at this stage in the technology.

That being said, there is, in fact, some auto-deletion of precomputes going on under your very nose! But, as it should be with all automatic functions that cause you to lose things, it is very conservative and you may not even notice. The only auto-deletion of precomputes occurs when you are making creative decisions quickly and removing or changing effects. If you are rendering the effects one at a time and then quickly deleting them, there is at no time any opportunity for the system to save the sequence with those effects in place. There is no record that they will be needed in the future because they have not been saved. Saving happens automatically at regular intervals and, when you render effects in a bunch, they are saved as the last step before allowing you to continue. Every time you close a bin, you also force a save of the contents. That is why the software auto-deletes precomputes only when you are rendering one at a time and quickly removing or unrendering them by changing and tweaking. If a save occurs while an effect is in a sequence, the precompute is not deleted automatically. Every little bit helps to keep the drives unclogged, but you still must evaluate the amount of precomputes and delete them on a regular basis. This is really not so hard even though it is a little intimidating at first because it involves deleting material that someone (probably the editor) may need if you get it wrong. This chapter will deal with the isolation of precomputes when we look at efficient deleting strategies.

THE IMPORTANCE OF EMPTY SPACE

Remember the Media Database? Any file that keeps track of all the media is a concern if you overfill your drives. That file must be allowed to enlarge to deal with the many files you add during the course of editing. If there is just not enough room for the media database file to update and grow larger, you may have media file corruption and eventually drive failure. A good rule of thumb is to leave at least 5% to 10% of each partition empty. This limit is flexible, but if you detect slow performance in the form of more underruns or dropped frames then consider moving media to an emptier drive. It is also a good idea to erase media drives completely after a job has been completed. Don't get initializing a drive confused with low-level reformatting! That is only a very last resort to save a dying drive. However, reformatting causes a drive to lock out any bad sectors. Those same bad sectors that have may have been giving you problems will be eliminated from future problems

BETWEEN SESSIONS

If there are multiple shifts using the same system, then the changeover period may be as simple as shutting down, turning some drives off, turning others on, and booting back up. You want to minimize the actual disconnecting and reconnecting of SCSI cables (they are very fragile and subject to data errors if twisted). If you find yourself moving drives very often then you may want to change to removable drives or a shared media solution like Avid Unity MediaNetwork.

If the media is moving to another system, consider putting the project files on the same drives with the media. The project can be picked up and continued wherever the drives go. Remember that drives used for transport or temporary storage don't need to be fast enough to play the media. They only need to be big enough to hold it all and then copy it to the destination system.

There will come a time when a project must be moved from one system or facility to another. It may be important to keep only part of a job and free up space for something new. These two situations require a better understanding of the sophisticated media management tools of the Avid editing systems and a few more steps on a regular basis to keep things running smoothly.

CONSOLIDATE

The best tool you have for moving media while inside the software is Consolidate. Consolidate can be used for two main purposes:

- Moving media from multiple drives to one drive
- Eliminating unneeded material

Take whatever you need to move, either the media relating to an entire bin or a finished sequence, and consolidate it to another drive.

There are some choices when you consolidate that may make things less confusing. First, you need enough free space on your drives equal to the amount of material you wish to consolidate.

You are able to specify a number of drives for consolidation. Using the list of drives means that, even if you run out of space on one drive, the next drive in line will take the overflow material. The second drive will take the material until it is almost filled, then the third drive, and so on until the sequence is finished consolidating.

These are the two major reasons to use the Consolidate function, although as a quick troubleshooting tip you may choose to consolidate a clip that is not playing back correctly. If the clip plays back better after consolidating to another drive or partition, you may have a drive problem or you may have captured the clip to a drive that was too slow to play back that resolution.

Consolidating a Sequence

The beauty of Consolidate is the advanced way that it looks at everything that is needed in a sequence and copies only that. There are, after all, other ways to copy media, but it is very difficult to tell at the operating system level what clips are really necessary to play a sequence. The Consolidate function will search all your drives for you, gather only the bits you need, and then copy them to the desired drive.

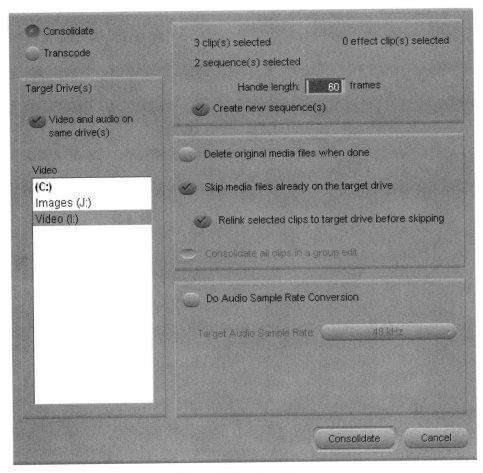

Figure 4.3 Consolidate

Consolidating will break the sequence into new individual master clips and copy just the material required for the sequence to play. This creates shorter versions of original master clips because you are copying only the bit that is needed. You can then selectively delete unused media.

Consolidate is especially important at the end of projects when the final sequence has been completed and it is time to back up. Instead of backing up all the media, you consolidate first and back up only the amount of media that was actually used. It is also useful before sending audio to be sweetened on a digital audio workstation. Delete the video track from a copy of the finished sequence, then consolidate this sequence and make sure the audio media files go to a removable drive.

After you select the sequence and consolidate it, the system looks at the original master clips to determine exactly what is necessary. If you have an original master clip that is five minutes long, but you used only 10 seconds of it,

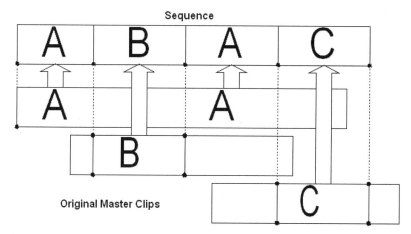

Figure 4.4 Consolidate Chart 1: Original Master Clips Link to the Sequence

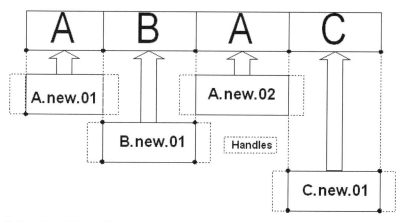

Figure 4.5 Consolidate Chart 2: New Consolidated Master Clips with Handles Link to the Original Sequence

then only that 10 seconds will be copied. The new 10-second-long master clip will have the original name and a ".new.01" if it is the first time in the sequence that shot is used. If you use another five seconds of the same original master clip, then you will have a second consolidated clip with ".new.02" and so forth. The Consolidate process also allows you to specify handles, or a little extra at the beginning and end of the clip, so you can make some little trims or add a short dissolve later. If the handles are too long, you will throw off your space calculations, so be careful; however, if the handles cause two clips to overlap material, then Consolidate will combine the two clips into one new clip rather than copy the media twice.

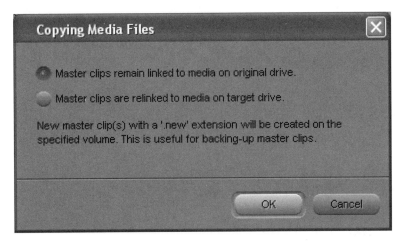

Figure 4.6   Copying Master Clips with Consolidate

Consolidating Master Clips

If you want to move media from multiple drives to a single drive, you can consolidate master clips. This will take all of the media linked to the clips in a bin and copy them in their entirety to another drive. Notice the difference between move and copy. You are really just copying and then must decide whether or not to delete the original. The copy and then the delete are the two steps that make up move.

Consolidating master clips is a fantastic method for being able to clear off multiple drives and put everything from one project onto a single drive or onto a series of drives so you can remove them or back them up. Consolidating master clips does not shorten material. It is just convenient and takes a complicated media management task and makes it one step. Be careful not to overfill any one partition.

If the idea is to move media from multiple drives to one target drive, then you must ask yourself, "Have I used this target drive before for capturing this project or a previous consolidating?" If the answer is yes, then you may already have some of the material you need on the destination drive. You don't need to copy the media twice! Be sure to check the option Skip media files already on the target disk. You would use this only if you were consolidating master clips and not sequences. Using the media already on the drives is perfectly fine. When the Consolidate function finds the long, original media file already on the target drive, it will leave it alone, untouched. This is the best setting for moving media from multiple drives to a single selected drive.

There is a secondary choice that becomes available to you only when you check Skip media files already on the target disk, and that is the somewhat confusing Relink selected clips to target disk before skipping. You want to make sure that when you skip the media that is already on the drive that the consolidated master clips are linked to it. When the system makes consolidated masterclips that link to existing media it doesn't add a .new to the end of the masterclip name. It will add a .old to the masterclip in the original bin, however, since it needs to distinguish between the two clips.

Consolidate Summary

If you are consolidating sequences:

- Choose the drive(s).
- Determine the handle length.
- Uncheck Skip media already on the target drive.

If you want to delete the original media after you have created the shorter consolidated clips, then check Delete original media files when done. If you are feeling prudent, go back and delete the original media in the Media Tool after consolidation. This is especially important if you are consolidating multiple sequences and they all share media. Do not delete the original files until you have consolidated all the sequences that use the media! If you are working with multicam, you should choose to consolidate all the clips in the group if you want to continue to have all the camera angles available in the consolidated sequence.

If you are consolidating master clips:

- Choose the drive(s).
- Choose Skip media already on the target drive.
- Choose Relink selected clips to target drive before skipping. Do not choose this if you are continuing to work on the project and someone else will take the consolidated media away and work on it elsewhere.

You probably do not want to delete original media here and will proceed with more sophisticated media management later. If you do not delete your media now, you will have a mix of .old with original clips on the original drive and .new with the new clips on the target drive.

Fixing Capture Mistakes with Consolidate

Let's take the example where you have captured too many tracks. Perhaps you (or someone just like you) weren't paying attention when you were capturing and you captured a video track with a voice-over master clip. Delete the unneeded video and keep the audio through consolidate. You can take this master clip and make a subclip of the entire length, making sure that you turn off the tracks that you don't want. If you have captured voice-over and accidentally captured (black) video, turn off the video track while it is in the Source window before you subclip the master clip. The subclip will be audio only. Highlight that audio subclip in the bin and choose Consolidate under the Clip menu (or right-mouse/Shift-Ctl-click). The Consolidate function will copy only the media you want to keep. You will have a new subclip and a new master clip with only the audio tracks and can delete the original master clip to free up disk space.

There is even a faster method to do this particular technique. You can find the clip in the Media Tool and delete just the media file that you don't need. You are given a choice of whether to delete the audio or video media, so for the video with voice-over problems, pick the video media for deletion. Then unlink the clip and choose Modify. Change the masterclip to reflect that it is now just audio. You will not be able to change the number of tracks of the master clip unless you unlink first. Then relink the clip back to the audio media. This method also ensures that you don't end up with strange media management problems down the line when batch capturing or restoring from an archive.

Subclipping Strategy with Capture

You can also use the subclipping then consolidating method just to shorten your master clips after you decide what part of them you really need for the project. In fact, many people like this method as a general strategy and capture master clips that are quite long, maybe an entire scene or the full length of an already edited master tape. This creates fewer files on your drives for the computer to keep track of, fewer objects for an object-oriented program, and can speed up performance. Then subclip all the sections you will actually use, consolidate the subclips to create new master clips, and delete the rest.

USING THE OPERATING SYSTEM FOR COPYING

If your goal is to move an entire project to another drive, you may be better off working at the desktop level. If you have been using MediaMover (www.randomvideo.com), your job is a snap. MediaMover will search all your media drives, find all the media from a specific project, and then move that media into a folder with the project name. Copy the entire folder with the name of your project from each of the affected media drives. It is easy to do a Find File (Windows-F on Windows or Cmd-F on Mac) and copy every folder found with the project name if you have a dozen drives.

The situation may be complicated if there are two different resolutions to keep track of in this project. You may want to copy only the low-resolution material and leave the high resolution on the drives or vice versa. Planning helps here and, if you are on a Macintosh, I recommend that you use the Labels function provided by the Macintosh operating system [this is for OS 9.x only]. If you go to the Control Panel for labels, you can change what the colors represent. Change the blue label to Project X Low Res and the green label to Project X High Res. Select all media files immediately after you have captured them and change their label color. You can sort the media files by label color and copy or delete only the files you want. This also allows you to track down those stray media files that always manage to escape even the most careful herding.

Whether you choose to use Consolidate or Copy on the Finder level, you need to keep track of all media. It must be searched for, backed up, copied, or deleted. The number of objects on the system should be checked periodically, and unneeded rendered effects or precomputes must be deleted on a routine basis. If your drives fill up, the session stops.

DECIDING WHAT TO DELETE

A place where an administrator or an assistant must be very careful and yet very efficient is the deletion of unneeded material. Sometimes this can be agreed upon mutually with the editor and you can eliminate all of the material for Show 1 when you are well under way with Show 2. Many times different projects share the same material and you must be careful not to delete that which is needed by both. The Media Tool and MediaMover can both be used efficiently project by project, but this may not be good enough. You must find some other criteria to sort or sift by, protect certain shots, or change the project name of the material you want to keep.

USING CREATION DATE

Creation date becomes a very important criterion to look for individual shots, and it is often overlooked by many assistants. It does not work like the modified date on the desktop, which updates to reflect the last time someone opened and modified the file. The creation date is stamped on the clip when it is logged, so if the shot is logged and captured on the same day, this becomes a useful heading to eliminate material that was captured at the beginning of a project.

I especially like to use creation date as a heading in my sequence bin. I duplicate my sequence whenever I am at a major turning point or even if I am going to step away for lunch, dinner, or elevenses. When I duplicate the sequence, creation date time-stamps it so I know that I am working on the latest version. Then I can take the older sequence and put it away in an archived sequence bin (or several archive bins as the bins get too big). I have control over the exact times that I have stored a version, instead of leaving it up to the auto-save, and I can keep the amount of sequences in any open bin to a minimum.

Sequences will be the largest files you will work with. In an effort to have the bins open and close quickly and not use up too much RAM, I try to keep the bins small and keep only the latest version of any sequence.

If you are trying to determine which sequence is the latest version and the editor is not present, creation date is the best tool. There is always the chance that the editor duplicated the sequence and continued to work on the old one, but you should have an agreement with the editor about how to determine this crucial fact.

CUSTOM COLUMNS

There are many other criteria you can use for sorting and sifting. If you plan it well, you can create custom columns to give you an extra tool to work with. Some people will create a custom column with an X or some other marker to show whether a shot has been used. One way to use sifting powers of the bin is to change the sift criteria to "match exactly" and have it search for a blank space in a custom column. This way, any shot that has *not* been marked is called up. The Media Tool is the only way to search for media that is online across bins. You can create a Media Management bin view, which can be used in the Media Tool. Your custom columns will show up there, too.

BASIC MEDIA DELETION USING MEDIA RELATIVES

Another way to find out if a shot has been used within a project or sequence is to use the Find Media Relatives menu choice that is in every bin. This is the best way to search across bins to clear unneeded shots, other than consolidating and deleting the old media.

     The most useful way to use media relatives is with sequences.

1. Open the Media Tool.
    You have a choice whenever you open up the Media Tool, to show master clips, precomputes, or individual media files for all the projects on the drive or just for this project.
2. Show master clips and precomputes for this step.
    I like to show "All Projects" when checking for media relatives because many times the bins I am working with came from another project.
3. Open all your relevant sequence bins for this method to be accurate.
4. Put all the sequences that you want to work with into one bin.
5. Select all the sequences and then choose Find Media Relatives from that sequence bin.
    The system searches all open bins and the Media Tool, and highlights all the master clips, subclips, and precomputes that you will need to keep.
6. Go back to the Media Tool and choose Reverse Selection in the pull-down menu.
7. After reversing the selection to highlight all the unused media, press Delete.

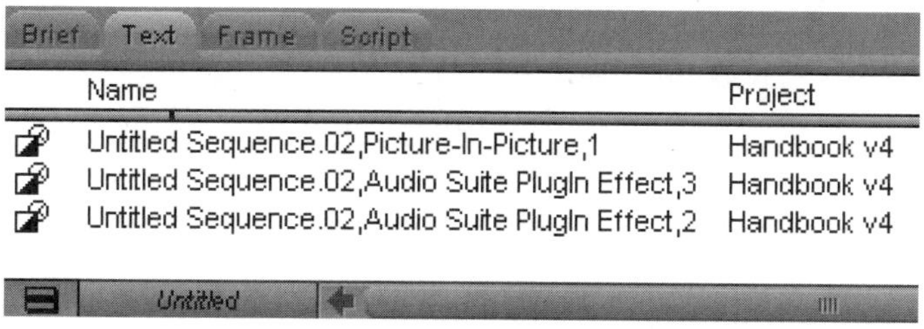

	Name	Project
📄	Untitled Sequence.02,Picture-In-Picture,1	Handbook v4
📄	Untitled Sequence.02,Audio Suite PlugIn Effect,3	Handbook v4
📄	Untitled Sequence.02,Audio Suite PlugIn Effect,2	Handbook v4

Brief   Text   Frame   Script

Untitled

Figure 4.7   Showing Precomputes as Media Relatives

Keep in mind that Find Media Relatives does not unhighlight something, so before you begin this operation, make sure nothing in the Media Tool is already selected. Also, give everything selected for deletion one final look. This is the last time you will see these files, and you don't want to trust anything or anyone except yourself at this stage—there is no undo. Don't do this when you are tired! Not all of us are "Morning" people!

You can choose to show only precomputes in the Media Tool. You will see all the rendered effects for a project with all the effects on your drives. You can use the previous method, Find Media Relatives, to track down the precomputes used for the sequences you've selected and then, remembering to reverse the selection, delete the unneeded precomputes. This process can be done every day if you are working on intensive effects sequences, like in a promotions department. It doesn't need to be done unless you need the space or memory requirements are climbing into the hundreds of thousands of objects. There is always the chance you might delete something you want to keep, so be careful.

Occasionally you will want to find exactly where a specific file is and for some reason, Consolidate, MediaMover, or the Media Tool are not sufficient to perform the task you need. You can use a function called Reveal File, which will go to the desktop level and highlight the media file associated with a specific master clip.

LOCK ITEMS IN BIN

If I have achieved my goal of making you slightly paranoid about deleting material during a project, then you will approach this task with the proper amount of stress. There is a great way to relieve some of this stress that does not require a doctor's prescription. Many years ago I spent a week at a major network broadcast facility observing their operations. They needed someone who knew nothing about the individual projects being edited to come in late in the day and clear drive space for the next morning. On longer jobs, this person would be familiar with the exact needs of the editor. In this case, they were working on as many as five or more projects a day. At the end of the day, some material needed to stay—material that had been grabbed from their router and therefore had no timecode—but the majority of the material needed to be deleted. This is where the idea for Lock Items in Bin originated.

It was supposed to work like this: As the editors gather shots and use them, they decide that they need some shots for the continuing story tomorrow and clone the master clips (not the media). They Option-drag and clone the master clips into a bin named Stock Footage. At intervals during the day, the editor goes to the stock footage bin and selects all (Ctl/Cmd-A). The editor chooses Lock Bin Selection from the Clip pulldown menu (or right-mouse /Shift-Ctl-click). A lock icon appears in the Lock heading that is displayed as part of the view for that bin. The file is locked automatically on the desktop level so that if some intrepid assistant decides to throw everything away, they see a warning that certain items are locked and cannot be thrown away. Any good assistant, however, knows that on the Mac if you hold down the Option key when emptying the trash, you can throw almost anything away and cheerfully ignore these pesky warnings. On Windows, you need to go to the file, right-click the icon, and unlock the file under Properties. You can select many files simultaneously for locking or unlocking, but if you drag them to the Recycle bin you will get a warning that the file is read only. It may allow you to delete at that point, but since you are going to the Recycle bin you will get the chance to retrieve it. You can see whether a media file is locked on Windows by looking at the Properties to see if the file is read only.

A happy side effect of the Lock Items in Bin command is that it can also be applied to sequences. Subsequent copies of that sequence are also locked automatically when the sequence is duplicated. This helps avoid the all too common problem when people have mixed clips and sequences in the same bin and they decide to delete files. They select all and delete and then—Hey, where did my sequence go? Certain types of objects are selected automatically as choices to be deleted when you select all in a bin. The sequence is automatically *not* checked for deletion if it is mixed in a bin with clips. If a bin has only sequences then they will indeed all be checked for deletion. Deleting a sequence by accident was usually someone's first visit to the Attic, only to recall the version of the sequence they worked on 15 minutes ago!

There are ways to outsmart the Lock Items in Bin command if you are determined. You can duplicate the locked clip and then unlock the duplicate. Because both clips link to the same media, you can delete the media when you delete the second clip. You can also unlock the media at the desktop level and throw it away without even opening the editing application. But you can't delete the media until someone, somewhere, unlocks it. Using these methods is really the honor system for media management. If you need more security for files you may consider using Avid MediaManager with a Unity system. There is a set of permissions that must be granted to delete media for a particular project. When media is deleted MediaManager searches all uses of the media and if someone is still using it, the master clip will disappear from the project but the media will remain on the drives. When the last use of the media is deleted, then the media will be deleted.

LOCKING BINS

You can also lock an entire bin. This becomes activated only when sharing projects over a Unity system. An editor who is busy at work on an important sequence or is organizing a large and complicated project doesn't want another editor to start changing things in the bin; however, the other editor may be working on the same material and need access to the latest version of the sequence or the latest subclips and custom headings.

So what is a good compromise? The editor who owns the bins can decide if anyone will get write access or read-only access. Write access allows anyone to add things to the bin or make changes to *anything* in the bin (like duplicating, deleting, or trimming sequences). No other editor can get write access to the bin while it is being used by another editor with write access. The write access is displayed by the green unlock icon at the bottom of the bin.

Read-only access is the limited ability to work on the bin without making any permanent changes. The read-only status is displayed by a red lock icon at the bottom of the bin. This means you can view clips and, if you really want to make changes, you must copy the clips or sequences to another new bin. Now you are forced to work on your own separate version of anything. Any bin can be opened at any time by Alt-clicking to open the bin.

Restricting all changes would unfairly hobble fast-working professionals. This would mean you could view but not mark points or add locators. Flexibility was added so you can make some minor changes to the read-only bin, but the changes will not be saved when the bin is closed. There is a warning when you close the bin (and at the first auto-save) that your work will not be saved. If you have made changes in the read-only bin, cancel the bin closing and drag your work to another bin. This flexibility is a trade-off with stupidity. This short-term capability is a convenience, but it should be clearly stated to any inexperienced user the full consequences of working in a read-only bin as a standard procedure. Sure, do something just to look at it quickly, but if you like the change you must drag the work to another bin—you cannot change the original files without permission.

Obviously, it is important to give write access only to qualified people who have the responsibility to own up to the changes they make. Ronald Reagan said, "Trust, but verify," so there is a log of any user who has dirtied a bin. When a user dirties a bin, the asterisk appears next to the bin name in the header bar across the top of the bin. This is a hint that there are unsaved changes (on the Mac this is a diamond icon). The log is accessed by right-clicking the mouse on the colored lock icon at the bottom of the bin. To lock a bin from write access by other editors:

1. Select the bin or all the bins in the Project window.
2. Under the Clip window choose Lock Selected Bins.

The bin name will now have an asterisk next to the project name in the locking editor's bin. In all other editors' bins the bin name will appear in bold and name of the person who has the write access will appear in their project window next to the bin name. If you want write access to that bin, then give the owner a call and ask permission for them to release their control.

CHANGING THE PROJECT NAME

There is another way to organize shots for easy media management: Change a master clip's project name. There can be confusion about exactly which project is associated with which media. It is so easy to borrow clips from another project that you may not even know that certain shots are from another project unless you choose to show that information in your bin headings. This is why I create a custom bin view just for media management. It shows project name, lock status, creation date, tracks, video resolution, disk, and any other customized headings that relate to media management. If the clips have been borrowed from another project and brought into the new project, even if you recapture them, they retain their connection to the project in which they were logged.

Project affiliation makes a big difference when it comes to recapturing at a higher resolution. Because all the shots from one tape are captured at once, if you are trying to recapture from Tape 001 in Project X and the clip is really from Project Y, the software considers these two different tapes. While batch capturing, the system will ask for Tape 001 twice since it really should be two separate tapes if it was logged correctly. There are several steps to changing project affiliation. The first thing to keep in mind, as mentioned in Chapter 2, is that project name is part of the tape name. The trick to changing the project name is to change the tape name to a tape name from the right project. All tapes have a Project column in the Select Tapes window when assigning tape names. This allows you to see which project each tape is from.

In Capture mode, whenever you show the list of tapes that have been captured and you check Show other projects' tapes, you might see multiple Tape 001 from other projects. Use the Scan for Tapes button in recent versions to make sure the system has updated all the tapes captured and used on the system.

Here's the best way to change the project name of a master clip:

1. Open the new project.
2. Open the Media Tool and show the original project media files.

3. Consolidate the master clips into a new bin in the new project.
4. Do not skip files that are already on the drive.
5. Select all clips in the bin and choose Modify from the Clip menu.

If the tape name you want has not been used before in this project, then you can create a new tape name. You can choose an existing name if you made an earlier mistake and need to rename these clips so they all come from an existing tape in the new project. As you finish with the modification process, you get a series of warnings that are important if you are working with key numbers in a film project since the new clips need them reentered after you modify the tape name. A film project will not allow you to change a source name of a clip without unlinking first. If you are not using key numbers, just check OK.

Now the clips are associated with the new project as far as the Media Tool and the headings in the bin are concerned. The change has not really affected the actual media file at this stage, just the master clip in the project. This is not enough for MediaMover to recognize there has been a change because it looks only at the media file and not the project. The project information about a media file is actually recorded in the media object identification (MOB ID—it has nothing to do with gangsters). The MOB ID is attached to the media file when it is captured, and the media file itself must now be changed to update this information. Here is another use for Consolidate: You need the space on the drives to copy the files you want to change.

To change the project name of the media files, select all the clips and Consolidate with the following options checked:

- Have the old name link to the new media.
- Do not skip over the media if it is already on the drive.
- Delete the old media when you are done.

If you want two sets of the media—one set of media in one project and one set in the other—do not delete the original media after you consolidate.

BACKING UP MEDIA

Now that you have eliminated the unneeded material by deleting, whether it is captured media files or precomputes, you can evaluate the choices available for backing up. As very large media drives become a reality, there may be a room full of gigantic backup drives in your future. Clearly, this is the fastest way to copy media and, if it is done over a very fast network like Fiber Channel, those drives could be anywhere in the building or, eventually, the world. Perhaps there is a future business for someone to set up thousands of drives in a warehouse somewhere in the desert and have people rent out the space for backing up over the Internet.

But that day is a few years away and for now you probably have to deal with digital linear tapes (DLT) or another form of archival tape backup (digital audio tape [DAT], advanced intelligent tape [AIT], or linear tape open [LTO]). Until recently, 8 mm drives were considered along with DLTs, but the size of the 8 mm capacities has not grown as fast as the size of the DLTs, although 8 mm drives are cheaper. According to the advertisements for different kinds of backup products, it may seem that an 8 mm drive is more than adequate for your storage needs. Maybe you can pocket the extra money and use it for that industrial cappuccino machine. This is because most advertisements feature the amount of storage they are capable of holding when the data is compressed. But you have already compressed; you are working with Motion JPEG and Avid's compression schemes that have already squeezed the heck out of it. These files are not getting any smaller, which means you should always choose to avoid any further compression as an option during your backup session. Further compression takes extra time and adds a layer of unneeded complexity to already very large and complex files. You may be courting data corruption, and that's not a cheap date!

The linear part of tape backup is what most people object to when using them, plus the fact that they can be rather slow to back up and just as slow to restore. Just like hard drives, they, too, get faster all the time. Consider that you may need to back up hundreds of gigabytes or many terabytes. Almost anything is inadequate now, but this technology will continue to double in speed and capacity every 18 to 24 months.

Many people use DLT magnetic tape drive or the cheaper, slower DAT to archive material that has no timecode. If they were to delete and recapture or reimport nontimecoded material, they would have to reedit it in by eye. If the material is graphics, they may have many layers to reedit and potentially some very complex sequences to modify. I still recommend dubbing nontimecoded material to a timecoded format, but sometimes, especially for fast turnaround projects or lots of graphics, this is just not practical. You also lose the quality benefit of keeping the images completely digital and uncompressed until they are imported unless your timecoded tape format is also uncompressed (not likely). Dubbing nontimecoded material to tape with timecode is still the simpler and more reliable way to get material back on the drives and get it to link to your sequence. With batch import, reimporting and relinking graphics and animations is now a one-step process if they are in file form and not used as a tape source.

Many times it is actually faster to recapture a shot or two than rely on a slow and cumbersome restoration process, which is why people don't back up their media as much as their project information. The exception to this would include video that has been carefully color-corrected during capture using an external color corrector and the color correction information has not been saved. Also, there are many instances where the original tapes become unavailable because either they are sent to another facility or, working without a net, they are recorded over. Again, the question is not if your hard drives die—it is when. What is truly important is not how easy it is to back up, but how easy it is to restore.

The professionally paranoid know that something is really out to get them and you can never be too prepared. For people like this, it is usually a good idea to do a full backup of all material after the first day of capturing. From that point on, incremental or normal backups can be scheduled every night to store only the material that has been added throughout the day. You may even want to consider backing up over a fast network so that you can back up all the editing systems from a central location. With the size of storage on even one machine easily overwhelming the capabilities of a smaller capacity DLT, this may be just wishful thinking. You may need a DLT for each of your important systems. Figure out how much it would cost you to lose that information for even four hours and then compare it to the cost of a DLT.

The only reason that a DLT or a DAT is not a foolproof way to back up media files for archival purposes is that over time, there may be a significant change in the structure of media files. This may occur because of a software change, as in versions Media Composer 7.x, when the structure for the media files changed from MFM to MSM to be compatible with NewsCutter and the EditCam. In the process the media became OMFI native. MSM ensures that all media files are server-ready, no matter which edit system you are using. If you upgrade to that version of the software, then all your media files must be either converted to the new format or recaptured. Avid made the switch from the NuVista + media, which was captured at 640×480 square pixels to the ABVB 720×486 non-square pixels. No pixel-based images are improved by expanding and distorting. The way to take advantage of the new technology was to fall back to the old technology and recapture from tape. In the short run, or if you are not planning to upgrade any time soon, the DLT is your best answer. A DLT will allow you to keep entire projects, with all the rendered precomputes and nontimecoded material, all in one easy-to-restore package that takes up very little shelf space. Avid has committed to moving all of its editing systems over to using AAF as metadata and MXF (Material Exchange Format) for media. Avid systems should be able to play both the OMF and MXF media into the near future, but at some point the older format, OMF, will no longer be supported natively. There probably will be a method to convert or import the old media and change it into the new format, but that conversion is usually time consuming for large amounts of media. It will probably be faster to recapture from original sources and, since the compression schemes will probably be better, this will result in a better looking image. Backing up media should be focused on quick retrieval in a near-to mid-term reuse, not long-term archiving.

ARCHIVE TO VIDEOTAPE

If you have a digital tape deck like a Sony Digital betacam you may want to consider backing up your source material to videotape using the Archive to Tape function under the Clip menu. This function on Xpress and above allows you to select a sequence and play out all of the media that was captured originally through the Capture Tool. This is an excellent way to store a project for a few days if you have run out of room on your hard drives. Be aware that digital betacam has excellent compression but is not completely lossless. If you do this several times to the same media you will begin to see some artifacts. This would be perfectly acceptable for offline quality media though.

Archive to Tape is also an excellent way to make a selects reel for a project where the sources were lots of different formats. By archiving to videotape you take a project that was on DV, Hi-8, VHS, or U-Matic and put everything on a nice neat digital betacam for future restoration. DS Nitris also has an archiving function that saves all the rendered files (caches) as well as the imported graphics, but only to one tape. If you have imported material you will have to back up those original files along with the project data to another format like CD-ROM or DVD-RAM.

BACKING UP USER SETTINGS

You should be able to reproduce User Settings in about five minutes, but some people resist it like a trip to the dentist. There is no way to lock your User Settings to keep unauthorized folks from making a few "improvements" or just accidentally changing them, so it is a good idea to back them up somewhere safe and hidden. Many freelancers carry their User Settings on a floppy or a USB memory device (sometimes called a SANdisk or thumb drive) that can hold over 256 megabytes of data. Other more intrepid instructors or field support technicians carry their own portable hard drive containing everything they could possibly need.

When software versions change, especially substantial version changes, you should always remake your User Settings. I know this sounds tedious, but recreating them solves many unusual and unpredictable problems, especially if your customized keyboard is complex. You may be trying to access menus that have been moved or deleted! Or there may be more subtle problems that don't immediately appear to relate to User Settings. One of the first things Customer Support asks you to do if you are getting unusual behavior, like a common feature suddenly not working, is to create a new project or User Settings. If the software version has not changed recently, having a backup of your User Settings may be enough to fix the problem, rather than having to recreate them from scratch.

NETWORKS

Any good administrator must evaluate how a network can improve their operation. Almost everyone should be using some form of a network today so plan your investment accordingly. One of the smartest broadcast engineers I ever met told me he was studying networks at night school, and that was in 1995. Just as "plastics" was the secret answer in the 1960s, networks may be the not-so-secret answer in the next decade. Networks, as they exist today, are relatively slow, but just like hard drives and DLTs, they are getting faster all the time.

As you may suspect, the more money you invest, the faster you can go. Consider Ethernet. It may not be the fastest network choice out there, but it sure is easy. All systems currently shipping from Avid are Ethernet ready.

Ethernet can easily be expanded by the addition of an Ethernet hub or upgraded to gigabit Ethernet with a PCI card. For very little money you can tie all your editing and logging systems together on one local network. Almost no one still uses plain vanilla Ethernet, technically known as 10 Base-T, although it is still adequate to share project information and small graphics. You can use it to send the occasional media file, but there are better options for that (we will cover those later in this chapter).

The next step up is 100 Base-T Ethernet or 1000 Base-T (also called gigabit Ethernet) and, as you have probably guessed, it is faster and uses a thicker cable for the connections. It also requires a small addition to your CPU on older models, so you might need an extra slot inside your computer for a circuit board that will allow you to accelerate the Ethernet. The more recent CPUs are shipping with faster Ethernet already installed. This allows you to consider larger graphics, captured audio, and the occasional video media file sent over the network. It is also a good way to back up just the contents of the internal hard drives with project, bin, and script information every night. Many people use this to tie into their digital audio workstation. If you can, you should skip over 100 Base-T and upgrade straight to gigabit Ethernet.

FIBER CHANNEL

A fiber channel-arbitrated loop is so fast that you will find yourself not copying files across this network, but actually using media files from a central group of fiber channel drives that everyone shares. Fiber channel makes it possible to access large media files at speeds that will equal or exceed the amount of time necessary to play these shots back in real time. There is no server required, no IDs, 128 devices per loop, and up to 30 meters of copper cable between drives or workstations. Optical cable will go much farther. Fiber channel is still not quite as fast as the fastest SCSI connections, but more flexible.

SERVERS

Every administrator should consider setting up a server. Simply, a server is a basic stand-alone computer fast enough to handle large transfers of data. It is connected as a central storage place for resources important to everyone working on the project. It may be a place to store the final version of any graphic, logo, or script, so the chances of someone using an older or unapproved version of something are reduced. Many times people end up using the wrong version of something because it was changed on one system and not copied to all the other systems. A good policy to enforce and maintain a level of version control compliance should exist.

On one project, we were so concerned about individual editors changing the tape-named bins that everyone was asked (repeatedly) to capture and edit from bins that were on the server, but not on their system. In other words, they would access all bins over the Ethernet network. The server icon would appear on their desktop like another drive, and they could open any bin stored there just as easily as any bin actually on their own system. If editors wanted to customize or modify a bin, they were forced to copy the bin to their individual system. From that time forward, it was their own responsibility to keep track of that bin and name it for themselves. This maintained a pristine copy of every original bin in the project in a place where all editors could get it at any time.

A server is also a good place from which to base a facility-wide backup or just to connect a Zip drive or CD-RW to back up central bins every so often without disturbing anyone else. It is also a good place to connect a printer for everyone to use. An assistant can be printing out bins nonstop and never interrupt any editing or logging and the server conceivably could be used as a logging station in a pinch. The server can be a good place to do database searches with a program if you have been exporting bins as text to a database program and MediaLog or to make EDLs with EDL Manager. You don't need a computer full of fancy circuit boards to open a bin! Besides, what else are you going to do with that old G3 or PIII?

AVID UNITY MEDIA NETWORK

Avid has developed a new way to share media and metadata (projects, bins, sequences, etc.) that takes advantage of the speed of fiber channel and a standard server. Avid found that going through a standard server configuration was too slow to get multiple users sharing the same media at the same time. Many shared storage solutions share drives, but Unity really shares media. It means multiple people can be using the same frame of the same master clip at the same time. To do this they created their own file system and moved the server out of the way between the editor and the media. The editor connects to a fiber channel switch and then directly to the drives. This eliminates the most serious bottlenecks but still preserves a server in the network to track all the usage and serve as the system administrator. With the continuing speed of fiber channel switches and drives growing quite quickly, it is only a matter of time before we are sharing high-definition files across a Unity Media Network.

With Avid Unity you can continue to add more drives to the network without disrupting the existing configuration. An administrator can change the size of any partition at any time without affecting the other projects on the drives. This means that you can get a much more efficient use from all of your storage. How many times have you had one room with extra drive space and another room that needs it right now? How many times has a project moved from one suite to another? Imagine if that changeover was as easy as assigning permissions to another editor for the same media. So even if you do not need to have multiple editors using the same media at the same time, the ease of assigning drive space to a project or moving a project instantly from suite to suite makes Unity a huge advantage to a busy facility.

The idea of sharing the media and metadata will eventually change post-production more than any other single aspect of nonlinear technology. Having editors working in parallel with sound designers and graphic artists will improve efficiency, collaboration, and allow facilities to offer new services. That is just the beginning. Systems that can be configured as render stations, capture stations, view and approve stations, and compression and DVD mastering stations are happening now. Any new facility should design all the workflow around central storage and begin to really get more value from owning multiple systems.

SNEAKER NET

Finally, the last, cheapest, and worst choice for networking: the Sneaker Net. If you have invested in the removable drives, this is not so bad. You just shuttle the media from room to room by hand (sneakers are optional). It means stopping work on two systems until the transfer is complete. If the goal is just to take everything to another location, it is simple and easy. Just don't trip! (Shoelaces should be tied!) If possible, don't fall into the common practice of disconnecting standard, fixed drives and moving them from room to room. You may get away with this for a while, but eventually you end up with bent pins, kinked cables, and a few error messages. There are some extra complications with Windows NT 4.0 and striped drives, so carefully follow the Disk Mounter procedures outlined in Chapter 13 if you are still on NT. Make sure the person doing the Sneaker Net transfer knows everything there is to know about SCSI configurations; this may end up being you!

RELINKING

All master clips, subclips, sequences, and graphics must link to media in order to play. When you log a shot into a bin, you create a text file. When you capture the shot, you create a media file on the media drive. The relationship between the master clip and the media file is considered to be a link. If the master clip becomes unlinked from the media file, it is considered offline. The media may be on the media drive, but if the master clip is unlinked, it is considered offline and unavailable. In order to link, you must highlight the clip and choose Relink from the Clip pulldown menu. The ability to link, relink, and unlink is very important to sophisticated media management. We will revisit these concepts over and over again in this handbook. Following are 10 rules that govern this complex and confusing behavior.

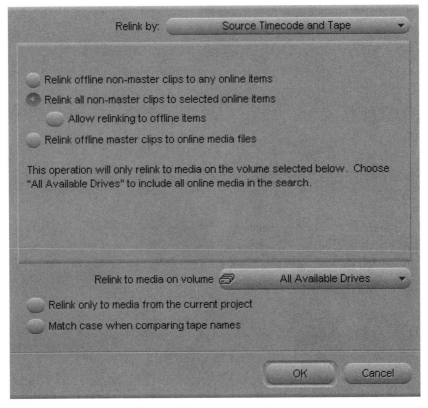

Figure 4.8 Relink to Selected

Rule 1 — Tape name and timecode

All linking between a master clip and a media file is based on identical tape name and timecode with media in the 5.x, 6.x, or OMFI MediaFiles and the MXF MediaFiles name of the future. Symphony Universal has a relink by key number for picture only, which is excellent for linking media from a new film transfer to a finished sequence. There will also be a choice in all models to relink based on resolution. This will allow you to force master clips and sequences to link to media based on the highest quality resolution available so that you can instantly go from low resolution to high resolution without deleting or hiding media.

If you have media offline that you know is on a connected drive you can simply Refresh Media Directories under the File menu. If this doesn't work then you will need to try relink. You can relink media when you have a master clip, subclip, sequence, or graphic that is offline. Highlight the object in the bin and choose Relink from the Clip menu. Your system searches the active MediaFiles folder for media that uses the same tape name and timecode.

Rule 2 — Tape name and project name

Just because the tape name looks the same doesn't mean it is the same tape. There can be only one tape logged per project with a particular name. Every time you add a tape as New in the Tape Name dialog, you are creating a unique tape that is associated with that project. You can use two different Tape 001s in the same project, but they are logged in separate projects and are always considered different tapes.

The project name is displayed as a column next to the tape name in the Tape Name dialog window. You can choose to show other project's tapes if you think the tape you need is in another project. To reduce confusion, all tapes for a project should be logged in the same project or in another project with the exact same name.

Having said all that, there is a way to bypass project name. In the Relink dialog, uncheck the Relink only to media from the current project. Then only the tape name is important and not the project. You can choose whether the media is from the same project and you can make sure that you will match the case of the tape name. The main use for unchecking the case of the tape name is when you are trying to link to an EDL that has been imported. You can link to these clips if the tape name is the same, but most EDL formats change the original Avid name for the tape to something with all capital letters. An EDL may also truncate a long tape name, which is a good reason to use short or numerical tape names, as recommended earlier. If these two choices are checked, then the first time you try to relink nothing may happen; however, if you are relinking to a terabyte or two of media, you will appreciate these choices when working with the tape names of dozens of projects online. Uncheck the choices and try the relink again.

Rule 3 — Size does matter

A master clip cannot link to captured media that is more than a few frames different from the master clip's start and end times, even though it has common timecode and tape name. For various reasons, this rule became looser in later versions of the software, but only by a few frames.

Rule 4 — Subclips are less choosy

A subclip will relink to media that is longer than the subclip. This is true even if the subclip is exactly the same length as a master clip that will not relink. Subclips are programmed to link to more media than the subclip start and end times.

Rule 5 — Sequences are really subclips

A sequence may relink when the individual master clips that are contained in it will not. Think of a sequence as many subclips.

Rule 6 — Multipart files make things more complicated

A sequence or a subclip will not relink to a media file that is shorter than the media it needs unless it is a multipart file. In recent versions, master clips were able to have multiple media files because they were split among several drives during the capturing. This new feature allowed users to avoid the Macintosh two-gigabyte file size limit and use disk space more efficiently. It also made relinking a bit more complicated. If only one of several media files links to the sequence or subclip, then there will be black in place of the missing media.

Rule 7 — Relinking master clips is different than relinking sequences

In the Relink dialog, the system will gray out inappropriate choices. If a master clip and a sequence are both selected, then you must choose which one you really want to relink.

Rule 8 — Relinking a sequence to selected master clips works only in the same bin

In the Relink dialog box, the Relink all non-master clips to selected online items button can be used only when you have highlighted specific online master clips that you want to relink to a sequence. In older versions, this was called Relink to Selected. (See Fig. 4.9.) There is no way to unlink a sequence, only forcing it to relink to other media, taking the original media offline or using the decompose function (discussed in Chapter 9).

This Relink to selected online items option works only within a single bin. It is used primarily to force a sequence to relink to media at another resolution or from another project. To relink to clips from the Media Tool, the clips must first be dragged from the Media Tool into the same bin with the sequence. Everything in the bin to be relinked must be highlighted. Do not check this function unless you are specifically linking To Selected, or nothing will happen.

Rule 9 — Media databases affect linking

If all of your efforts fail to produce a relinking, make sure you are working with the version of the bin that was actually captured (i.e., the same project). With the latest version you can uncheck Relink only to media from the current project and try again. If this still does not work, you will have to rebuild the Media databases and desktop file [Mac OS 9 only]. Your drives may be too full to rebuild the Media databases correctly, or you may have too many small objects and not enough RAM. You should delete unneeded precomputes or media to create space and force the Media databases to rebuild.

Rule 10 — Media management demands updated Media databases

When using any of the recent versions of the software, be very careful when copying files at the desktop level. Media databases are called msmOMFI.mdb and msmMac.pmr files on Mac, or msmFMID and msmMMOB on Windows. They must be the most recent files in the OMFI MediaFiles folder. If they are not, you may have to force them to update.

Choose the Refresh Media Directories command under the Avid File menu to force the system to look at all the media in all the MediaFiles folders and to update the database if necessary. Opening the Media Tool will do the same thing. If this still does not work, you may have corrupted databases and must throw them away. The system automatically will make new ones when you relaunch the editing software or click back onto an open window in the Avid software.

Although you needed to force the rebuilding of the media databases occasionally under the earlier releases, you will need to do it more often now — basically every time you copy something from one drive to another. This is because in a shared media environment (like a SAN or Avid Unity) your system would stop and rebuild the database every time someone else did something with media to any of the other systems sharing the drives. This would slow all editing to a halt.

If media is still offline after throwing away the Media databases, try the Relink command. Overall, you must be more responsible for choosing when to update your Media databases yourself since, in the interest of speed, it will not happen automatically under all the same conditions. As long as you understand that the Media databases must be updated (thrown away and recreated) or reloaded whenever you have a serious media management problem, you will be fine.

UNLINKING

Unlinking is one of those powerful, dangerous, useful, and poorly understood functions that people know they should use but don't really know when. Sometimes a link must be deliberately broken using Unlink. In the Clip menu, highlight the desired clips, hold down Shift-Ctl on Mac or Shift-Alt on Windows, and Relink becomes Unlink. Any media that has been captured for this master clip now becomes offline. The system considers this master clip as never been captured and is not linked to any media *or to any sequences*. The sequence linking is important because otherwise, every use of this clip in any sequence will be changed. You want to change this one master clip, but you probably don't want every use of it to change, too. That is why Unlink is required as a safeguard.

You then can modify duration of the clip, but you must recapture all the unlinked clips. Do not unlink media that has no timecode and do not modify the master clip because you will be unable to batch recapture.

Unlink is extremely useful for multicam projects. After batch capturing Reel 1 from Camera 1, you can duplicate all the master clips, unlink them, and change the tape name (Reel 2, Reel 3, etc.). Now you can batch capture all the other camera angles. Just be sure to duplicate the original clips using Ctl/Cmd-D and not Alt/Option-drag to another bin. You must be working with a true duplicate and not a clone of the master clip before you unlink.

There is no way to unlink a sequence using the Unlink command. Sequences have a loose link to media that allows them to change resolutions easily. The best way to unlink a sequence is to duplicate and decompose. You can throw away all the new decomposed master clips and just use the sequence. Because a sequence is loose about linking, you don't really need to unlink most of the time. You can just force the sequence to link to new material (Relink all non-master clips to selected online items with both media and sequence in the same bin), and it will automatically break the links to the old media.

USE COMMON SENSE

Even though there is a lot to be in charge of when administering an Avid system, much of it is common sense and taking advantage of existing computer peripherals and software that make your job easier. Make sure the policy you decide on is followed uniformly. Make sure that all members of your staff are educated on the correct procedures as well as just a little troubleshooting. Then they can deal with those questions themselves during the night shift. You may want to consider creating a media management policy in writing and making sure your clients know it, even by getting them to sign it when beginning a project. If set up right, you will significantly reduce downtime and make it easier to diagnose and solve technical problems and missing media.

5

Long Format and Improving System Performance

Many of the long format performance issues have been solved by faster computers and drives but it is still true that whenever you push a system to the limits of its capabilities, different rules of operation may apply. Although the Avid system is remarkably more responsive, renders are faster, and more effects are in realtime, there will always be those who are looking for the edge of moving even faster. We have come a long way from the Macintosh IIci when I first started teaching Avid courses, but any time you work with extremely large files on a computer, you want to minimize the wait times involved with the extra work all the hardware must do to keep up. Since the Avid system has evolved from these slower systems, there are many techniques for optimizing performance when the length of a project becomes more demanding. As with all programs that give you an advanced level of flexibility, the Avid system allows you to work only on the specific parts of the project you want to change. If you are working on a sound mix, you may want to minimize the video or vice versa. This chapter describes various techniques developed specifically to increase performance and to cope with situations that occur only when working on longer format pieces.

Any sequence with a duration of longer than a few minutes could be considered long format. Only as the number of drives, master clips, and the length of the sequences creep up will you experience any benefit from many of these procedures. Many of these techniques will speed up the response of the systems no matter how long the project. If you experience a sudden slowdown in performance, you may have encountered a simple technical problem that you can troubleshoot and solve quickly. Generally, a sudden slowdown occurs if lots of very large bins are being opened and closed or other pieces of software have been launched and quit throughout the day. This is just normal RAM fragmentation on a Mac OS 9.x and can be solved easily by shutting down the computer and restarting. On every other system, a reboot is probably the first and easiest action you can take to restore the system performance. Most people do this before the beginning of a shift, and some, working with lots of media and bins, also shut down at lunch and then restart. Dismounting drives that hold unneeded media can instantly reduce the number of objects the system needs to account for. This can be done on the Macintosh by dragging the icons to the trash (or highlighting them and using Cmd-Y on Mac OS 9 to put away or Cmd-E on Mac OS X to eject). You may want to rebuild the desktop (Mac OS 9 only) and the Media databases as well. On Windows you can dismount drives using the Disk Administrator. Details on those procedures are covered in Chapter 13.

The absolute first thing that must be considered before starting a long project is how much RAM is in the system. Over 512 megabytes of RAM is not too much. The more objects or captured master clips, precomputes, sequences, subclips, and so on, the more RAM you need to handle them. As I have mentioned before, any cost for RAM more than pays for itself as the project gets longer and more complex.

BUILDING THE PIPES

Every time you press Play, there is a preloading process where all media in the sequence must be scanned and analyzed. If there are mismatched resolutions in monitored tracks or mismatched sample rates in the monitored audio, you may get an error message and the sequence will not play. You may see the video slate Wrong Format when trying to play any resolution that could not mix with the resolution of the very first shot of the sequence on Meridien or ABVB systems.

The benefit of this preloading or "building the pipes" is the ability to play the entire sequence backward or forward at any time. The drawback is that the system is working very hard to provide access to all your media, although you may want to see only the next 30 seconds. When the sequence gets quite long, there may be a noticeable delay after pressing the Play button.

There are three methods to reduce this delay. The first method is to work with shorter sequences. This has many benefits, including not having to worry about sync 20 minutes down the timeline when you just want to make a few quick changes to the opening montage. Focusing on a specific trouble section and massaging it without worrying about affecting the rest of the sequence cannot help but speed up the editing process, even disregarding technical playback limitations. Many projects lend themselves to being broken into smaller pieces anyway. Films break easily between scenes, episodics break between acts, and corporate programs and documentaries break at location or subject changes. When it becomes necessary to view the entire project at once, it is a simple process to Shift-click all the smaller sequences and drag them to the record monitor where they instantly become a composite master. Perhaps a fade to black or dissolve needs to be added between these segments, but otherwise you have a quick, rough assembly of the entire show.

After the approval viewing or the digital cut is over, go back to the individual segments to make the changes and archive the composite master. There is nothing worse than making changes on a short individual segment but then showing the older composite master sequence without those changes!

In fact you may want to create a naming scheme that incorporates three stages of any sequence you are building. Create a sequence bin for every day of work on a multiday project. Keep only that day's work in the bin to minimize the number of sequences in open bins at any one time. At the beginning of the day, take the finished sequence from the day before and copy it to the new bin. Name it something like PreComp or StartSeq so you know it represents a starting point for the day's work. Cut out the part you are going to work on that day and call it something obvious for the day's work in progress. At the end of the day stick it back into a copy of the morning's starting point sequence, paying particular attention to sync issues if you have changed the length or the number of tracks. Name this end of day assembly as PostComp or EndSeq. It will be what you output to tape or compress to DVD for viewing. It is also where you will start tomorrow. If your producer insists on seeing everything together, you can easily track down the EndSeq or FinalVersion of all the acts and slap them together into a ScratchComp or TempSeq. This naming method should assure continuity of versions and avoid the mistake of working from a temporary viewing copy of any sequence.

The second method is to only Play to Out (Alt-Play or Option-Play). The system builds the pipes only up to the marked-out point. This approach will be necessary at a later point in the project, after all the segments have been combined or if there is no way to break the sequence into smaller pieces. If you get into the habit of quickly jumping down the timeline a minute or so, marking an out, and then using Play to Out, you get a faster response from the system.

PLAYLENGTH

The third method to reduce playback delay is called Playlength. This is a button in the Command Palette (PL) under Play and can be mapped to the record side of the source/record mode or mapped to the shifted function of the Play key. When you activate Playlength, the play button changes the normal black icon to a white icon. When active, Playlength restricts the building of the pipes to one minute in front of the blue position bar and one minute after the blue bar. When it reaches the end of the one minute, it stops playing and you need to press Play again. Because the preload is so much shorter, the response time of the system for much of its operation (including trims) is noticeably snappier.

Playlength does not affect the Digital Cut. This prevents a possible embarrassing situation when you go to tape, or as some very brave souls have tried, straight to air. Since the use of the Playlength button defaults to two minutes, you may want to change it to make it longer or shorter. When you launch the application the next time, it will always go back to the two minutes, but it will stay at the new length for the duration of the session. You can set the length of the Playlength feature through the Console. The Console is powerful, but potentially dangerous because it is designed primarily for the programmers to debug the program or run various tests. Here is the obligatory warning: Any command typed into the Console is completely unsupported by official Avid policy. Under the wrong conditions, you can permanently change the way your software operates and you may have to reload it from the original disks to get it to behave correctly again. Don't mess around with the Console unless you know what you are doing!

Despite all these warnings, to change the length of pipes that are built using the Playlength button:

1. Open the Console under the Tools menu (Ctl/Cmd-6).
2. Type "Playlength 5". Notice the space before the number.
3. Press Return. On newer systems, you will get a message like this confirming what you have done: "Play Length interval = 5 minutes forwards or backwards."

If you want to return Playlength to the default of two minutes you can go back to the Console and retype "Playlength 2" (see Fig. 5.1). If you haven't entered anything else in the console you can press the F3 button, which will reload the last command. You can then change the custom Playlength number back to the default. Or you can just restart the application and the next time you activate the Playlength feature it will be two minutes and it will always return the system to normal Play behavior.

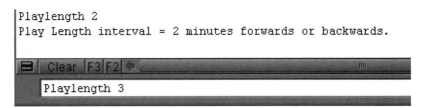

Figure 5.1   Specifying Playlength in the Console

PRE-FILLED FRAMES AND DESKTOP DELAY

On Xpress DV, a User Setting called Pre-filled frames helps you find the balance between response time and the ability to play more complex effects. Located under the Video Display Setting, it will determine how many frames are loaded into RAM to preprocess any realtime effects before playback. The system will delay a maximum of 10 seconds before it plays. If the system delays less than this, it may be because you don't have enough RAM to prefill that much. Ten seconds is a very long time to delay; you should play around with a much smaller delay to figure out the best compromise for getting complex realtime effects to preview in realtime.

The other response setting, Desktop Delay, is meant to synchronize your desktop playback with the client monitor. Because of the inherent lag in recompressing DV25 back to standard, baseband video, the client monitor can be many frames behind the audio and video coming from your computer monitor. You can turn off your computer speakers and point everyone to look only at the client monitor for playback or you can tweak this setting. By finding the precise amount of delay, you can get the two monitors to play back at the same time but at the cost of responsiveness. Your system will always wait that maximum of 30 frames every time you press Play. You may want this setting to be greater than 0 only for important approval playback sessions. This is an excellent setting to save as a Client Playback user setting

RENDER ON THE FLY

Other ways to improve performance are to utilize techniques that reduce the amount of media that must be retrieved, played, or rendered. There is a menu choice under the Special (Media Composer and Symphony) or Clip (Xpress) pulldown menu called Render on the Fly. Render on the Fly is a temporary render of one frame performed in the RAM when you place the blue position bar over the effect in the sequence. Nothing actually goes to a media drive. With Render on the Fly turned off, when the blue position bar stops over an unrendered effect, the system does not try to composite that single frame for viewing. The system shows the highest numbered track that does not contain an effect. This means you can scroll through the timeline and stop in the middle of an unrendered effects sequence without the delay involved in previewing all the layers. You can move quickly to perform tasks that do not require seeing these composites, including audio mixing, trimming to marks based on audio, and moving segments around the timeline.

Editors who turn off Render on the Fly on a regular basis end up mapping it to a function key. The only drawback to disabling Render on the Fly, besides not being able to see every layer, is that some objects, like titles or imported matte keys, are considered "effects with source." This means that they must be rendered on the fly to be seen. If Render on the Fly is disabled, you see only black when loading a title into the source window. Ironically, you can view them as soon as you press Play because then they are rendered in realtime. Leaving Render on the Fly disabled has resulted in a few calls to Avid Support, so remember to turn it back on before handing the system over to the next editor.

Render on the Fly does not affect non-realtime effects until you release the left mouse button (mouse up). This noticeably speeds up scrolling through the sequence with the mouse since the nonrealtime effects (blue dot) will not Render on the Fly until you release the mouse button.

There is a checkbox for enabling Render on the Fly in Trim mode under the Trim setting. It defaults to being off, but once turned on, it allows you to see multiple unrendered layers and effects while in Trim mode. This slows trimming considerably but may be necessary to make content-based trims with those layers in view. If you need this ability on a regular basis, make two separate Trim settings and switch between them in the Project window. Use a Workspace to make this switch quickly as part of a layered or effects mode. Use the "Activate settings linked by name" to link the Render on the Fly trim setting to the mode you use for working with effects.

MOVING THE VIDEO MONITOR

Another way to improve performance while working with effects is to move the video monitor icon to a lower track. This means that since you are not monitoring those tracks, the media does not need to be retrieved and played. Again, you can move quickly through a multilayered sequence without waiting for unneeded images to be recalled.

A User setting in the Timeline setting under the Edit tab allows you to turn off Auto-Monitoring. Auto-Monitoring is when the video monitor icon follows whatever you are doing in the timeline. If you patch to a higher channel, even if you have turned the video monitor off, it will turn the video monitor back on for that top track and force a longer Render on the Fly. With Auto-Monitor turned off, you are responsible for moving the monitor where you want it to go, but it will always stay where you leave it. This setting defaults to on so you will have to disable it manually.

To move even faster, turn the video monitor off completely. Sound mixing and rearranging of large chunks of the sequence will speed up remarkably. This is also a good way to isolate whether playing back a video track is causing a particular error. If you stop monitoring the video track and the playback problem disappears, you have isolated the problem to that track.

RENDERING FOR SPEED

By occasionally rendering realtime effects during a break, you can make it easier for the system to cope with playing many short clips. The system is always trying to display all realtime effects by doing the compositing in the CPU. Playing back a precompute — a rendered effect file — takes less effort and can result in faster response time. If you are having problems playing a particular section of fast, short cuts, you can apply a submaster to the track above and render it (use Add Edits to mark the area on the empty track and drag the submaster effect into the defined area). This puts all the short media files into one rendered precompute so the system needs to play back only one single file when you monitor the track with the submaster. If you must make an EDL, avoid the top video track with the submaster and still generate a clean, accurate list.

VIDEO MIXDOWN FOR SPEED

If you don't need an EDL, you can create a video mixdown of the section. A video mixdown combines all the separate media files and precomputes and makes them one new media file and a new master clip. This also plays back very quickly with the added benefit that, if the segment has several layers, those media files required for the layers can be completely eliminated. If all the effects in a sequence have been rendered, then making a video mixdown is as fast as copying a file. This is an insignificant amount of time compared to the continued benefit of faster access time.

Even after you render an effect, the layers below the top track continue to cause media to be retrieved in case it is needed. With video mixdown, there is just one layer, one media file, and subsequently, faster retrieval time. Don't use a video mixdown if you need an EDL or you need to match frame back to an original source in the segment. A video mixdown has no timecode and retains no links with the original media; it is purely a convenience for simplifying a complex segment.

Always make a duplicate of the sequence before cutting a video mixdown over the original layers so you'll have a nonmixdown version of the sequence to come back to for revisions. Or you can take the original layers and put them onto higher tracks (collapse them down using the collapse button and move the submaster vertically if you have limited tracks; see Chapter 7 for details). Then monitor only the lower tracks for normal work and monitor the higher tracks for match framing if you need to.

MINIMIZE TIMELINE REDRAW TIME

There are ways to minimize the screen redraw time of the timeline. This is especially important when many effect icons need to be updated constantly. You can go into the Timeline fast menu and start disabling the display for various icons. By turning off effect icons and dissolve icons, you lose that visual information, but if you are positioned over the clip, you can still see the effect in the record window. Effect Contents can be disabled so that, if there are nested effects, they show up only as the single outside effect icon without trying to show all the layers inside. If you have not seen Effects Contents at work, it is because it is visible only when you are zoomed in closely on the Timeline and you have enlarged the height of that track (Ctl/Cmd-L).

Never work with the audio waveforms displayed unless you specifically need them. There is a Timeline User setting that shows waveforms only between mark in points and out points and this might be more useful than trying to display waveforms for the entire sequence.

Any of the modes that require images to be drawn in the timeline should be disabled, like Heads View or the Frame track, and in general, any display you can work without temporarily should be turned off. Although you can't turn off the top Frame track on Xpress or Xpress DV, you can minimize it by clicking on the space between the track panels and dragging it smaller than the other tracks. I always leave sync breaks on, but I might turn them off for audio when I am editing primarily video. You should save a Timeline view with all these display features turned off so you can get to it quickly when you really want to move fast. You may also want to Link the Timeline view to a Workspace called Long Format, Effects, or Fast Navigation.

An instance where redrawing images on the screen can slow performance occurs when using the segment editing arrows. As soon as you begin to drag a segment in the timeline on Symphony or Media Composer, the four screens showing all the transition frames must be drawn on the screen. When moving especially fast, this redraw can delay the system enough so that it perceives a click and drag as a double-click on older CPUs. The system waits for you to release the mouse so it can display nesting, the command it has received from a double-click over an effect in the Segment mode. Then it does not allow you to

Figure 5.2 Four-Frame Display

drag at all! On these systems the four frames can be disabled as part of a User setting under the Timeline setting (Show Four-Frame Display). On Xpress and Xpress DV this display is not an option.

With the Timeline User setting in Media Composer and Symphony, there is also a command to turn off the segment arrow double-click that displays nesting. This setting does not exist on Xpress or Xpress DV. Step In and Step Out are in Xpress Meridien 4.5 and Xpress DV 3.0 and later so you have an alternate way to look at nesting. Although faster CPUs and OGL graphics cards take away some of the benefit of these Timeline settings being turned off, they are still welcome additions for the speed freaks out there who are always working a few steps faster than the system can display.

On Windows systems, another little trick gets the screen to update faster and makes the display a little cleaner. On Windows XP go to Control Panels/Display/Appearance/Effects and uncheck Show window contents while dragging. This will now give you just an outline of any window while you are holding the mouse down and moving it. Since the system doesn't have to redraw so much information, it can give you control and update the screen faster whenever you move windows. If you have trouble seeing where your window is going, you might want to turn it back on.

Working with long projects means you will always be trying to squeeze all the extra speed out of your system. It means stripping down the features to the bare necessities to minimize screen redraw time and wasted time accessing images and effects you don't need at that moment. Many of these speed-based User settings can be turned on or off using the Workspace functions and mapped to a single key. In the end, you are working with the most flexible, customizable system for media management and maximizing the use of the drives and CPU. No other system comes close.

6

Importing and Exporting

Graphics come in many shapes and sizes from artists, ad agencies, the Internet, scanners, digital cameras, and a wide range of other graphics programs, but very rarely will it be perfectly prepared for video. In general, graphics tend to be the wrong size or the wrong resolution either through accident, ignorance, or repurposing. This chapter will help you learn how to compensate for that.

Avid has built in some system intelligence to deal with all the different graphic formats and, if the graphic format given to you is recognized, the system automatically imports it. This is possible because Avid has incorporated the technology called HIIP (Host Independent Image Protocol). There is auto-detection of file type using HIIP in current versions; over 25 formats for import and export exist.

Also, the new drag-and-drop capabilities of the last few releases speed up workflow and simplify basic tasks. The user creates an export template based on the requirements outlined in this chapter and is assured every export will be consistently correct. By dragging a sequence or a master clip to the desktop level, even complicated exports can be done by beginners. The same HIIP technology allows users to drag a graphic straight from the network and drop it on the open bin. By creating preset, named, copied and carried, import and export templates, you can make interoperability with other software a one-step process.

Import and export procedures depend on the end result. Do you want the highest quality? You must follow some basic requirements for format, size, and resolution. Do you want to play the material on a CD-ROM? You must make some quality tradeoffs to get a decent throughput of the images and audio on slower computers. You must, above all, educate your subcontractors and graphics department (even if that is just you!) and find a consistent procedure for acquiring the types of graphics and animations you demand for your clients and productions.

IMPORT AND EXPORT BASICS

Some basic issues should be understood when working with computers, graphics, and video. Because of the way computers have developed, with their reliance on RGB color for their screens and memory, and the way video developed, beginning with black and white analog, the two mediums have never had a particularly easy coexistence. One of the hopes for high-definition television is that some of those issues will be resolved, but the addition of more incompatible formats has rarely made things simpler.

COLOR SPACE CONVERSION

Computers and video work with different color formats and different kinds of scanning and scan rates. We commonly refer to methods of representing colors as color space. It usually is represented graphically by a cube with white at the top and black at the bottom. The range of colors possible within a color space makes up the height and width of the cube. Different color spaces use different methods of distributing those colors inside the cube shape.

Computers traditionally work in Red-Green-Blue (RGB) color space and digital component video works in YCbCr. RGB is easier for displaying images on an RGB computer screen and working with computer memory. RGB has a more limited range of colors to choose from, but it is very dense, with many more possible colors within that color space of perceptible colors than YCbCr (we really care only about perceptible colors!). The ironic part about camera-originated video is that it begins as RGB and is then converted by the camera to a different color space called YCbCr (or commonly and incorrectly referred to as YUV, which applies only to a composite signal). This new YCbCr color space spans beyond the range of visible colors and has fewer colors within the visible range. Thus a conversion between YCbCr and RGB can force colors to change because a true exact match does not exist or because colors are out of the range in the new color space ("out of gamut").

The change of colors when the video camera's RGB values are captured and then converted to YCbCr as the signal is recorded to tape is a greater source of color change than anything that happens later inside the computer. The amount of color information is cut in half as you move from a 4:4:4 RGB signal within the camera to a 4:2:2 YCbCr signal on tape. So yes, there is potentially some change to your video colors during the capture process on an Avid Broadcast Video Board (ABVB) as it converts the signal to RGB for editing. Most likely you will never see it on a standard video monitor. The conversion never affects monochrome values no matter how many generations you go using the ABVB. The Meridien video subsystem, DS Nitris, Media Composer Adrenaline, Xpress

Pro, and Xpress DV all work in YCbCr, so this is no longer an issue during capture. They will occasionally switch to RGB color space when the results are better in that color space (e.g., during color correction), but the realtime effects are YCbCr for better performance. If your colors or video levels are changing, it is more likely you are doing something wrong like exporting or importing into the system using the RGB levels choice instead of 601 levels. This chapter will discuss the details of these choices later.

The Meridien Video Subsystem, the Nitris and Adrenaline DNAs, and the software-only systems like Xpress DV are all ITU-R.bt.601-compliant. This is a SMPTE recommendation governing the use of nonsquare pixels and the distribution of bits to represent luminance and chrominance values. It is the predominant specification for digital standard definition video today. It used to be called CCIR-601, but SMPTE changed the name to reflect another committee, which makes things a little more complicated because most computer graphics programs do not conform to this video standard. At the same time, it opens up some interesting possibilities. The recommended standard for high definition video is ITU-R.bt.709 and specifies square pixels.

The reason it is more complicated is that the early video boards (including the first Avid board: the Truvision NuVista+) and "prosumer" video boards do not capture the entire range of the possible video levels. A true ITU-R.bt.601 video signal can use a signal that is as black as possible without going into the sync section of the signal. It can also go above the brightest luminance level that should be broadcast, giving some extra range for digitizing bright signals without chopping out detail. The other, older design boards use a more limited range, which in NTSC is from 7.5 IRE, the blackest you can legally broadcast, to exactly 100 IRE, lower than the brightest levels to be broadcast. In PAL, the range can be shifted, but it usually goes from 0 volts to 0.92 volts. This made it impossible, for instance, to play proper SMPTE color bars, which are designed to exceed those levels for test signal monitoring purposes. Why did the video board manufacturers pick these values? Because they are the limit of RGB color space! This is a really serious mistake.

Not allowing a video signal to utilize the full 601 video range means you are shortchanging yourself in many ways. The most important is that you have a much more limited range to start color grading. If you want to get more detail in the blacks or the highlights of an image you are out of luck. If you want to see more clearly the face of someone in shadow or bring back the details outside a window you will have lost that part of the signal. Those levels have been clipped during the signal ingest process through the video board. Those details are gone—never to be seen again! A minor tweak to make the video look better (usually at the client's request) just results in adding noise to the signal. Now you may be thinking that chopping this part of the image is a type of de facto "legalizer," which will make sure you conform to broadcast standards, and although this is true, it is missing the point. The RGB color space limitations are actually *less* than legal broadcast levels, so you are cheating yourself. Ultimately, you can always make the image look better if you have control over the extra detail in those parts of the signal. You can run the signal through a third-party legalizer later, after the program has been color graded. But the best solution is always to keep an eye on the waveform and vectorscopes like online professionals do to ensure quality. Getting the levels right through careful monitoring and control throughout the process is the best way to reproduce all the detail and color of the original images.

Clipping a signal to under broadcast levels during the ingest stage also assumes that you will be outputting to video only. What if you are going back to film? You have seriously degraded the quality of the final film print. And it gets even worse if you are using systems where the video board and the software are from different manufacturers (*nonturnkey* is the technical term for cobbling multiple proprietary systems together). The video hardware will clip your signals, then the editing software will stretch them back again to the full range only to be clipped again by the hardware during output! Some very popular video hardware clips video levels during rendering of effects so you will have jumps in your video levels all over the final master sequence. All of these nonturnkey issues compound the original problem and seriously degrade a signal that may already be marginal for broadcast or printing to film because of compromises made on location during the shoot.

If you can't use the video levels all the way down to 0 IRE (called superblack), you cannot use some graphics in a luminance key. If the only way to use the graphic is from videotape and there is no mask, matte, or hi-con, you need to key the graphic with a luminance key. If you have black in the graphic that needs to stay opaque (not keyed out) then you need to set the key clip using the levels on the tape that are blacker than the blacks in the graphic. Without the superblack for luminance keying you will find yourself hand-drawing mattes around the graphics.

Another characteristic of the ITU-R.bt.601 requirements is that the frame aspect ratio is different from most graphics computer programs. With standard computer monitors, most people work with square pixel graphics with a frame size of 648 × 486 or 720 × 540 pixels in NTSC. This is fine if you are going to one of the older, non-ITU-R.bt.601 video boards. But Avid standard definition video products require the NTSC graphic to be 720 × 486, a nonsquare pixel size. If you are importing a graphic that is 648 × 486, this is not a terrible mistake, but it means that the import has to resize the graphic for you. All of the current versions of Avid editing programs do an excellent job of changing square to nonsquare so this is not really a problem except that the resize may take longer during the import stage if you have many graphics. If you have a choice, ask for the images or the animations in the proper frame aspect ratio, 720 × 486 NTSC or 720 × 576 PAL.

The final characteristic of graphics and video that must be taken into consideration is a video field. There are two fields to every standard definition video frame since a standard definition television broadcast signal scans the screen twice. This is called interlacing (or abbreviated as "i" after the frame rate like "25i"). Generally, whether you are PAL or NTSC, you must use interlaced scan lines—every other line is part of a different pass drawing the image on the screen. This has been the standard definition requirement for almost all material shot or transferred to video. Progressive scan cameras are coming onto the market, but you must be able to edit in a progressive resolution to take advantage of them, like with Avid's "p" resolutions. There are many common high-definition formats that are progressive, like 720p or 1080p. Some cameras will enable you to capture at 480p, which most people do not consider high definition, but which comes with the benefit of progressive scan.

Broadcasting an interlaced signal takes less bandwidth because only half of the image is transmitted at any one time, but it complicates things when you are taking an interlaced image to a noninterlaced medium like the computer. A computer uses a progressive scan for display on your monitor, which means that the image is drawn on the screen as a single frame, not two fields. If you export an interlaced frame from the Avid editing application and there is some kind of horizontal motion in the frame, you see a difference between the first set of scan lines, field one, and the second set, field two. Although they are only a 60th or 50th of a second apart, you see jagged horizontal displacement of the image every other line. If you are working with interlaced images you will need to consider when and how to de-interlace them when exporting to graphics or animation programs.

With the ease of basic desktop editing and the combination of graphics that go straight from one computer graphic format to a computer video format, you have some quality challenges. All of these image type mismatches can be dealt with if you are careful when converting formats. These graphics have too much fine detail to reproduce well in the relatively low-resolution, interlaced world of standard definition video. Consequently, the images buzz, flicker, and give us unpredictable results if played back with little compression. A thin line may look fine on a progressive scan monitor, but the moment it moves to an interlaced scan medium like standard definition video, that line may be only one scan line wide. This means the line is drawn on the screen only every other field, causing a disturbing flicker. Images that originated with a video camera can never be recorded with that kind of problem. And since, in the past, most graphics were seen through video monitors as they were being created, they could be adjusted on the spot so that the design could take into consideration the limitations of what looked good on video. Colors were toned down and detail was blurred until the image was acceptable, and then it was put to tape.

Until all graphic workstations can figure out how to approximate what the final product will look like after being interlaced and reduced in resolution for SD broadcast, you must be able to tweak the graphics after you receive them. The most common adjustments are to open the graphic in a graphics program on the editing workstation and add a little blur to areas that are buzzing with too much detail. If done with a little skill, the blur will never be noticed. In fact, a very slightly blurred image looks better than one that is too sharp. The Avid painting tools allow you to do this without any third-party programs, but they must be rendered. Another common adjustment is to lower the saturation of a particular color or, in a worst-case scenario, all the colors. Again, this can be done easily and safely using the Avid color effect or by running all the images in a batch through a program that allows you to create a desaturation macro like Equilibrium's DeBabelizer or Adobe Photoshop 4.0 or later.

IMPORTING REQUIREMENTS

There are three important factors to get right when importing images. They are size, resolution, and format. If you get these three correct, then any graphic should import correctly. With recent versions of Avid software, most of these requirements have become much simpler.

- Most common file formats are recognized automatically (over 25 types).
- The optimal resolution for images should be 72 pixels per inch (ppi), but are imported correctly even if they are not. This is important since higher resolution will not result in a better looking video image.
- All graphics levels will be mapped to ITU-R.bt.601 levels.
- Images will be resized correctly during import. The correct frame sizes are as follows.

Resolution	Frame Size	NTSC/PAL
ABVB:		
Resolutions	NTSC	PAL
AVR 12–77	720 × 486	720 × 576
AVR 2s–9s	720 × 243	720 × 288
AVR 2m–6m	352 × 243	352 × 288

Meridien, DNA, and software-only systems:

Resolutions	NTSC	PAL
DV	720 × 480	720 × 576
Two field	720 × 486	720 × 576
Single field	352 × 243	352 × 288
Multicam	288 × 243	288 × 288

All of the resizing is done automatically and at high quality. You can work completely in a square pixel format and the system will take care of the correct nonsquare size during import (more on that later). As progress marches on, the process has gotten simpler and more foolproof without reducing choice.

Aspect Ratio, Pixel Aspect: 601

When correctly importing a graphic, several important choices must be made. The dialog boxes are designed to reflect the type of graphic to be imported. The system will import graphics correctly, assuming that accurate information is given about your graphics. The next few sections will show you the differences between settings.

Size and resolution are related. For all Avid editing systems, the size of the standard definition graphic must be 720 × 486 pixels for NTSC and 720 × 576 for PAL. The exception is when you are working with NTSC DV native material, like in Xpress DV, where the frame size is a slightly different 720 × 480 to allow for better compression. But if you are working in a square pixel graphics program like Adobe Photoshop, then your starting frame size should be 720 × 540 NTSC or 118768 × 576 PAL. When you import the graphic, the system will resize the

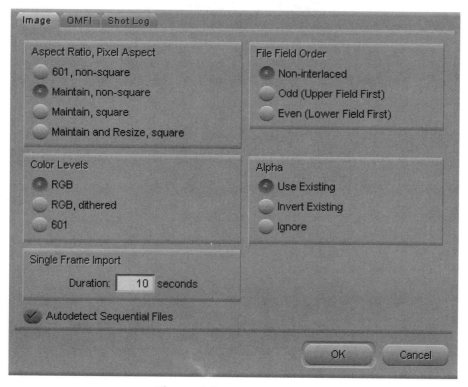

Figure 6.1 Import Settings

image correctly to non-square pixels when you choose the Aspect Ratio, Pixel aspect: 601. The system will take any frame size and try to fit it to the correct 601 frame size. If your graphic is tall and thin or otherwise nonstandard-sized you need to choose one of the other three types of aspect ratios.

Importing Nonstandard Frame Sizes

Maintain, nonsquare; Maintain, square; and Maintain and Resize, square are the three settings for graphics that do not fit the standard video frame size. Figures 6.2 through 6.5 show examples illustrating what happens when importing an unusual size graphic.

Most people would consider the choice Maintain and Resize, square to be the best choice since the import process has enlarged the graphic to fit within the maximum vertical space. If the image comes from a nonsquare source like an exported image from an Avid system after it has been ingested and converted to 601, you will want to use nonsquare. Notice what happens when you take a graphic created in a square pixel graphic program and force it to be

Figure 6.2   Original Source Graphic 215 × 900

Figure 6.3 Imported Graphic: Maintain and Resize, Square

nonsquare during the import. There is an unnatural distortion, which is slimming in NTSC and flattering to some humans, but it squashes the image vertically in PAL, which does no one any good.

Maintain will keep the original aspect ratio, which is why you will end up with black around the edges if the frame size is nonstandard. Use Maintain if you want to bring in a small logo or bug and stick it in the corner of the frame. If sized correctly by the graphic artist, don't mess with it.

Resizing Images and Safe Areas

When preparing graphics for video that will be resized you must previsualize how text elements may be placed. All important visual elements in video for broadcast must be within the safe picture areas. "Safe Picture" or "Action Safe" is everything in the outer 10 percent of the frame and will probably be at least partially cut off on most home TV sets. If that information is important, you must compensate for it before importing by pushing important details away from the edge of the graphic. With natural curving of glass tubes and the normal drifting of correct horizontal and vertical positioning of many home television sets, only a subset of the safe picture, called the safe title area, will be completely readable. All text must be within the safe title area (only if you want to read it at home of course!). In the Avid editor, the safe title overlay can be superimposed over the image in the source or record monitor using the grid button under the FX section of the Command Palette or under the default fast menu. You may want to make your own overlay guide to use in graphics

Figure 6.4 Imported Graphic: Maintain, Nonsquare

Figure 6.5 Imported Graphic: Maintain, Square

programs or find a template premade on the Internet. You can also squeeze the image smaller after it is imported by using a picture-in-picture effect, but this is not recommended because it will seriously degrade the quality of the image.

Figure 6.6 is an example of the safe picture area on the outside grid and safe title on the inside grid.

Figure 6.6　Imported Graphic with Safe Picture/Safe Title Grid

16:9 Anamorphic Graphics

Many projects are now completely 16:9, so the graphics must be handled correctly to match the way the video is unsqueezed inside the Avid while editing. A standard definition 16:9 image is recorded anamorphically squeezed to fit the 4:3 video aspect ratio. A 16:9 monitor then unsqueezes the image during playback to make it fit the wider frame size. High-definition format is always 16:9, so you may have downconverted footage that has been anamorphically squeezed to 4:3 for offline editing. You may want to create the graphic or animation in the original high-definition resolution and then make another one for the standard definition offline with the HD version being true 16:9 and the SD being 4:3 anamorphic. Many animation programs will let you output two versions in a render queue. During the batch import in the final high-definition online, you can point the Avid system to the correct graphic.

If you are creating graphics or animations in NTSC:

- Create a composition in the animation program with a size of 864 × 486 square pixel. Crop or resize your graphic to this size with the graphics program.
- When you output from the animation program, use the output module's Stretch function to resize the animation to 720 × 486. You can also nest the composition within a 720 × 486 composition. Use the Avid Codec for rendering.
- When you import your single graphic frame, choose pixel aspect ratio 601, nonsquare.

In PAL:

- Create a composition in the animation program with a size of 1024 × 576 square pixel. Crop or resize your graphic to this size with the graphics program.
- When you output from the animation program, use the output module's Stretch function to resize the animation to 720 × 576. You can also nest the composition within a 720 × 576 composition. Use the Avid Codec for rendering.
- When you import your single graphic frame choose pixel aspect ratio 601, nonsquare.

Resolution

Graphic artists may insist that the higher the resolution, the better the image will appear in the final product, but this is not true for standard definition television above 72 ppi. The image will never look any better in standard definition if it is created at a higher resolution than 72 ppi, but the system will import the graphic correctly nonetheless. There are several plug-ins available (including the Avid Pan and Zoom effect) that will allow you to import a graphic at a higher resolution. You can then pan and zoom that graphic inside the frame with animation keyframes without reverting to standard definition 72 ppi until the animation is rendered. This is known as a modal import and will not affect the graphic in the same way as, say, using the resize effect. By working with a higher resolution image throughout the process you will end up with a better looking end result. This will allow you to create the equivalent of a motion controlled camera without the cost, and the results generally are excellent.

You should always work at the highest resolution possible when creating the graphic since you will have more flexibility with cropping and enlarging a section of an image. You should keep a high-resolution version as the original graphic and resize it before sending it off to the video session. You could even use a macro to do this resize as a final step.

Format

The last characteristic that must be taken into consideration is format. With the adoption of Avid's HIIP technology, most formats are automatically recognized and imported. However, the size of a 35-millimeter scanned slide does not exactly match the aspect ratio of video so even though you may be able to import something directly, it might not be a good idea. The slide may also have black borders left over from the scan. The system will squeeze the sides if you pick 601, nonsquare as it attempts to make the aspect ratio fit. You really must decide which part of this image to discard, or live with black bands (especially if it is a vertical composition!). Choose what must be discarded from the image and crop in a graphics program.

Choose a format that has little or no lossy compression like TIFF with None as the compression choice. You want the graphic to have as much original information as the original and lossy compression can create visual artifacts. However, if you are working with a graphic that is much larger than video format size, you will be shrinking that graphic anyway, so compression is less of an issue. Compression really becomes important if you are working with small graphics that must be resized up or with graphics that have lots of detail and are about the right size already.

You can now import graphics in the Adobe Photoshop format and retain the original layers in the Avid editor. Each layer in Photoshop could be text or other graphic elements with an alpha channel that you don't want to merge with the background image. When you import them into the Avid the graphic becomes a multilayered sequence that you can use as a source. Color correction layers do not come across and there are some other minor restrictions since not everything in Photoshop can be recreated in the Avid. All types of blending options will not transfer except Dissolve or Normal and will be ignored during import. You may want to flatten or merge those layers in Photoshop before importing if you want to keep that effect. All layers need proper alpha channels to key correctly, but all the standard selection tools easily create alpha channels in Photoshop. This is a wonderful breakthrough that allows you to continue to manipulate the different parts of a multilayered image with video transitions and DVE moves. You can build an image onto the screen by bringing in the layers separately and can create animation for an element that otherwise would be static.

When you import the image you will be given the choice to flatten the image or keep all the layers intact (up to 24 layers allowed in Symphony or Media Composer; fewer in Xpress). You can choose to discard layers at this point and the system will remember the layers for later batch importing and bring it in correctly later during the online. Not only does this allow you more creative freedom later in the process, but it is a great way to handle changing versions. If you will be using the same graphic in different videos, you can have the graphic artist include all the versions as different layers in a single multipurpose graphic. You can then import all the layers and use only the ones you want or choose the layers for each version during import.

Alpha Channels

An alpha channel should be considered seriously as the method of choice for keying images. It is a fourth channel after the red, green, and blue channels and turns an RGB graphic into an RGBA graphic. An alpha channel contains grayscale shapes that determine what part of a graphic will be opaque and what part will be transparent when it is superimposed. It is the changing of the alpha channel from white to black that determines how the image actually is blended with the background in the Avid system.

There are two types of alpha channels: straight and premultiplied. Only the DS Nitris can handle the premultiplied type; for the other Avid editing products you should specify straight. This is especially important for animations because many animators are used to working with premultiplied alpha channels.

It is a bit of work, carefully drawing around edges to create a good alpha channel. It easily could be handed off to someone more skilled at drawing or who has a dedicated computer just for graphics. Once you become good at it, however, you may find it easier and faster to just whip one off yourself. Depending on your image—perhaps it is high contrast already—and your skill with the graphics program, you may be able to use some automatic selection functions (like eyedroppers) to create your alpha channel. If you find yourself creating alpha channels on a regular basis it is well worth the investment to purchase a pen and tablet for your editing workstation.

When saving in a graphics program you must always choose to save a graphic with an alpha channel as a 32-bit graphic (each channel contains 8 bits of data). If you save your graphic with 24 bits you will unceremoniously be deleting your alpha channel. You will also want to be aware of the way transparencies are preserved. If the alpha channel blends from opaque to transparent how will that opacity change be represented after importing? Although there are many formats that can support an alpha channel, it is the TARGA format that handles it best when using Photoshop. This is because the TARGA format does not blend the fill with another color. When the fill retains full opacity and does not blend with another color (like black or gray), the blending of the alpha channel is most effective and closest to the original graphic look.

Here's a quick tip if you don't have a graphic program for cutting mattes or creating alpha channels. You can use the matte cutting tools in the Animatte effect in the Intraframe option in Media Composer and Symphony:

- Apply the Animatte effect to draw a shape in the Effects mode.
- With the object selected, choose Export Matte PICT from the File menu and export to the desktop.
- Reimport the image to a bin and you have a realtime key.
- Step into the key once it is in the sequence, and replace the static video with moving video.

You can also stay in the Animatte effect and be able to change that matte over time with keyframes.

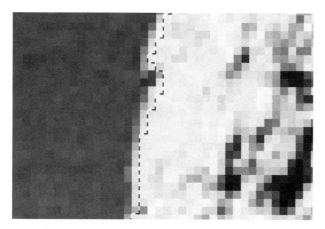

Figure 6.7   A Selection Choked by One Pixel

Working with a graphics tablet makes it easier for the fine adjustments to a matte selection to be made in a graphics program. Make sure the edges you create are always anti-aliased since this gives you the better keying edge. Some people insist that anti-aliasing should never be added to an alpha channel because it gives a little unwanted border around the outside edge. This is a result of the three-pixel-wide edge that is created in a straight alpha channel when you anti-alias. It is part of why an alpha channel gives a nice, smooth edge that blends with the background from full opacity to full transparency. The three pixels cause the edge to extend one pixel beyond the edge that you have so carefully drawn. Luckily, it is very easy to adjust the edge of the selection and "choke" it back by the necessary one pixel using Modify Selection. All professional-level graphics programs should allow you to choke the selected edge by a pixel. This clears up any border that is a result of anti-aliased edges of a straight alpha channel.

The standard in television graphics on dedicated graphics workstations has been to make any mask or hi-con area black where the keyed graphic was to be transparent and white where it was to be opaque. Most video switchers in tape suites inverted this if necessary, but nonetheless this was common practice in the video world. Unfortunately, the Avid import default is the opposite. This means that, for simplicity's sake, all graphics should be created with an alpha channel that has white for transparent areas and black for opaque areas. Of course, if you know that the graphic was created in the older style, you can invert the alpha channel during import. Confusion arises when graphics coming from one source are one way and graphics coming from another source are the opposite. This is something that should be specified as part of any facility's production guidelines. Do one test import to see if you have background and no foreground in your keyed graphic. If so, time to use the Invert Alpha Channel during import!

RGB VS ITU-R 601 FOR IMPORT

We need to refer to the earlier discussion of ITU-R.bt.601 (CCIR) levels versus RGB levels when bringing in our graphic. Chances are, because the graphic originated in a graphics program, it was created with RGB values. This means that the levels of red, green, and blue at 0, 0, and 0 equal the blackest black possible in that program. Conversely, 255, 255, and 255 are the brightest. Now comes the dilemma: How dark and bright should they be in the video program? In the Avid software, you are dealing with the digital ITU-R.bt.601 values. During import, you are mapping the RGB values from the graphics program to another set of digital video values, the 601 specification. These values are the same for both PAL and NTSC.

The original graphic values, RGB, import the 0 levels from the graphics program to digital level 16, broadcast black of 7.5 IRE, or 0 mV. The 255 levels will be mapped to digital level 235, which is the brightest broadcast white (both PAL and NTSC). This happens automatically during import when you choose RGB as your import choice.

RGB Import

Most of the time your graphic will be RGB so this is generally the best choice. If your image was exported from a video program it may be 601 already. If you import a 601 image as RGB then you will have compressed your signal a second time. The blacks will appear to be gray (at 32 instead of 16) and the dynamic range of the image will be compressed.

RGB, Dithered

When working with 8-bit graphics, occasionally a gradation from black to white or "ramp" can look like it has horizontal bands running through it. This is because certain types of digital graphics have more detail than can be recreated with 256 luminance levels. The jump between the missing levels is called banding. Although this can't be eliminated completely without working at 10 bits (1024 luminance levels) or higher, there is an import choice that can make it better. This technique, called dithering, is the insertion of a certain amount of random noise into the image. You can't really see the noise, but it does make a difference in smoothing out these banding problems. The next time you see a graphic that appears to have bands running through it, try this import option. You probably don't want to use it every time or use it multiple times on the same graphic, because eventually you will begin to see the noise.

601 Import

Importing using the 601 option takes the 0, 0, 0 RGB levels of the graphic and maps them to 0 digital. The darkest part of the picture is mapped to the darkest possible video level. The highest RGB value, 255, 255, 255, maps to 255 digital, the highest video level. Nothing is changed from the graphic's levels to the video levels. You want to use this exclusively when you are importing a graphic that began as an export from a video program that supports 601 levels. In other words, you have a graphic that is using 601 levels for correct broadcast black and broadcast white. You actually may have a graphic created in a graphic program where the artist was careful to stay in video safe levels (16 to 235), but this is very unlikely.

If the graphic has been combined with exported video and has also been modified with the full 0 to 255 level, you will have illegal levels when importing using the 601 choice. The only way to avoid this problem is to run the graphic first through a third-party, safe-levels filter that makes sure that everything in the graphic is safe both in video level and color saturation. Sometimes these filters can change the hue as a by-product of this change.

Sometimes you don't care if the video levels are safe, you just want them to match something already captured. Clearly, importing at RGB levels changes your levels and importing the 601 levels does not, so after you have finished tweaking a shot you will want it to fit seamlessly into the existing video. The 601 levels option would be the way to go. 601 is of absolute importance if you are working with a frame of video exported that has values below or above the legal broadcast ranges. In other words, you have exported a frame of video from the Avid that has illegal video levels and you need to match the graphic seamlessly. Some editors would say you should just recapture the video again and get the levels within the proper specifications, but let's face it, that often is not a real-world scenario. The amount you are over or under RGB levels may be tiny, but still visually noticeable if the match frame from the graphic to the video is not perfect.

Another reason you might not care if your video levels are safe is if you want to use the superblack for luminance keys. Using superblack allows the graphic or logo to contain black in the design and still be keyed using a luminance key because the level of black that is used to key out is at 0 IRE. Importing 601 and preserving 0 black allows any 7.5 IRE normal black levels to key in or remain visible. To get a good key, make sure that nothing in the image you want to key is even close to the 0 black of the background that will be keyed out. Between 11 IRE and 15 IRE in NTSC on graphics is about right for normal blacks if the graphic has subtle shading. So, if you want something to be imported exactly the way it was exported or if you want to make use of parts of the video signal beyond normal broadcast ranges, use the 601 values choice during import.

EXPORTING

There is certainly a joy to working on a general-purpose computer instead of a dedicated graphics workstation. You always have the ability to quickly export a frame from the video application, tweak it in a graphics program, and import it back. You no longer need to call the graphics department at the last minute to make the minor changes that are inevitable as the deadlines get closer. It makes it easy to take any frame and use it as a background, do some simple rotoscoping (painting on the video frame by frame), or isolate parts of a frame with an alpha channel. The trap, of course, is that you will always be counted on to do this once you have shown how easy it is!

Export Templates

Since exporting usually is done with only a handful of the potential formats, most people find the export formats that suit them and ignore the rest. There is also a pattern to the type of exports; users find the video format or the graphic format they prefer and stick with it. It makes sense then to take the settings that are used in exporting most often and save them as templates. You can create

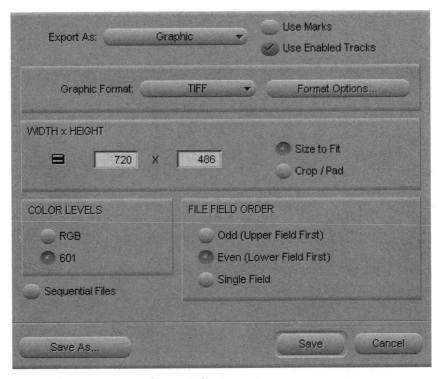

Figure 6.8 Export Settings

and save these templates as User Settings so you can take them with you from job to job with the assurance that you will always get the export right if you make the templates right. It is also great for experienced editors to make them for their less experienced colleagues or assistants, who can use them with an extra level of confidence knowing that they will be correct.

Export Basics

Exporting has its own set of choices, but most of them are decided by the ultimate use of the image. If you are exporting an image to work on and then you re-import back, you want the export and import to be exactly the same. This means choosing 601 video levels and the native frame size. In NTSC that native frame size is 720×486 and in PAL it is 720×576.

If you change the frame size during export, two things happen. First, it takes much longer to export since the application must do more work resizing. Second, and more seriously, the scan lines are disturbed from their precise, standardized relationship. If you export a still at a small size for use in a document or for a Web page, the scan lines are not as important because the image is not going back to full-frame video. A video imported back to the Avid system needs the interlaced scan line information to reproduce the image exactly. This means you should not de-interlace the image in the graphics program if you are going to re-import to the Avid system. If you resize the image, even to a square pixel size like 720×540 or 768×576, when you bring the image back to a nonsquare video playback like on the Avid system, the import process does not know exactly how the original scan lines were laid out. The image will be degraded.

During import, the software puts the lines in a slightly different order from the original if the size has been changed. Forget trying to match back seamlessly to the original video if the scan lines are in slightly different places. You will also experience a loss of resolution because some scan lines are doubled to make up for missing ones. If you are exporting to re-import, always choose the native frame size and do not resize the image at all in the graphics program.

EXPORTING METADATA

An important trend in interchange between systems is the ability to export with rich metadata. Avid was critical in the development of metadata exchange by creating OMFI and implementing it as OMF 1 and OMF 2. This allows users to retain more of the creative decisions when they move to a third-party program. The metadata creation is so important you can think of much of the editing process as metadata management. The interoperability with other devices in the distribution chain is critical to new applications for video. This is the reason Avid created the MetaSync feature and added a metadata track to the timeline.

You will be able to program points in your sequence that line up with actor's individual dialog in scripts, multilanguage subtitling overlays, and interactive television applications not yet invented.

There are several steps to making the metadata exchange work correctly. First, the user must export from the Avid system using the preferred format. Up until now that has been OMF 2, but is currently AAF. The third party must be able to import the metadata format and then, most importantly, the program must know what to do with the data! It is no good importing a rich metadata format like AAF and then stripping out all the information except for cuts and dissolves just because you don't know what to do with the information! Be wary of products that claim to import OMFI and AAF but only reduce it to an EDL once inside the program. You might as well just make an EDL!

Some manufacturers incorrectly made the assumption that OMF 2 was a closed Avid format. This left the door open for products like Automatic Duck's Automatic Composition Import. This product takes the OMF 2 sequence metadata

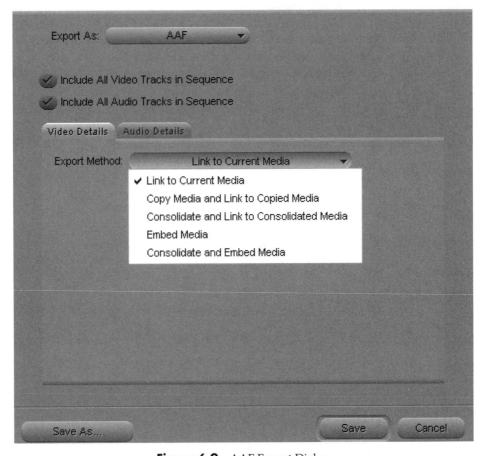

Figure 6.9   AAF Export Dialog

and opens it inside Adobe After Effects version 5.0 and later. Common effects are transferred to After Effects like picture-in-picture, matte keys, speed changes, and collapsed layers. The user can link the metadata to the media already on the drives or you can export the OMFI metadata with the media embedded. You can quickly use the advanced compositing functions of After Effects and then render as a QuickTime movie with alpha channel using the Avid Codec.

AAF has been codified by the AAF Organization (www.aaf.org), which is made up of major manufacturers and broadcasters. It is largely based on OMF 2, but all aspects are controlled by this independent body to make sure that proprietary information is handled correctly. A common misconception is that once everybody starts exporting and importing AAF, there will be transparency between applications. The AAF specification allows for the use of hidden or "opaque" data that benefits certain manufacturers that choose to share the ability to unencrypt this proprietary metadata. And even if the manufacturer chooses to make certain information public to all, there will be varying degrees of success as third parties try to incorporate the data correctly into their programs.

Avid has implemented the export of AAF already for several releases. This will enable Avid to share data better with Avid DS and Digidesign ProTools as well as any third party that correctly implements the standard. Currently there is an extremely high degree of information shared with Avid DS that can be used for conforming high-definition masters from Avid offlines. This interchange saves approximately 10 minutes, conforming each complex effect that was created in offline compared to recreating the effect by eye. Multiply this time savings by the typical number of effects in a one-hour prime-time program and you can see that using AAF between Avid applications can save hours per show during the high-definition online.

There are several choices when choosing to export metadata, described next.

Link to Current Media

This allows any program that opens the metadata to know exactly where all the video and audio media is on the hard drives. It will automatically link without having to copy the media to a new location. This can be valuable when using a third-party program on the same system or working on a very fast network like Unity MediaNetwork. However, if you are working with uncompressed images they are too big to play back over a standard 100 base-T Ethernet network so don't try to link to Current Media.

Copy Media and Link to Copied Media

This option assumes you need to move the media to another system. You may want to copy the media to a slower large drive just for transportation and then copy it again to fast drives once you get to your destination. If you are not careful, you may break the links between the media and the composition, so reduce the number of times you need to copy this media before you open the metadata in the destination application. If you break the links by accident you can use the Relink command (see Chapter 4). This choice is excellent if you want all of the original media of all the master clips used in the sequence.

Consolidate and Link to Consolidated Media

By consolidating first you reduce the amount of media that must be copied and moved. You will reduce the length of the original master clips to a length only long enough to play the sequence and some user-defined handles. If drive space is low, copy time is critical, or you are moving media over a slower network, you will want this choice. You will not have the freedom to recreate the project from scratch because you will not have all the original media, but that may not be important at a late stage in the project.

Embed Media and Consolidate and Embed Media

Embedding the media in the AAF or OMF export is a way of simplifying the transport of both metadata and media by combining it into one file. However, that may be a very big file. If you go over the 2-gigabyte file size limit you may have problems elsewhere. Although modern operating systems can handle files bigger than 2 gigabytes, many applications do not. Besides, the danger of file corruption, the slowness of any import or export and other inconveniences when dealing with such a large file may reduce this choice to being used only for very short segments or highly compressed media.

AvidLinks

This choice under the File menu is meant to simplify exporting metadata to other Avid products. It eliminates some of the confusing choices and makes it easier to get something you can use. If your needs are not complex, this may be the best way to export metadata for interoperating with ProTools, AudioVision, Avid Illusion, and Avid DS Nitris.

Exporting to CD, DVD, and the Internet

There are many different kinds of compression codecs on the market today. Some are proprietary like Sorenson, Radius Cinepak, or Windows Media, and others follow defined standards like MPEG (Motion Picture Experts Group) or Motion JPEG. Some companies have been able to add their expertise to the industry standard compressions to make them higher quality or compress faster. They may be worth the extra money if you plan to do much high-quality compression.

MPEG-1 is best used for smaller playback frames (about 352 × 240), progressive images, and video bit rates around 1.1 Mbps (megabits per second). This makes it an excellent choice for CD-ROM or Internet playback. MPEG-2 was the next improvement, and not only could it handle higher bit rates, it could also encode high definition signals for broadcast. Many broadcasts today, including satellite and HDTV broadcasts, are heavily compressed using MPEG-2. MPEG-4 improved on MPEG-2 by making it more robust, handling lower bit rates and some other features that have yet to gain wide acceptance. MPEG-4 is catching on much more slowly than MPEG-2 although big things may lie ahead for this format. It adds compression for both better quality low bandwidth streams and extremely flexible formats for high quality (like the Sony HDCAM SR 440 Mbps format). Since many broadcasters and production facilities are looking toward a multiresolution world, the ability of a single compression format to handle web streaming, desktop proxy editing, high definition editing and effects, broadcast transmission, and archive is quite attractive. H.263 is a compression format that you may not have seen before because it was designed originally for the extremely low bandwidth needs of teleconferencing. However, it has become the basis for Windows Media encoding and if the quality and speed of Windows Media 9 encoding is any indication of the future, this may become a commonplace format.

Compression for the Internet focuses mainly on the bandwidth of the connection. Many Web sites give the user a choice of video quality for either downloading or streaming depending on the speed of the connection. You may find yourself making multiple compressed copies for such clients and this is a significant benefit to creating batch compression queues.

CD-ROM used to be the preferred method to distribute video electronically, but nowadays, MPEG-1, DVD, and MPEG-2 are quickly replacing both CD-ROM and VHS. Very soon you may find yourself making DVDs instead of VHS dubs at the end of every session for client approval. If this becomes commonplace in your market, I highly recommend investing in a second computer for compression tasks. You may go so far as to invest in an MPEG compression board that will do the job in realtime. Some users simply play back the video stream out of their editing system's SDI output straight to the video compression board. This is a significant cost, but will give you fast, high quality results. Although not good enough quality for final delivery, if you want a DVD to replace VHS for approval dubs, there are stand-alone DVD recorders than can input component video in realtime for less than $500. For everyone else, however, creating a QuickTime Reference or AVI Reference movie is an excellent choice; we will explore this later in the chapter. Export a reference movie of the sequence, send it over the network to the compression computer, start the batch compression processes, then burn to DVD. Create a few templates based on client requirements and it all becomes very automatic at the end of the day.

DVD for commercial distribution may need more care to prepare, and some high-end facilities hire full time "compressionists" for just this task. Not only will you be required to know a little more about creating the interactive menus of the start page, you will have to evaluate the material to be compressed on a scene-by-scene basis. Some high-end facilities prefer that you put your final project to a high quality format videotape like Sony DigiBetacam or Panasonic D5. They can then run the signal through a DVNR system (Digital Video Noise Reduction) to eliminate problematic noise, which will interfere with good compression.

Like most compression types, MPEG-2 makes decisions based on the redundancy of information in the single frame (intraframe). It will throw away pixels that can be encoded in a digital shorthand. MPEG-2 has a temporal compression that attempts to save frames as only the pixels that are different from the frames before and the frames after.

Most DVD MPEG-2 compression uses a scheme called "long GOP" or IBP frames. GOP (Group of Pictures) is a unit of compression that begins with a reference frame (the I-frame or Intraframe) and then the difference frames (the B- and P-frames or bidirectional predicted and predicted). This means that not all of the information about a frame resides inside that frame. You may have to go back to an I-frame to get the full information about all the pixels in a particular frame on a DVD. This makes for a high quality playback at low bandwidth, but it is not so great for editing. Currently the only common form of MPEG for editing is Sony's IMX (SMPTE standard D10), which is a fixed resolution I-frame-only format (generally 50 megabit). IMX is used mostly for newsgathering. Making every frame an I-frame means that you can edit anywhere and not have to worry about whether you have chosen a difference frame for your cut point.

You will have to decide whether you want to compress your MPEG-2 file using CBR (Constant Bit Rate) or VBR (Variable Bit Rate). CBR encodes all of the images at the same bit rate regardless of the content. This takes up more room on the DVD, but ensures that the bit rate will always be high enough to give you a good image; one without the blockiness (sometimes called tiling), banding, or "mosquitoes" (small, moving compression blocks that swarm around titles). If you are putting a relatively short piece to a standard DVD you may be able to stick with a CBR; otherwise, you need to venture into the murky world of VBR.

VBR takes skill to use or a very smart compression algorithm since every scene must be evaluated. VBR also takes longer because the compression program must analyze every frame in the context of the surrounding frames. This analysis makes it a nonrealtime solution. If the project has many scenes of talking heads discussing important topics like life, death, and beauty, it will be easy to compress to a lower bandwidth. If your characters solve their problems by stealing cars and crashing them into each other, however, you will require a higher bit rate to capture all the quick changes from frame to frame. VBR can save space on the DVD by choosing the compression bit rate that fits the material for the best quality playback. VBR will probably be necessary for a full length motion picture on an inexpensive, standard DVD.

DE-INTERLACING

As I mentioned earlier, there is a slight difference between the image in the first scan line and the second scan line a 60th or 50th of a second later if there is motion in the frame. This is usually unacceptable in an exported frame if the end result is a print or is designated for Web use—some kind of a still. One field must be eliminated, leaving only half of the resolution. Since every other scan line must be doubled or interpolated to fill in the empty space left by deleting half the frame, the image appears to be somewhat softer. Photoshop has a de-interlace filter that nicely removes your problem. Photoshop version 4.0 and later allows you to program this de-interlace as an Action or macro. For Web use, you are going to be reducing the quality of this image anyway in order for it to load across the Internet with better than glacial speeds, but enough third-party programs out there blend scan lines, reduce video noise, and limit the color palette so that you should always plan on running your graphics through an extra, final step before their end use.

Another reason to deal with interlacing is when you are working on compositing or rotoscoping software. If you take the interlaced video exported from the Avid and de-interlace in the third-party software, you will reduce image quality problems. Then you re-interlace before rendering and exporting back to Avid.

If you use the 24P standard on the Universal Editing and Mastering options it means that you no longer have to worry about interlacing. Since you are working with a true progressive frame, you can work only on full frames and avoid interlacing artifacts. In NTSC you will save 20 percent disk space, eliminate all 2:3 pulldown frames, and always export sharper stills. If you are working on film originated material or other progressive formats you should seriously consider editing with an Avid progressive resolution. Even with Xpress Pro, you will be able to export true 24P material for DVD creation when working with advanced cadence pulldown 24P material or with material ingested using a Film Composer or Media Composer Adrenaline in a film project (see Chapter 12 for more detail).

SAFE COLORS

There is a danger that anything you do to the image in the third-party program may extend the video and chroma levels beyond the legal broadcast range. The first answer is to use only colors and brightness levels that are already in the image through the eyedropper or any other gadget you have in the graphics program for sampling part of the image. Remember that digital-safe levels are 16 for black and 235 for white. There is usually a window in the graphics program that you can open to allow you to read the RGB levels of any individual pixel if you are suspicious.

This technique is not going to help you if you have applied a special effect filter to your image because the filter is not really restricted by safe colors or brightness levels for video. In fact, most people believe the wilder looking the image, the better! This will be a good way to identify the look of the late 1990s, but for now it poses the problem of checking each and every image before importing back into the Avid. Both Adobe Photoshop and Equilibrium's DeBabelizer have safe-level filters. Many safe filters just lower the brightness level of the image, but since DeBabelizer works with both chroma and brightness levels, it is the better choice. Importing with Symphony's SafeColor enabled for imported graphics will fix any over or under levels during the import stage.

All of this works pretty well if you stay consistent and keep your expectations realistic. You will never get a great still for print from captured standard definition video that even comes close to a real photograph. Never plan to get your print advertising from freeze frames exported from the Avid unless you plan to run serious filters over them for a look that has nothing to do with reproduction accuracy.

IMPORTING AND EXPORTING MOTION VIDEO

The next step for import and export is using moving video. Usually, people want to import animations created by a 3D animation program or an effect sequence rendered from a compositing program. There is also the demand to export for Web pages, CD-ROM, or for material to be used in a compositing program like After Effects. All the already mentioned procedures apply, including resizing and safe colors, but with slight differences. Also, you must decide between two choices for formats: movies or sequences of stills.

Let's look at importing video first. The choice of whether to render as QuickTime, AVI, or PICT/TIFF sequence depends on several factors. The first and most important is: What format does your third-party application have as an export choice? Usually less complex programs on the Macintosh have only one choice—QuickTime. This does indeed make things simpler.

QuickTime

QuickTime has earned a mixed reputation because most people experience it as a postage-stamp, pixelated image that drops frames to stay in sync with the audio. Within the last 10 years, the performance of QuickTime, along with most other streaming media formats, has increased dramatically in quality while using the same bandwidth. Unfortunately there are still inherent sync issues in QuickTime that show up as "long frames" created during capture, and third parties must write special software to compensate as best they can. This is because the system may take longer to actually play certain complex frames and will stall playback briefly, making certain frames stay on the screen longer than other frames. This is great for streaming over slow modems so pay attention during capturing or playback accuracy, but you are not going to playback QuickTime! Not yet anyway. All you want to do with QuickTime is use it as a transfer medium from one program to another. Since it allows conversion without any further quality loss, it has become the de facto standard for transferring video in the Macintosh world. Try to play this uncompressed movie and you might get giant frames that update when they feel like it, but copy it to a drive or send it down the network and it imports clean and shiny, with no loss of quality, into your editing software.

A QuickTime movie is a single large file that must stay under the Macintosh operating system file size limit of two gigabytes (no longer such an important issue with OS X and later). Two gigabytes seems pretty large, but if you are working with high-quality video—this means between 250 and 675 kilobytes per frame—you will reach that size between around 4.5 minutes and 3.7 minutes (approximately 1.65 minutes with uncompressed NTSC). Usually when a project requires a long QuickTime movie for playback, it is a much reduced frame size and resolution or you should consider breaking the sequence into smaller chunks.

The Avid Codec

The key to getting a QuickTime movie rendered at high quality that imports quickly into the Avid is to use the Avid codec. A codec (compressor-decompressor) is a system extension that any third-party program can access because it is in a central location on the host computer. On the Macintosh OS 9 and earlier it is in the System folder along with other extensions and in OS X the codecs go in System:Library:QuickTime, where System is the name of your boot drive. On Windows XP, it is installed automatically in the C:\Windows\system32 folder. The Avid Codec is a .qtx file; it allows other programs to compress and decompress a rendered QuickTime movie using the Avid media file format. This means that even though you are rendering a QuickTime movie, you really are creating an Avid media file that is inside the QuickTime format. Technically, this is called *encapsulating* or *wrapping*. To work best, use the native frame size and do not mix resolutions in the exported sequence. You can change the resolution of the Avid Codec based QuickTime movie during import if you must, but it will slow down the import significantly. This is especially important if you have only one resolution of an animation and you must use it for offline and online. With the Adrenaline and software-only-based Avid editing systems you can mix and match resolutions in the same sequence, so create the animation at uncompressed quality and keep it that way when you import.

Make sure that you load the Avid Codec into any Macintosh or Windows system you are using for graphics or animation; you will be given Avid resolutions when you choose the quality of your rendered QuickTime movie. Send the

Codec to your graphics people or to anyone who is subcontracting graphics for you. And by all means, make sure it is the most recent version! You can get that information by calling Avid Support or by downloading from the Avid web site (www.avid.com); be aware that the version of the Codec may change on a different schedule from the software itself. The reward of using the Codec and native frame size is the almost realtime import speed as you come back to the Avid.

QuickTime Reference Movies

A new option has appeared designed to save you time and money when exporting if you have set your network up correctly. A QuickTime reference movie is a series of pointers, a small amount of metadata, about where the original media is on the drives. When you export a reference movie you send this small amount of metadata to another computer on the network. The other computer can use it for compression using a program like discreet Cleaner or Canopus ProCoder, or a hardware-based compression system like Anystream or Telestream. The compression system loads the QuickTime reference movie like it was a real Quick-Time movie and the compression software looks for the media in its original location. It then grabs the frames and compresses the media without slowing down or disturbing the main editing system. This now becomes a very fast background task to create a movie for the Web or for a DVD.

Size

Size of the frame is still a consideration for importing QuickTime, for all the same reasons. Refer back to the chart for proper frame sizes for the different resolutions. With a still image, maintaining the correct frame size is important because of the scan-line relationship, but there is now another consideration—import speed. If you render at anything other than the native frame size, you have to double lines to make up for the difference or throw away resolution and waste rendering time. You are also wasting import time since the Avid must resize each frame on the fly, which adds a considerable, unacceptable amount of extra time to the process.

The other choices involved, when rendering to import to the Avid, have to do with field order and field rendering. To get the absolute best-quality rendered movie, choose field rendering if your application makes it available. Field rendering takes longer to render, but it is worth it if you are working with complicated video and lots of detail and you want this to be a finished product. This is because you want the movement of your animations to be as smooth as possible and if you render using only frames then you are giving up half the motion resolution. With field rendering you get all 50 or 60 fields available to you rather than 25 or 30 frames. The extra fields smooth out motion of moving objects by giving you more discrete images within the same amount of time.

If you are working in a 24p or 25p project then you are working in progressive frames so a field rendered animation won't help much. In all other projects, however, you will gain significant quality improvements by taking the extra time to field render.

Importing with an Alpha Channel

With current versions of Avid editing software, you are able to import Quick-Time movies with an alpha channel attached. With the addition of a 32-bit QuickTime Codec, Avid has added a new choice for rendering and importing QuickTime movies with third-party compositing applications. Since 24 bits are used to represent the color and detail of the foreground, the extra 8 bits are used for the 256 levels of luminance for the moving matte or alpha channel. So a 32-bit QuickTime Codec allows you to render the alpha channel and the fill or foreground as a single movie and import it as a fast import. This makes Quick-Time movies easier to move around (although bigger files) and simpler to use since the effect comes in as a premade realtime matte effect.

AVI

The Avid AVI Codec does not support an alpha channel so it is less useful for rendering animations in third-party programs. It is still useful when exporting to a compression application or creating video to be placed on a Web site or Power Point presentation. In many ways Windows Media will begin to replace this format since it is extremely fast and high quality for highly compressed videos.

Importing Sequential Files

Most people choose QuickTime for animation because it is just one large file that is easy to keep track of and because it is the early universal choice of most animation programs. There are times when the animation program may originate on a computer platform that doesn't support QuickTime, such as a Sun Workstation, or you are choosing to work in a graphics program to affect each individual exported video frame. You may also want to avoid the 2 gigabyte file size limit of QuickTime and AVI. In these cases you want to work with a PICT or TIFF sequence.

There is a little twist involved in getting these stills sequences to import to the Avid, which has to do with file naming. Any PICT or TIFF sequence must be recognized as a sequence and not as a series of unrelated stills. The way to be sure that you have a PICT or TIFF sequence is with the filename, which must follow a strict format of: Name.001, Name.002, Name.003, and so on. The name must be followed by a period and a number in older versions, but not in current versions. With the older versions, you must get the PICT filename right or you will be faced with running every file through a renaming program like Knoll Renamer, or creating a macro in DeBabelizer. These tasks are not extremely time consuming, just tedious and distracting from the work at hand (especially if you must track down one of these renaming programs while the clock is ticking).

The first frame chosen for the import will be the first frame of the sequential animation. The end of the sequence will be the last consecutively numbered file. This is different from older versions where the system will look for .001 to be the first image in the animation. If you have started with .000, then that frame will be skipped when the sequence is created. On older systems, if there is no .001, the system does not see the rest of the images as a sequence. Today the system will autodetect sequential files if a series of files with the same name begins with and has sequential numbers in the filenames. Occasionally, if you have a series of images captured with a digital still camera they will be numbered automatically by the camera and look like a sequence to the Avid. If you do not want the system to assume these images are a very short sequence, you can turn this option off in the Import Settings (or rename the file). The finished PICT or TIFF sequence should come in as one video frame per image.

The main benefit of working with PICT or TIFF files to import into the Avid is that these file formats support an alpha channel. Remember that the alpha must be inverted from what most programs create as a default. Remind your animator or invert on import. Either way, make sure you know whether you have white on black (need to invert) or black on white (import without adjustment) before you import so you don't do it twice! Import one frame first to check the setting before importing the entire sequence.

This method is simple and relatively problem free unless the animation takes up thousands of stills. The Macintosh operating system will slow down to a point where it will be impossible to open a folder with too many images inside. You may want to break any animation into smaller sequences if you are concerned about filling a folder on the desktop with thousands of files.

Consider how easy it is to set up a network between an SGI and your Avid as an Ethernet or faster TCP/IP. TCP/IP is a protocol that you can use between two computers in your facility (technically, an intranet if done this way). This simplifies the transfer between computer platforms admirably and you no longer need to worry about a removable drive, compatible hard drive formats, or the time to copy and restore. Besides, most SGI workstations don't use a floppy drive.

The original source for your PICT or TIFF sequence may have come from the Avid. In this case, you are dealing with many individual PICT files and may be using these to go to an SGI that doesn't use QuickTime. Or you may want to paint or draw on each individual frame and paint out wires, boom microphones, or ex-spouses. This is time-consuming because of the huge amount of work involved in 30 or 25 hand-painted frames per second. It is the ultimate "fix it in post" and last resort for fixing problems created in the field, but the results can be magical if done right. The Avid intraframe tools are adequate for this task, but other third-party animation programs may be better if you need this often.

The Avid plug-in format, called AVX (Avid Visual eXchange) simplifies the filter process even more. You can buy a version of your favorite filter that has been converted to AVX and apply it in the Avid like any other effect. There are plug-ins like Profound Knowledge's Elastic Gasket that makes many third-party After Effects plug-ins available on your Avid. You cannot directly use the standard Adobe plug-ins. There are also excellent title animation and DVE programs available from Boris Effects, and wild creative "looks" from GenArts Sapphire collection.

Using OMFI for Pro Tools

When moving a project with media to ProTools or AudioVision there are few elements that are critical.

- All sample rates must be the same. You cannot mix 44.1 kHz and 48 kHz in the same sequence. If you have mixed sample rates you should create a copy of the sequence and convert the sample rate. Select all the sequences and choose Change Sample Rate under the Clip or Bin menu. If you are consolidating the media during the export you will have the option to convert the sample rate in the export dialog.
- Macintosh systems are not compatible with WAV files. If you have WAV files and are moving to a Macintosh or an older ProTools or AudioVision you may have to convert the files to AIFF-C. You will have to do this while embedding the media into the composition. When you choose OMF 2 export and embed you will be given the choice to convert the file to AIFF-C. You must embed the media when you export OMF or AAF in order to convert it from WAV to AIFF-C.
- Macintosh systems cannot mount drives striped together from a Windows system. A Macintosh system may have difficulty playing from any Windows formatted drive. You may use a Windows formatted drive for transport, but the media will be copied to faster HFS or HFS+ formatted drives once the media arrives at the audio studio. You can mount HFS drives (even stripes) on a Windows system using third-party drive mounting software like Mediafour's MacDrive, but again there may be performance issues and the media might have to be copied. Find out the platform, format, and the sample rate preferred by your audio facility.

If you are moving the audio to a Digidesign Pro Tools session, there are two methods to consider. The first method requires less drive space and is faster, but the second method allows more flexibility.

The first method is to hand over the drive with the audio media files on it and create an OMFI file that is composition only. You may want to consolidate your sequence before you do this if you want to put all the audio media on another drive for transport to the audio workstation. This allows you to keep working with the audio files you have. You might want to lock your audio tracks in the sequence so you don't accidentally change something while the mix is going on.

The second method is to make an OMFI Audio Only file. Consider the second method if you are going to use software that does not recognize the Sound Designer II format that Pro Tools and Avid Macintosh editing systems use. This creates an intermediate file and converts the audio to another, more widely used audio format, AIFF, along with all the edit information. Once this very large file is moved over to the digital audio workstation, it can be converted back to an SD II file for use with the Macintosh Pro Tools software. The intermediate OMFI file is opened in the OMF Tool that comes with Pro Tools and converted to a Pro Tools session. The original OMFI file can be deleted after the conversion.

The AvidLink for ProTools will allow you to choose either method. The AvidLink for AudioVision assumes you have the correct format audio file and saves only as an OMFI composition. You will need to copy the audio files to another drive manually or through a standard export dialog. Since AvidLinks is simplified workflow, if you have not created the audio files in the correct format you may have to use the more complicated method as outlined earlier.

To send the audio mix back to the editing software, you must "bounce" the audio tracks to a continuous audio track. This realtime process changes the audio file, which now has subframe edits, to a frame rate that can be used by the Avid. Alternately, you can output to a digital tape format, recapture it into the Avid, and line it up to the beginning of the sequence. Having some sort of synchronizing beep tones with a countdown or flash frame simplifies this final sync.

ADOBE AFTER EFFECTS

When preparing a composition in Adobe After Effects, you should always use certain settings when rendering for the highest quality:

- Always use 29.97 fps when exporting from Avid and rendering from After Effects (AE) in NTSC (not 30 fps). Of course, PAL is 25 fps and 24P is 24 fps.

- Graphics or other elements should be created at 720 × 540 (NTSC) or 768 × 576 (PAL). This is an accurate square pixel representation of the TV screen. Graphics are then resized in AE to the 720 × 486 (720 × 576) D1 pixel size.
- Whether one chooses to work in AE at 720 × 540, 720 × 486, or 648 × 486, it is vital that the final render takes place at 720 × 486. For PAL, the proper comp output is 720 × 576. This can be done by creating a comp in the final correct size, dropping your animation into it, and using Scale-to-Fit (Ctl-Alt-F/Cmd-Opt-F).
- If you need to work at 16 × 9 you can use the widescreen selection when choosing the pixel aspect ratio under new composition. This will keep the project the correct frame size for anamorphic standard definition. If you are moving material from the Avid to After Effects and then back to the Avid, using the widescreen pixel aspect will keep the material from being scaled twice.
- If field rendering in AE, choose upper field first when going to an ABVB system or a PAL Meridien system. If going to an NTSC Meridien system, it should be field rendered lower field first. All Xpress DV systems require lower field first.
- If working in 24P, do not field render! This is a progressive format and does not use fields.
- Using the Avid Codec is currently the best conduit for going back and forth between Avid and AE.

If rendering a graphic for compositing in an Avid, this is the best way to deal with the alpha channel:

- Always render a QuickTime movie with an embedded alpha channel because batch import allows for more control of the files once they're in the system.

When you render in After Effects:

- Select the comp in the render queue and choose Add Output Module from the Composition menu.
- One output module should be set up to save the RGB+, and set your color to Straight (Unmatted). Only DS uses premultiplied mattes in the Avid product line.

Importing and exporting video, audio, and graphics has many variations, formats, and choices. With this flexibility comes complexity, so any production company should find the processes that work best for it, simplify them as much as possible, and be aware that you have many tools at your disposal.

CONCLUSION

Learning to import and export graphics, animations, and metadata correctly from your Avid system means you position yourself as the hub of the creative process. You are in more control of the final result of any project and can confidently maintain optimal quality at every stage. You can collaborate better with graphics artists and animators as well as properly preparing your material for distribution on the Web or DVD. This makes you indispensable as both a technician and an artist.

7

Introduction to Effects

The effect capabilities of Avid editing systems are surprisingly deep. Even with the more limited layering features of the Avid Xpress, wonderfully complex and professional results can be achieved very quickly. One of the leaps forward in capabilities of the past few years is that the speed of CPUs and hard drives along with greater PCI bus bandwidth has made realtime effects more common. Many of the old restrictions of numbers of video streams and realtime effects have been shattered by the move to software or host-based Avid systems. However, some of the reliability of the hardware-based systems is gone as well. Now the capabilities of your system rely on the configuration of off-the-shelf computer parts rather than custom built Avid hardware. Numbers of streams and realtime effects may vary from system to system based on the host computer and not on Avid's hardware expertise. We will explore the capabilities and implications of this brave new world in this chapter.

With a Meridien system you can have two streams of video with one realtime effect, a color correction effect on each stream, and an uncompressed title all play at once in realtime—along with eight channels of audio. Your system will be able to create between 8 and 24 video tracks. Even with fewer video tracks, you can nest, collapse, and mix down to extend and expand any system's potential.

The trick to maximizing your system's resources is to do creative work in realtime and then render for superior quality. The last time I checked, time was still money and the best way to use the time with a client present is to show them multiple versions and make changes quickly. The lines between cheaper/slower and expensive/faster are blurring, especially if the preview quality is good enough to make important decisions about the final version. Incorporating faster CPUs and hard drives means that effects done on Avid systems can compete with much more expensive workstations.

There is much to be covered to deal completely with effects—too much for this book—but some basics can get you past the beginner stage. There is nothing like the experience of a hands-on class, and Avid offers several. This chapter can only hint at some of the techniques you will discover with enough time to experiment.

As you become more confident with effects, you will also become serious about nesting. Nesting is the feature that gives you more levels of video layering than you could ever practically use except for the densest of graphics sequences. Nesting gives you incredible power with an extra level of complexity. This chapter will discuss nesting after discussing the basics.

Adrenaline Systems

Over the last few years the speed of the PCI buses has increased and CPU, or host, processors have increased in number and speed and, finally, they have reached the stage where they can outperform most dedicated video boards for standard definition video. Without the restrictions of a pipeline architecture where the video streams are limited by hardware, you can get many more streams of realtime video and effects. However, there is no guarantee of any particular performance, everything is dependent on the types of effects, the speed of the drives, the complexity of the video, the resolution of the stream, the resolution of the images before and after, what other processes are going on in the background, and so on. In other words, it is very easy for the demands on the system to be so complex that most humans cannot predict whether something will always be realtime.

The Avid host-based system can make recommendations for rendering through colored lines in the timeline. In practice, all effects become conditional. But if the system shows a red line, then something absolutely has to be rendered. If it shows yellow then you may be pushing your luck. The benefit of the nonguaranteed realtime effects is that when you are being creative you can drop a few frames here and there to previsualize the timing and the overall appropriateness of the effect. You can tweak, play, tweak, play, and so on without the intermediate step of rendering. Of course, when it comes time to play out to tape for a master or to compress for DVD, the system will conservatively offer to render everything that is questionable. You can trust the system to understand that something is better off being rendered before final output.

So why do you need hardware at all? Input and output will still demand compression and decompression and conversion between formats and sampling rates. Although the Avid Adrenaline system can mix and match any compression type along with uncompressed standard definition, you may want to convert an analog signal to digital and back to analog at the end. It still requires hardware to do this with high quality if you don't have a digital video deck that will play analog formats.

TYPES OF EFFECTS

There are three kinds of effects: realtime, nonrealtime, and conditional real-time. On a Meridien or ABVB system, realtime effects can be handled by the compression and video boards without rendering as long as there are only two sources of video playing at the same time or there is only one effect at a time. Color correction can be realtime on each of two video streams without using up the one effect restriction on Meridien systems. If more streams of video or multiple, simultaneous effects are desired then rendering is necessary. The exception to this is the ability to play a realtime effect and an uncompressed title. These titles are put into a downstream key (DSK) and are separate from the standard playback restrictions. The only drawback is that in order to scale them up or down in size, move or crop, you must disable the DSK checkbox on the title in Effects mode.

With the host-based systems, realtime is constrained only by the quality of the video. If the images are highly compressed, they take up less bandwidth and you should be able to get many video streams even above four or five. Ironically, it is easier to play effects on multiple streams of uncompressed video than on high-quality compressed video. This is because the computer's processor must work harder to uncompress and recompress the images in the normal course of using them for composites. However, the uncompressed video needs the faster drives, either fiber or SCSI, to really get the full capabilities for the Avid system. The ability to play a realtime effect is the combination of the effort of the computer's processor and the amount of media traveling across the PCI bus.

When an effect is not capable of being played back in realtime on a Meridien or ABVB system, the effect will not play at all. Since the number of video streams is limited by the video boards, the system will play only that maximum number of tracks. The system makes an analysis from the bottom track up (V1 and higher). As soon as the capabilities of the system are reached, the higher tracks simply do not appear until rendered or until the user stops on a frame and the system "renders on the fly." With the host-based systems, more tracks will attempt to be played. If you run out of CPU power or bus bandwidth, the system will begin to stutter or hold frames until it can update the image again. This is extremely useful for quick previsualization of a complex effect. Don't worry about the dropped frames; when you render the effect you will get them all back!

NuVista, ABVB, and Meridien systems are all dependent on DVE hardware manufactured by Pinnacle Systems. Avid created the user interface and integrated the hardware into the video flow. The Adrenaline and Xpress Pro and Xpress DV systems now rely on the Open GL graphics card that shipped with the system. You can continue to upgrade your OGL card for very little money year after year to get better quality resizing and edges. The Nitris systems have an Avid custom designed DVE that is focused on maximum quality as well as offering many more realtime streams than ever before.

Realtime effects have an orange dot over the effect icon when unrendered. On a Meridien system they can take advantage of a fast-render scheme that allows them to render at between eight and six times realtime using the video board. One of the main advantages of a host-based system is the ability to use the faster CPUs and large amounts of RAM to speed up any rendering. A faster PCI bus and an operating system that takes full advantage of all the bus speed available will also make a difference. In practice, all realtime effects become conditional on an Adrenaline or host-based system. This is because realtime is always determined by the capabilities of the computer and the context of the effect in the sequence.

A nonrealtime effect is too complex to be dealt with so quickly and sports a blue dot once it is in the sequence. With the host-based systems, nonrealtime effects tend to be some AVX plug-ins, and certain types of motion effects. The system will always try hard to play something, but the results may be unpredictable. Even fancy wipes that you shouldn't use anyway (a.k.a. "weasel wipes") will play a realtime preview. Clearly, if a realtime effect does the trick, it is preferable and your effect design should take this into consideration. You may want to substitute a realtime effect as a temporary replacement for the final nonrealtime effect just to get the timing correct with a realtime preview. It is easy enough to replace one effect with another after all the multilayered video and audio timings are perfect.

A conditional realtime effect is an effect that, under certain conditions, may switch from being realtime to an effect that must be rendered. It is not a nonrealtime effect, however, and the colored dot is green, not blue. The green dot indicates that the conditional realtime effect can still take advantage of the fast rendering scheme on a Meridien system. They are not nearly so slow to render as a true nonrealtime effect.

Realtime effects also become conditional when there are two effects on the same master clip. If you put a dissolve on a segment effect, the dissolve becomes a conditional realtime effect. It may or may not play depending on the segment effect. Defaulting to the transition effect for the conditional effect ensures that usually it is the shorter effect that needs to be rendered.

Certain effects that reposition the image like the PIP or the 3D warp have two settings for quality on the host-based Media Composer and Xpress Pro. This is because the resize algorithm can be fast and OK or slow and beautiful. You need the fast version when you want to view the effect in realtime and it may be good enough for the final effect. Or after the effect is in place and the timing is correct, you will choose the slow and beautiful setting and render. This is the nuclear green HQ button. You can change the quality when you render the individual effect, or when you "render in to out." There is a

Render Setting in the Project window that allows you to change the quality to standard, highest, or as set in each effect. You may have a mix of some effects that have been set to HQ in the effect interface and others that have been left in standard quality. You can choose to leave them that way to render everything more quickly or you may just want to take a long lunch and switch them all to high quality.

The one other time an effect becomes conditional is on ABVB and Meridien systems when two 3D effects have completely different shapes; say one is a page turn and the other is a water drop, within five seconds or less of each other. One effect will need to be rendered.

When layering effects on top of each other vertically, the material on the top track always has priority. This is not a true multichannel digital effect device in the way most people think of standard DVEs. It is more like having many single channel devices. Each video track can be considered a separate channel of effects and, with nesting, much more than that. It means that the separate video tracks do not interact with each other because they are each like separate sequences. This modularity allows quick exchanges of shots when it comes time to modify an effect. It also means that if moving objects are going to change their layering priority on the screen, then the lower object must be moved to a higher video track.

In this CPU-dependent world with so many shades of realtime even the Digital Cut dialog has a choice called Effects Safe. This checkbox is the last chance to make sure everything will play out to tape. As you would expect, this setting is a little conservative. If the system can figure out the minimum to render before being absolutely positive there are no dropped frames or any other problem with the many layers of realtime possible today, you should let the system take over. Think of the system as using a "manumatic"

Figure 7.1 Digital Cut with Effects Safe

method of looking ahead and saving you time by doing the right thing before a digital cut.

EFFECT DESIGN

Good effect design tries to achieve the most spectacular effects with the simplest use of layers. The fewer layers used, the fewer problems with trimming and rendering in a track-based effects model. Simpler design means most of the time you can modify faster because it is easier to figure out what is affecting what. If you can do something in fewer tracks, it looks better and renders faster.

Tree-based compositing is extremely powerful for creating graphic representations of the effect flow. This is the type of control offered by the DS Nitris system and some other third-party programs like Eyeon's Digital Fusion. You create branches by connecting effect nodes that could have mattes fed from one branch to another. Intermediate results (a traditional "work part" or submaster) can be connected to advanced controls without the restrictions of the Media Composer effect interface. The tree opens a whole new world for deep, complex effects that can be understood by this graphic signal flow. But all effects happen over time so there still needs to be a timeline and keyframe aspect that is tightly integrated with the tree. Consider signal flow, the order of effects and the way they change over time when designing any composite.

RENDERING

Rendering effects can be reduced significantly by using some basic strategies. In general, you render only tracks that are combined with nonrealtime effects. With so much realtime capability these days, you should try to see if something plays without dropping frames before you consider rendering. Whenever you render an effect, you are rendering a composite of everything below. If you want to play the tracks below by themselves you can move the video track monitor down to lower tracks or stripping off the very top tracks to make multiple versions. Otherwise, you can leave these lower tracks unrendered. One simple method to rendering only a top track is to put a submaster effect on an empty track above the effect sequence. Put Add Edits in the empty track on either side of the area to be rendered and then drag a submaster effect between them. By rendering the submaster effect you are assured of rendering only the top track, if there are many effects sequences in a row this can be a timesaver.

What confuses people is that many times there is not just one track available to render as the top track. The beginning of the show may have

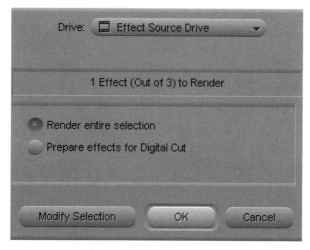

Figure 7.2 ExpertRender

a complicated layering section that has 10 layers. Then most of the show may not go above track three and the end has five tracks. The best solution is ExpertRender™.

ExpertRender

ExpertRender is a feature designed to let the intelligence of the system solve your rendering problems for you. If the main problem with rendering is that people render too much, then the best solution would be to make sure everyone feels comfortable with a minimal style of rendering. The reason people render too much is that they don't really know what will play in real time and what won't. This is because many effects are conditional and depend on what else is going on in the sequence. Don't take the time to step through a long and complicated sequence effect by effect and still, perhaps, guess wrong. It is easier to mark in at the beginning of the sequence, out at the end, turn on all the video and audio tracks, and use render in to out. The expert part of the feature will leave as much realtime as necessary and render only what is absolutely necessary.

If you don't use Expert Render then rendering in to out may be simpler, but there are some definite drawbacks. The first is the extra time involved in rendering effects that will play just fine in realtime. The other problem is that when you render graphics or titles that are using the Meridien DSK (downstream key for animations and imported graphics), you are taking an uncompressed image and bringing it down to the level of the rest of the sequence. Of course, if you are working at 1:1 resolution, then this will not create a noticeable difference, but at 3:1 or DV25 you may be able to see the difference with complicated images. Lastly, any time you want to go in to change any of the effects that could have been real time, you are unrendering them when you make the changes. You may think that you are all done and ready to go to tape, but your client may have other plans, and you want to stay flexible to the very last minute.

The Meridien system knows what it can theoretically play and what it cannot. When preparing to play a sequence, the Meridien system must "build the pipes" and allocate the hardware resources on an effect-by-effect basis. The Avid player must look at each effect and say, "I have a video stream available for this effect and a DSK free for this effect, but this one must be rendered." This all happens at a very deep layer of the programming code and is very basic to the functioning of the machine, so it is very dependable. Avid can reveal this capability to the user in the form of ExpertRender. By marking in and out, turning on all the video tracks, and choosing ExpertRender in to out, we tell the system to do an analysis before rendering. The results are then displayed in the form of a dialog box that says "X effects out of Y will be rendered," and it will highlight those effects in the timeline. You can now see the direct benefit of how many effects you do not need to render. They will now stay realtime with a guarantee that they will play. The results can be astounding with only a handful of effects out of hundreds really needing to be rendered.

There are times when you may disagree with the Expert. Specifically, you may have plans for a certain section and you will be adding more effects to a higher track when you are done with the rendering. In this case, the system cannot read your mind to know what you will do next and can return only results based on the existing sequence. You can then choose to Modify the ExpertRender choices. By clicking on Modify in the ExpertRender dialog, the Expert leaves all of the chosen effects still highlighted in the sequence. You can Shift-select or Shift-deselect as you see fit and press the regular Render button when you are done. There is no need for a render in to out again because all the effects are already selected.

Another time you may disagree with the Expert is when you have dissolves between titles. This is relatively rare and really should be treated as an exception. In this case, the system will realize that the realtime dissolve cannot be played in real time, but because of the order that it must allocate resources, chooses the titles for rendering. Again, the user can override this situation easily, pick the shorter answer, and render the dissolve only. Or you can dissolve titles using the Fade Effect button, which creates keyframes that do not need to be rendered. This limits you to fading up and down; if you want to dissolve between titles then the best method is still a dissolve.

The beauty of this automated analysis is that a vast majority of the time the choices are the shortest rendering answer. In reality, you will save so much time by letting the Expert do the job for you that even the occasional over-render is easily overlooked. How much time is wasted stepping through effects by hand? The guarantee that the system will be able to play the entire sequence after the ExpertRender is, all by itself, money in the bank.

Partial Render

Partial Render is the ability for the system to render only part of an effect at a time and then come back later and pick up where it left off. This allows you to start a render at any time, even if you know you don't have enough time to finish rendering the entire effect. By pressing Ctl/Cmd-period you can escape from the render and keep or discard what has been rendered so far. This is especially useful if you have a series of slow, blue dot effects and you can start to render a little more anytime you take a break.

The system will create a new precompute for each partial render and tie them all together to play the final effect. You can see how much of an effect you need to render by changing the Render Range in the timeline view to show Partial. This is the most useful setting and should be left on most of the time since it will show you only what is left of an effect that has started rendering, but not quite finished. This is the default setting on current systems. If you change the view to All, then it will show you all effects in the sequence that are not rendered. Although this can be useful under some conditions, it can be confusing if you are relying on ExpertRender to figure out what needs to be rendered. The visual feedback in the timeline of the red or partial red line across the top of the clip with the effects can clash with the information from dupe detection. I would strongly suggest that Render Range display and Dupe Detection not be turned on at the same time. If these two functions are important to you, they can be made part of a workspace and changed with a single keystroke.

The only other drawback to Partial Render is that with long-term complicated effects projects you will be generating more precomputes. If you have been cleaning up precomputes once a week to keep the system operating without the high, unnecessary overhead of too many small files, you may want to do it more often. Most of the time, however, you won't even be aware that Partial Render is at work. The best features, many times, are the ones that make you more productive without attracting attention. Long after the sizzle of the product demo is over, you will be making your deadlines with projects you are proud of, and you won't really care why!

SOFTENING

As layers are added to a sequence and there is more detail in the image, some extra compression of the image is required to ensure playback at that resolution when you are using the ABVB or the host-based systems (Meridien systems do not have a Soften option). If the combined complexity of all your images merged together exceeds the ability of your drives and computer's bandwidth to play back, you may have to reduce the amount of detail in the rendered

Figure 7.3 Adrenaline Render Setting

image. Minor amounts of compression are handled invisibly, but when the level of complexity gets too great, you need to soften the images in the effect.

If at all possible, never soften an effect on purpose. Not only is it unacceptable for finished quality work, it slows down the render on ABVB systems. On ABVB systems this difference is dramatic since the system must drop out of hardware rendering to use the (by now somewhat ancient) processor for software rendering. If you are cautious and up against hard deadlines you might want to set the Softening setting to Ask for Permission to Soften. If you choose Do Not Soften or if you choose not to soften when the Permission to Soften dialog box pops up, at a later date you might get an underrun error message. This is because the drives cannot keep up with playing back the level of complexity of that image. If you get the underrun error, you should do almost anything else to improve playback before softening. You can reduce the amount of audio tracks with an audio mixdown or render all the realtime effects around a trouble spot to help simplify that part of the sequence.

The Permission to Soften message you might get during rendering is a little conservative. Avid engineers are always planning for the worst-case scenario of a maximum amount of audio tracks, rubberbanded audio, and highly complex realtime effects when estimating the ability to play back a complex image. You may never get near that level of demand on your system in normal work and almost never at high compression rates like DV25. This means that if your sequence is simple, but the individual effect is complex, you have some wiggle room to push the machine beyond the prescribed limit for a short time. You might also want to set the threshold to 120 percent to push the system a little harder. This means you are alerted only if the effect is seriously over the supported limit of complexity. I personally always used to put this setting to Don't soften and almost never had a playback problem because of effects; however, if you have nonstriped or slower drives, this occasionally may cause a problem with high quality levels, like with 10-bit uncompressed clips.

If you have the time to stop and troubleshoot an area that underruns by trying some of these solutions, you usually will have overall better-looking programs than if you leave the system to soften automatically when needed. If you can never take the chance of an underrun during playback, you may want to make the quality trade off to soften when necessary in favor of dependable performance.

KEYFRAMES

Almost all effects can be manipulated by keyframes. The only exceptions are the color corrections and a few other segment effects like flip and flop. Keyframes are the method to change an effect over time, and always need at least two keyframes, beginning and end, if you want the parameters to change. At the first keyframe, certain values about position, shape, or color are entered, and the settings change to match the values on the next keyframe. The change, if there is any, is smoothly interpolated between the keyframes. Keyframes can be added in Effects mode on the fly while playing and pressing the Keyframe key (the apostrophe key is the default on Symphony and Media Composer; Xpress systems use the N key). Keyframes can be copied and pasted, dragged by holding down the Alt/Option key, and can be moved with the Trim keys. If you want the effect to just hang on the screen, with no motion, then you can copy and paste the same settings between two keyframes or highlight both keyframes when changing parameters.

Much has improved with keyframes and some of the old advice no longer applies. Avid significantly improved the keyframe model for certain effects and will, over time, migrate that functionality to all the effects. Now the user can choose to work in the "classic" keyframe mode or promote the effect to the new keyframe model. You now have a keyframe per parameter—the ability to change each parameter with its own timeline and set of keyframes. Using the new model, you need only one keyframe if you want the parameter to change from the default but remain the same throughout the effect. As we will see in the next section, you also have a wide range of choice as to how the motion between the keyframes is interpolated and how trimming affects the timing.

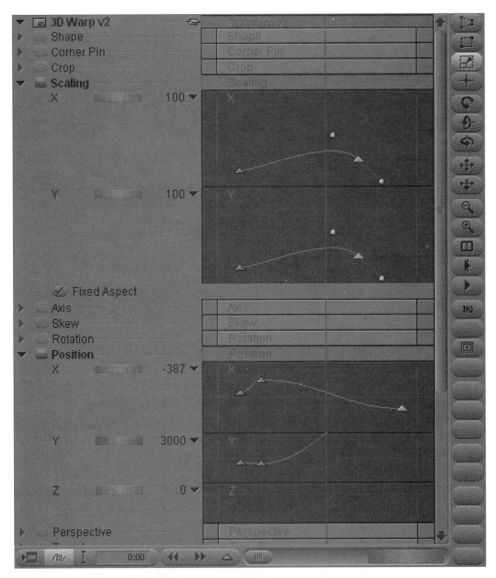

Figure 7.4   Advanced Keyframe Window

Advanced Keyframe Model

Avid keyframes were given a major overhaul in Xpress v5, Media Composer v11 and Symphony v4, which added power and complexity while preserving much of the old keyframe methods. You can choose to promote several effects to the advanced keyframe model as desired once you open the Effects editor in these later versions. Of course, you can keep the effect just the way it came to you from an older offline machine. But if you need to add some more sparkle to a project you can do much more while staying in the Avid program.

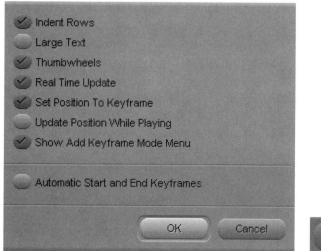

Figure 7.5 Effects Editor Setting

Look at the bottom of the effect in the Effects editor for this pink multitimeline icon. By clicking on this all of your parameters are preserved (except Acceleration, which is replaced by something more powerful) and the keyframes are moved into a "timeline per parameter" effects interface. You can now add keyframes only to specific parameters and can add different kinds of motion interpolation much more sophisticated than simple ease-in/ease-out.

Creating, Deleting, Copying, and Moving Keyframes

When you move away from the concept of a single keyframe affecting all the parameters in an effect at the same time, you open some interesting possibilities that are also more complex. To maintain speed you need a series of choices for basic keyframe housekeeping. The first is a series of choices for adding exactly as many keyframes as you need.

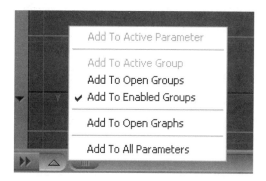

Figure 7.6 Add Keyframe Menu

Hold down the mouse over the Add Keyframe Icon in the Effects editor and look at the choices. Let's look at why you would use each one.

Add to Active Parameter

This is legacy choice of adding a keyframe to only the parameter that has been last activated by clicking on it. This choice is available if you have an active parameter chosen. It will add just one keyframe to this one parameter. If you have X and Y parameters of scale checked for Fixed Aspect, however, you will get keyframes on both parameters even with this choice.

Add to Active Group

This will add a keyframe to an entire group of parameters like Position. If Position is Active (selected and highlighted pink) then a keyframe will be added to the X, Y, and Z parameters simultaneously. This is quite handy if a specific effect needs all axes to line up precisely at the same time.

Add to Open Groups

Like the previous choice, this will add a keyframe to all parameters that are part of a group that has been opened (the small triangle has been spun down to display all the sliders).

Add to Enabled Groups

When you promote a 2D effect to a 3D Warp effect you get the enable buttons. These buttons allow you to adjust a parameter and then just disable it without resetting it to the defaults. This is a quick way to see multiple versions of the same effect since all those disabled parameters are still embedded in the effect, ready to be turned on to see the alternate version later. If several parameters are enabled you can choose to add keyframes to all of them with this choice.

Add to Open Graphs

This is a different way of thinking about convenience. You may have many parameters enabled in a complex effect, but to save screen real estate you have only the critical parameters showing the full keyframe graph. Rather than spend time scrolling up and down to make sure all unwanted parameters are disabled or closed, you can focus only on the open graphs where you are doing all the work.

Add to All Parameters

Not so sure about this Advanced Keyframe nonsense? You might want to go
back to adding a single keyframe for all parameters and worry about which
ones to tweak later.

Here are two sneaky shortcuts that we designed to make it even easier to
use the power of keyframes per parameter. If you right-click or Shift-Ctl-click
on the area of each parameter where the name of the parameter shows up in the
timeline, the name of the parameter will change to the choice Apply to Group.
This is a quick way to apply some change to the entire group without changing
a single default setting. This can be applied to the entire effect by right- or shift-
Ctl-clicking on the very top of the parameter timelines, where the name of the
effect is shown. This will change to Apply to All.

Deleting Keyframes

You can activate a keyframe and press the Delete button. You can also
Alt/Opt-click on the Add Keyframe Icon to delete any selected keyframes.
You can also right-click or Shift-Ctl-click on any keyframe and get a menu
choice for delete. You can even Shift-click to activate many keyframes and

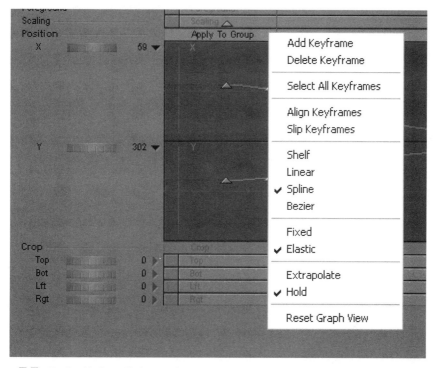

Figure 7.7 Right-Click or Shift-Ctl-Click on Parameter Timeline for Apply to Group Shortcut

delete them all at once through any of these methods. Which way do you remember the fastest?

Changing Parameters Over Time

With the new graphs representing parameters changing over time you need some modifier keys to control the direct manipulation of the keyframes. You can just click on any keyframe and drag it up and down to change the parameters, but what if you want to move the keyframe sideways to change the position in time? You have several choices. If you like the mouse you can hold down the Alt/Opt key and while dragging you will have complete freedom to move anywhere on the graph. However, you may now need to constrain such movement so the parameter doesn't change, just the placement on the timeline. In this case, you would hold down the Shift and the Alt/Opt keys and now move only sideways. If the parameter graph is closed (click on the small left triangle to close a parameter graph) then the motion of the keyframe automatically will be constrained to time changes only. You only need to use the Alt/Opt key. Finally, you can use the trim buttons that are mapped to the keyboard for a very accurate nudge. You can push all active keyframes one frame or ten frames depending on the trim key you use.

What is really interesting about this new method of displaying keyframes over time is that you can have keyframes before or after the effect itself. In other words, you can add parameters that will begin before the effect starts to play. This allows you to adjust timing with trimming of the clip with the effect. It is also a great way to have an effect match another effect by starting in the same place or same time and then syncing up later when both effects are later visible (landing at the same time or bumping against each other, for instance).

Aligning and Slipping Keyframes

Clearly there can be many more keyframes in each effect using the Advanced Keyframe model than ever before. Chances are that you will need many of those keyframes to start and end at the same time. You need to align keyframes from different parameters so that they have a common point. This might be the beginning, end, or somewhere critical in the middle (like on the drum beat and cymbal crash). This is what the functions Align and Slip are for.

Imagine that you have added a keyframe to the Position X parameter that needs to match the Scale X parameter. You need the motion to stop at the same time the resize begins. Unfortunately you have already created all the keyframes and realize that the effect is timed slightly wrong only after playing it back once.

The Position keyframe is in the right place for the timing so it becomes the reference keyframe. Click on the reference keyframe to move the blue position bar to that location. If you have Set Position to Keyframe unchecked in the Effects editor setting then you will have to drag the blue bar to the reference location. Align always uses the blue bar as the point in time to match up. Then right-click or Shift-Ctl-click in the Effects editor in the Resize parameter area and choose Align. The highlighted pink keyframes will move to the new position. If you want to align more than one parameter at a time (like the X, Y, and Z parameters of Position) then you can use the sneaky shortcut of right-clicking or Shift-Ctl-clicking on the name of the Parameter in the timeline above the graph or on the name of the effect at the very top of the timeline graph and Apply to All. This Apply to All is so powerful that you may end up affecting too many keyframes. Make sure that all the other parameters besides the ones you want to change have their graphs closed and no keyframes are highlighted pink. If you accidentally moved too many keyframes then undo the Align, go to those parameters, close the graphs, and Ctl/Opt-click on the pink keyframes to turn them gray. Then do the Align process again.

Slip is just like Align except that all the keyframes to the end of the effect are affected too. When the active keyframe moves to the position of the blue bar in the timeline, all the other keyframes in the effect stay in the correct relationship and shift the same amount. This makes sure that you don't change the timing of the rest of effect when you line up one keyframe.

Trimming Effects and Keyframes

With the previous keyframe model you had only one choice of behavior when trimming a clip with an effect to make it longer or shorter. The keyframes followed along to make the effect slower or faster. This could be an amazing timesaver since it meant that the timing of the effect automatically followed the length of the clip, but it was also limiting. Sometimes you wanted the effect to just stay at the end of its trajectory, landing in just the right place on the screen at just the right time, but you needed an extra beat or breath to absorb it before the cut. In this case, you don't want the timing of the trajectory or the moment of the effect landing to change when you make the shot just a little longer. This is why Avid created both Elastic and Fixed keyframes.

In a standard effect you can make any keyframe either elastic or fixed. You highlight the keyframe and then right-click or Shift-Ctl-click to get the menu of choices. If you make a keyframe elastic then it behaves like it always did and changes timing with the length of the clip. If you choose Fixed then it will stay put in the relative timing of the effect no matter how you trim the shot. Now that the keyframe is fixed you need to determine what happens to the effect with the extra material in the shot. Does the effect stay absolutely still and hold in position? Or does it extrapolate and continue to move in the same direction of the trajectory? This all depends on the effect itself. If the effect has come to rest on the screen then either choice is fine, but if the effect was a slow move off the screen and you trim the clip longer and extrapolate, it will continue to move in the same direction a little longer.

Controlling Motion Between Keyframes

By far the most powerful aspect of the Advanced Keyframe model is the ability to control the way the effect moves between the keyframes. There are now four different types of motion that improve upon the standard acceleration of 2D effects and Spline in the 3D effects. Let's explore each one and how they might be used.

Shelf

In many other programs Shelf would be referred to as Hold. When the user chooses Shelf it means that the effect stays in place until the time of the next keyframe and then it jumps instantly to the new position. This effect can be used to change a parameter when the object is hidden for a fraction of a second so that when it reappears on the screen it has changed. You don't have to worry about the overshoot or undershoot of other motion types. You can also use it to bounce an object around the screen very quickly, but mostly it will be used to keep a parameter the same over time with a minimum of complexity.

Linear

This motion type is usually associated with very mechanical types of motion. The object moves between keyframes at a completely steady pace. There is no speed-up or slowdown of the object, and this resembles the kind of unreal motion that only a robot could emulate. This type of movement is used when you have many objects moving at the same time, but they start and stop at different times and still manage to sync up. If objects are speeding up and slowing down at different times, it is very difficult to get them to land at the same time or combine into a

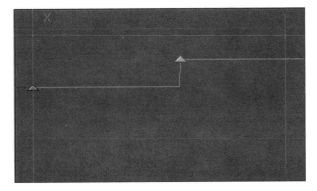

Figure 7.8 Shelf Motion between Keyframes

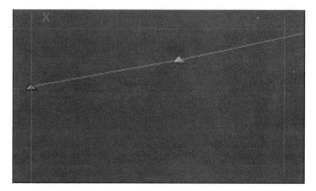

Figure 7.9 Linear Motion between Keyframes

graphic effect simultaneously. This motion type is also used when objects start and stop off screen. If you are moving large letters across the screen so that they spell a word, you want them all to stay evenly spaced apart even though each letter is a different layer and 2D PIP. You also don't really want objects to appear to accelerate onto the screen and then decelerate as they exit. The illusion is that they are just passing by and not grinding to a halt somewhere just out of sight.

Spline

Those who have used the older 3D Warp effect are familiar with the Spline control. This was defined by animators to reproduce natural types of motion. Spline in this sense emulates the smoothest natural motion between multiple keyframes. There is enough intelligence built into the spline effect so that it can look over three or more keyframes to determine the smoothest path through all of them. As you move the keyframes, Spline automatically readjusts. At the simplest, Spline creates the ease in/ease out effect that is basic to DVE moves. It is simple and effective to create basic smoothness without complex handles.

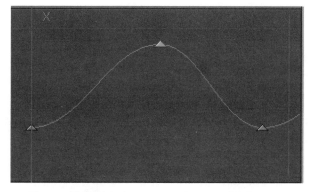

Figure 7.10 Spline Motion between Keyframes

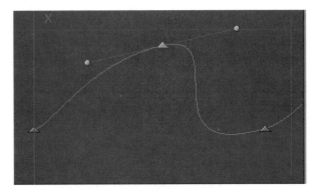

Figure 7.11   Bezier Motion between Keyframes

Bezier

Bezier begins as a Spline, but gives the user much more control — in some cases, perhaps a bit too much control, as it can create unexpected results from complex handles. There are handles on each keyframe that can be adjusted to control the amount of ease in/ease out, the speed of the velocity change, and make the effect behave differently on the other side of the keyframe. This is because a Bezier handle can be adjusted three ways: Symmetric, Asymmetric, and Independent. You can cycle between the three types by holding down the Alt/Opt key when adjusting the handle. The effect defaults to Symmetric. Hold down the Alt/Opt key and click the handle to change the mode to Asymmetric. You can adjust the handle freely in this mode without a modifier key. If you hold down the Alt/Opt key again you will cycle to Independent.

Symmetric

A Symmetric effect is created by pulling on both sides of the Bezier handle the same amount. This means that if the effect swoops and slows down to the keyframe, it will swoop away and speed up by the same amount.

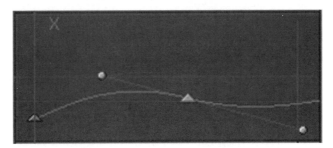

Figure 7.12   Symmetric Motion between Keyframes

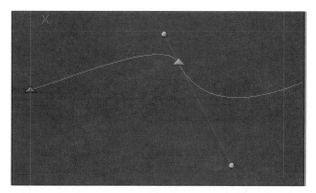

Figure 7.13   Asymmetric Motion between Keyframes

Asymmetric

You can adjust the handle to create a different speed curve before the keyframe and after the keyframe. By holding down the Alt/Opt key while pulling on a handle on one side you can change the speed at two different rates.

Independent

The Independent mode is sometimes referred to as breaking the cusp. This allows very different motion, sometimes quite extreme, on either side of the keyframe. Experiment with this mode for dramatic and unusual movement.

Keyframe Settings

Now that we have looked at all the new capabilities of the advanced keyframe model, we should take a close look at the settings that control the interface. Open the Effects editor setting in the Project window to see the new choices.

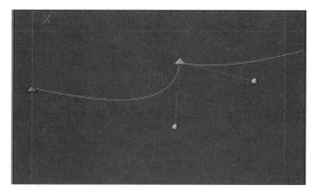

Figure 7.14   Independent Motion between Keyframes

These settings are a combination of controls for display and use of screen real estate along with performance enhancers for slower machines. The first three choices—indent rows, large text, and thumbwheels—make the text easier to read on high resolution monitors and the thumbwheels save valuable horizontal space that is used up by the classic parameter sliders.

The next four choices are ways of turning off displays so that they don't try to update the user interface during the creative process. If you have a good idea of what you need to change in the parameters, then having the computer slow down to try to display those frames is not that valuable. The Set Position to Keyframe control, however, can be turned off if you are doing a lot of align and slipping of keyframes. This is because you are clicking on keyframes for the alignment and can work faster if the screen doesn't try to update a complex effect every time.

Show Add Keyframe Mode Menu is also a time saver. You can set a default for the Add Keyframe button (mapped to the apostrophe key by default) or for the Add Keyframe Icon in the Effects editor window. If you find that you need the flexibility of adding keyframes to open groups and some of the other more advanced choices, you will want the Show Add Keyframe Mode menu to be on. A single click on the Add Keyframe icon brings up a menu of choices. Click twice to take the checked choice or change the choice depending on how many keyframes you want to add. The Add Keyframe keystroke will follow the choice made in Mode menu.

The final choice in the Effects editor Window, Automatic Start and End Keyframes, gives you a choice between preserving classic Avid behavior and moving completely into the more advanced new methods. The old Avid keyframe model always had two keyframes and, although this was at times comforting and familiar to the experienced Avid editor, it didn't follow the model of other programs. If there is no change in the parameters over time then why add a keyframe at all? The first and second keyframes would need to be highlighted to make any change that applied to the whole effect and, more likely than not, you would forget to select one and end up with some unwanted animation.

As soon as you promote an effect to the advanced keyframe model you don't need a keyframe to change the default effect parameters. Just change the parameters without one. If you add a single keyframe you are not held back by needing another one at the end of the effect. This simplifies basic parameter changes and makes simple trimming situations even simpler. I would uncheck this choice and move completely into the Advanced Keyframe world.

Timewarps

The traditional "Source Side" motion effects have been replaced by a much more advanced timewarp control for creating motion effects in the context of the timeline. Timewarps are now applied through the Effect Palette like all other effects. However, you do not have to be in Effects mode to use it. This small fact becomes very important later as we look at the interrelationship between time-warps and trimming. The important thing about the new motion effects user interface and the timewarp effect is that you can use the advanced keyframe model to control speed changes. You also have a wide range of techniques for complete control over remapping time. Additionally, you have a wide range of new types of motion, which elevates this particular technique to an art form needing just the right touch.

There are two types of control over the timewarp: speed and position. Most people will use speed since it makes the most sense for a wider set of circumstances, but both are very useful.

Speed

This pane allows you to add keyframes mapping speed to relative time. Although this sounds like a mind bending concept out of Dr. Who, it is really straightforward. You add a keyframe at the time you want to change and move it up or down to determine the speed. Add another keyframe and now you have a ramped speed change. You can start the speed at 100 and then ramp it down to 0 to have a smooth transition to a freeze frame. You can add as many keyframes as you like to change the timing and control the smoothness with the wide range of motion interpolation choices discussed in the advanced keyframe section.

You need an anchor frame for the speed change to really work. Fortunately you will always have the first frame of the effect mapped as the default anchor frame. This is the frame that doesn't move when all around it is changing. It is the frame that stays in exactly the same place as you change the keyframes around it. This is critical to having a motion effect that doesn't have the first frame change when you make it slower. You have picked the first frame and you are making the frames in front or behind it speed up or slow down. Keeping the first frame exactly at the beginning of the effect gives you a reference point you can count on.

You can move the anchor frame if you want to get more sophisticated. This would be critical if you wanted a shot to start at a specific place and go backward. You don't want to put the first frame of this backward effect at the beginning of the cut! Edit the clip in to the sequence by making the outpoint of the source clip the first frame of the reverse motion. Go to the last frame of the clip in the sequence (the starting point for the backward motion) and click on the Set Anchor Frame button. Then change the speed to –100. All of the frames in the effect are visible in the timeline; they just go in the opposite order. Experiment with the anchor frame if you want a specific point to be matched to audio cues and have all the frames around it update with speed change keyframes.

Position

Position is even a bit more mind bending. With Position you really are mapping a source timecode to a sequence timecode. This is an excellent choice if you have multiple points that have to hit exactly to a cue like a music sting or an explosion sound effect. You go to the exact frame of the source material and add a keyframe. Then move that keyframe so that it matches the chosen point in the sequence. Keep adding keyframes until all the points are mapped to the proper places.

Timewarp and Trim

Unlike other systems that allow you to keyframe motion effects, the Avid system is designed to keep you from destroying the rest of your sequence! Imagine that you have a full length project with lots of audio and video edits. You add one motion effect in the middle and decide to make it just a little slower. You certainly don't want to knock the rest of your sequence out of sync! But other systems seem not to care that whenever you change the speed of the clip you are changing the length in the timeline too.

Avid designed a different solution. Here when you change the speed of an effect you do not change the length of the clip in the sequence. You need to go to Trim mode and make the adjustment in a controlled, predictable way using the best trim model in the business. Don't leave the result to chance when you change speeds, especially when you have multiple keyframes and you are spending lots of time experimenting. The ability to adjust the speed of a clip while you are in Trim mode works quite well as a basic technique.

Formats

You can change formats of the image during the speed change. This is very important when working with 24 fps film in a NTSC 29.97 project. Since all 24 fps film has 3:2 pulldown when transferred to 29.97 videotape, these extra frames become very visible when the speed gets slower. Generally our eyes compensate, but as soon as you extend the amount of time the pulldown frames are on the screen the motion looks very jerky. The best approach is to remove the pulldown before creating the motion effect. Go to the Formats button and choose Film with 3:2 pulldown. The system will then try to detect the cadence of the pulldown since the beginning of the edit is probably not at the A frame or beginning of the pulldown cadence. The system is very good at detecting where the duplicated frames are and removing them. However, if it makes a mistake or is fooled by an animation that starts on repeated frames, then you can override the cadence detection and try it on your own. Choose progressive as the output to get the best results.

If you have shot progressive frames on video you can actually add in 3:2 pulldown for a film look. Choose Progressive as the input and Film with 3:2 pulldown as the output. The system will assume the first frame of the clip is the A frame and add in the duplicated frames.

Motion Effect Types

With so many different types of motion effects available we need a quick overview of what each is doing and when you would use it. Many of these effects are realtime within certain restrictions on the host-based systems.

Duplicated Fields
This type drops out the second field of each frame, which significantly softens the picture. However, this type is excellent for experimenting with motion since rendering Duplicated Fields is much faster than any other type. To get a really fast render you should experiment with the 3:2 pulldown still in the image. Once you have the motion close to the way you want, remove the 3:2 pulldown with the Format button and change to a higher quality motion type. Think of this as the draft mode for motion effects. You may receive offline sequences from older systems with this choice selected by the offline editor. Chapter 10 goes into detail about how to accommodate this choice when rendering in the online.

Both Fields
Although this method preserves both fields of every frame, it almost always looks choppy when used on moving material. It is excellent for preserving sharpness of a still image that you must extend to fill a sound bite.

Interpolated Fields
Although this creates smoother motion than the previous two choices, this method is also slow to render and somewhat soft. The system is making mathematical calculations as to which actual fields to combine together into a single frame for smoothest motion. Unfortunately you may get field one from frame one and field one from frame three combined into the same resultant frame. Since the frame never actually gets the information from field two, it will always look slightly soft. This combining of fields is rather random depending on the speeds you have chosen, but to keep the image from bouncing back and forth from soft to sharp, all of the frames are slightly soft.

VTR-Style
This method reproduces the type of motion effect a VTR would create when playing back each field. This method is sharper than interpolated and smoother than both fields, but isn't doing any fancy math to compensate for jitter. You

will see a little horizontal movement as you go between fields. Since so many people are used to seeing this from VTR playback, it may not even be noticeable under medium speeds. At very slow speeds, however, you will see a difference, and interpolated may be a better choice.

Blended Interpolated and Blended VTR

This modification of the previous styles tends to smooth out the motion even more. It averages the frames and performs somewhat of a dissolve between frames. Although it preserves the best aspects of the original motion type, the blending occasionally can call attention to itself as a "look." But if the image is looking jerky, try this twist.

FluidMotion™

Someday all motion effects will be as smooth and sharp as FluidMotion. This computationally intensive breakthrough in motion effects actually makes new pixels from combining original source pixels. If you want to make the smoothest possible motion effect you need to recreate the look of a high-speed frame rate. Because usually you have only a limited number of frames from normal film and video production, you need to manufacture the in-between frames that will eliminate the jerkiness of normal slow motion. Blending and interpolating will get you only so far; then you need to predict pixels.

FluidMotion makes new in-between frames by looking at the real frames before and after and tries to predict where each new pixel should be. It can combine any number of frames together as long as it properly tracks the individual pixels and in what direction they are moving (their motion vectors). The problem arises when the system can't predict what the next frame will look like. This happens with extremely fast motion where the pixels from the real frames jump huge distances in a single field. The pixels change so radically from frame to frame that the system cannot track them (a similar problem arises in the tracker controls in Symphony and DS Nitris). It also happens with *occlusion*, or when part of the image is covered up by a foreground object that is not moving, like a tree. This is because the prediction from one frame to another is interrupted if important pixels just disappear for a few frames and then show up with no history. If a pixel pops out from behind a tree you can predict where it is going only by looking into the future. The basic algorithm of looking into the past as well as the future to make a weighted decision about predicted in-between position is disrupted. FluidMotion doesn't know about trees, only pixels moving through time and ones that don't. This also applies to objects entering and leaving the frames.

When the pixels go astray they appear to morph objects into each other. There are some fascinating controls to help correct for these pixel prediction problems. The user draws around the problem area on a frame-by-frame basis and forces the area to have a certain vector or direction. This is done through a basic paint tool and an eyedropper.

Stop on the frame where the image is morphing incorrectly. Click the paintbrush icon next to the FluidMotion choice in the Timewarp interface. You are shown an analysis of the vectors in the image—what direction each object, really each collection of pixels, is supposed to be going. The direction, or vector, is represented by a color. The colors are mapped to the points of a standard vectorscope like directions of a compass. If an area is yellow then the FluidMotion effect is predicting that the pixels in that object are moving to the left. If this pixel prediction is incorrect, use one of the drawing tools to draw a selection around the yellow object. The selection will turn gray, showing that it has "zeroed out" the vector. While the rest of the image will have its motion predicted, this section will act more like a blended interpolated effect. This may solve the problem. Feather the edge of the selection and render. If it doesn't then you need to grab the eyedropper over the color selection tool of Set Vector mode. With object selection still active, grab a color from somewhere else in the image that matches the correct vector. If the object should really be moving up in the picture then grab the color of something moving in that direction. Up would be red on the vectorscope/compass. The selection will turn red and you can feather the edges a little to make sure it blends correctly. Render the effect and see if it does the right thing.

FluidMotion is an excellent choice to hide that fact that there is a motion effect at all. However, it does have its own look and can be used quite effectively to make something look unique.

SAVING EFFECT TEMPLATES

After all this work making the effects just right, you can save them as effects templates so they can easily be applied over and over again. Any effect can be stored in a bin by clicking the effect icon in the upper left-hand corner of the Effects editor, and dragging and dropping the effect icon in the bin. If a bin that holds an effect template is open, that effect will be available in the Effect Palette. The effect template can also be dragged back from the bin and applied to the timeline. When you apply the effect, it looks slightly different from the original effect unless the new clip you apply it to is exactly the same length as the original clip, or you have chosen fixed keyframes for all the parameters.

There is a sneaky trick very few people know for applying just part of a template. Using the Effects editor, open a specific parameter. You can drag and drop an effect template directly onto the single parameter. The open effect will take on only the parameter that came from the template, not the rest of the effect. This is very useful for matching drop shadows or border color and width. This also works with the color correction mode if you just want to repeat a hue adjustment but nothing else.

If the Alt/Option key is held down when dragging the effect icon to a bin, then the effect template is saved with the video (segment effects only). This is an "effect with source" and can be edited into the sequence like a master clip. If you add an effect to a title, then just saving the effect template for the title is always "with source," so you don't need to hold down a modifier key. If you want the title effect template to be just the keyframes alone so the effect can be applied to another title, hold down the Alt/Option key when saving it. Here's how to remember it:

- Alt/Option drag the effect template for effects gives you an effect and the source clip. Use this like a subclip with an effect attached.
- Alt/Option drag the title template on titles gives you the title's keyframes and no source. Apply this template to another title to get a similar title move.

ADD EDITS

If an effect cannot be manipulated with keyframes, it can be split into sections using the Add Edit button. By splitting one effect into multiple effects, each one can be manipulated separately and then recombined by a dissolve. This is most useful for a color correction that can be used to change a color over time with add edits and dissolves. If you have a camera that moves from exterior light to interior or from bright sun to shadow you can mix the two color corrections seamlessly. This is much easier to do with a dissolve than with complicated keyframes since it is essentially two complete setups rather than selective parameter changes. The system will play both color effects and dissolves in realtime.

Creating add edits adds extra keyframes to an effect sequence. If you have an effect with two keyframes and you split it with an add edit, then you have two effects with two keyframes each. If the original effect had a smooth motion across the screen and now it has double the keyframes, you could have a problem with acceleration on a basic 2D effect. Acceleration is an effect parameter that smooths the motion of an object's path across the screen with an ease-in/ease-out speed change. Adding an extra keyframe causes the effect to slow to a stop at the new point and pause where there used to be a continuous motion. If your effect has an object in motion with acceleration, you should apply only an add edit at a point where a keyframe already exists.

NESTING

Nesting is the most complex, but the most powerful, characteristic of effects with the Avid editing systems. So far we have discussed building effects vertically, which, depending on your model system, may be limited by the number of tracks. Once you understand nesting, you can expand the amount of tracks dramatically. Nesting involves stepping into an effect and adding video tracks inside. More effects can be added inside the nest, and then you can step into those. It is a way of layering multiple effects on a single clip, but also much more. The only real limitations are how long you want to render and how much RAM you have.

There are two methods to view a nest. You can apply an effect and then use the two arrows at the bottom-left of the timeline to step in or out (these buttons are also mappable to the keyboard). Once inside the nest you can no longer hear audio, but you can focus on that level alone and work on it like it is a separate sequence. Within that layer you can add as many new video tracks as your model allows. Red numbers on the timecode track in the timeline will indicate how many layers you are nested in later versions.

The other method to view nesting is used on all models. Using the segment arrow to double-click on a segment with an effect, the tracks in the timeline expand to see all the layers inside at the next level. Continuing to double-click layers inside the first effect reveals those tracks as well. Tracks can be patched and edited in this mode, and the audio can still be monitored. It is a little easier to understand all the effects going on because the display is more graphic.

This mode of viewing nesting can frustrate people who open it up by accident and then are confused about what they are looking at. You can always close the expanded view by double-clicking on the original track again with the

Figure 7.15   The Up and Down Arrows for Traditional Nesting

Figure 7.16   The New Nesting Display

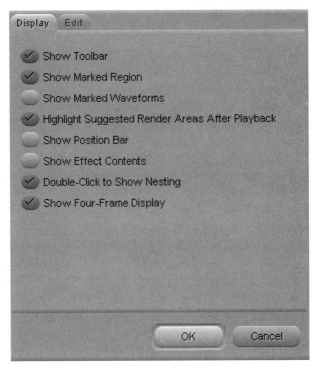

Figure 7.17 Timeline Settings

segment arrow or, in Media Composer or Symphony, Alt/Option-clicking on the down-nesting arrow. The view can be turned off in Media Composer and Symphony by unchecking the checkbox in the Timeline Settings called Double Click Shows Nesting. I recommend turning this feature off if you like to move very quickly and you have an older Mac.

Autonesting

Nesting as just described implies a certain order of assembly. Apply the outer effect and then step inside. You must apply the PIP and then step in for the color effect. But real life doesn't always work this way. Many times the nest is a secondary thought, used well after the first effect is in place and rendered. In this case there is autonesting:

- Select the clip with the segment arrow in the timeline.
- Alt/Option-double-click an effect in the Effect Palette.
- The second effect does not replace the first effect, but it covers it.

This adds the layers from the outside instead of stepping into the effect and building them from the inside.

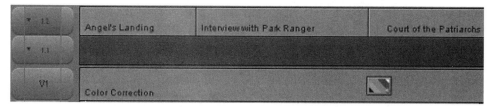

Figure 7.18   Autonesting

All these methods are for adding multiple effects to a single clip, but they are just as useful for adding one effect to multiple clips. If you want a color effect to cover an entire montage, then it is a waste of time and energy to put a separate effect on every clip. What happens when the effect must be changed? Now you need to change one, turn it into a template, and apply it to all the others. But there is a faster way that uses autonesting:

- Shift-select multiple clips in the sequence with the segment arrow.
- Alt/Option-double-click on the effect in the Effect Palette.
- The effect autonests as one effect that covers all the clips.
- Adjust the one effect and all the clips are changed.

If you want to change one of the clips and replace it with another shot, just step inside the effect and make the edit. You can also step inside the multiple-clip effect and add dissolves or other transition effects. You must render these inside effects, but you can leave the outside effect in realtime to allow for future changes. Of course, if you render the outside effect, it will create a composite of everything inside.

The main drawback to this method is that you will have to step inside the nest to trim the clips. But if the trimming stage is long over and you are tweaking and finishing this is not such an issue. With the color correction mode it is faster to save a correction as one of the four "buckets" so you may prefer to use this mode instead of nesting. By mapping the buckets to function keys you can move just as quickly through short sections. The Symphony's Program side color correction is even easier by using the Use marks for segment correction Color Correction user setting described in Chapter 12.

Viewing and Changing Nesting

Collapsing Effects

You can nest an entire effect sequence into one effect after it has been built using vertical layers. If keeping track of all the video layers becomes tedious, you can collapse them into a single layer. In order to nest effects, you must have an outside effect at the outermost level so collapsing places a submaster effect over all the layers and nests them inside. Select the area to be collapsed by marking in and out and highlighting the desired tracks. Then press the collapse button and watch the animation.

There is no way to really uncollapse an effect segment. Here is the best method to work around it:

- Step inside the collapsed effect.
- Mark an inpoint and outpoint around the entire segment.
- Turn on all the video tracks (except for V1 if it is empty).
- Use Copy to Clipboard button while inside the Collapse.
- Paste the Clipboard Contents into the Source window. On a Media Composer or Symphony you can use the Alt/Option key when Copying to the Clipboard and the pasting to the Source window happens automatically. The layered segment can be used as a subsequence.
- Cut the subsequence back over the top of the collapsed effect in the timeline or drag to bin.

This alternate method is actually even simpler:

- Create some new video tracks—the same number as in the collapsed effect.
- Expand the nest in the timeline by using the double-click with a segment arrow method to show all the tracks.
- Drag the segment up to the empty tracks using the red selection arrow, and the Control (Windows) or Command (Mac) key to make sure they don't slip horizontally.

You could collapse all the tracks except the top track, like a title, and leave it in realtime so you can continue to make changes without rerendering. Rendering a nested effect is simple; just render the top, outside effect, the submaster. This leaves all the effects inside unrendered, but it is sufficient to play as long as you are monitoring that outside effect.

Figure 7.19   Leaving the Top Track in Real Time

You don't need to leave the top effect of a collapse as a submaster. You can replace the submaster with another segment effect by just dragging and dropping from the Effect Palette. You can replace the submaster with a mask (to simulate 16:9) or a color effect. You can also step into a nest and render the top track inside the nest (or ExpertRender) thus leaving the outside effect in realtime. You can keep tweaking the effect on the outside of the nest if the dissolves inside are rendered.

Collapse vs Video Mixdown

Although the collapse feature is excellent for simplifying complex effects sequences down to one video track, a collapse can still potentially become unrendered. If you are sure that an effect sequence will never need to be changed, match framed back to an original source, or used for an EDL, then you can use video mixdown. Video mixdown (under the Special menu) takes any section of video between marked points, whether it has effects on it or not, and turns it into one new media file. This new media file has no timecode (which timecode would it use if you had 15 layers?) and breaks all links to the original media. This is why match frame to original source clips will no longer work and EDLs will no longer reflect the original source timecodes.

Video mixdown should be used only for finishing and for something that you will be using as a single unit over and over, like the graphics bed for an opening sequence you use every week. Once all the effects are rendered, a video mixdown is as fast as copying the media to another place on the drive, an insignificant amount of time. If the effects are not rendered before the mixdown, they will be rendered first as part of the video mixdown process so don't forget to count on the rendering time in your calculations. This workflow encourages rendering first and video mixdown later when everything is signed off.

A video mixdown will significantly improve the performance of the Avid during long sequences with lots of effects. Instead of forcing the computer to "build the pipes" for complex effects with many short media files, it just needs to find one master clip. This means snappier reaction time when you press play. Always make a copy of the sequence before you overwrite a mixdown over your timecoded original sources so when the client changes his or her mind you will have a fall-back sequence. Video mixdowns are very powerful and time-saving for a wide range of purposes, but don't use them for offline if you plan to recapture or make an EDL!

3D EFFECTS

In Media Composer and Symphony, all 3D effects come from one effect, the 3D Warp. 2D PIPs (picture-in-picture), titles, and imported matte keys can all be "promoted" to 3D. In Xpress there are some premade effects broken out into simple-to-apply subsets of the entire 3D functionality. Titles and imported matte keys can be animated in real time using the Promote button. On ABVB systems, having the extra piece of 3D hardware gives you both realtime wipes and real-time keys.

On Meridien and ABVB systems the 3D Effects option and "m" AVRs are completely incompatible because to enable four or nine screens to play at once for hardware multicam, the "m" resolutions are at a much reduced frame size. The 3D Effects option must be disabled before working with this unusual frame size. This is accomplished by launching the application and holding down the F and X keys. You will be given the option to enable or disable the 3D. On later versions of the Meridien systems you don't need to disable the 3D but you still cannot add 3D effects to a multicam sequence.

Clearly there are more things you can do with the 3D effects than just spin and tilt. On ABVB and Meridien you also have Stamp and Clear, which uses a realtime buffer to stack unlimited layers without any rendering or loss of quality. These stamped layers must be added while the sequence plays in realtime since they cannot be rendered and must fade out all at once, but the possibility for titles is very powerful. Remember to clear a stamp buffer or the effect will pop back onto the screen after you have faded it out. And the ever-popular trail

Figure 7.20 3D Effect

effect can be added to any object that moves by taking advantage of the same buffer used for Stamp and Clear. Trails also are realtime only; render the trail and it disappears.

There is also corner pinning, which allows you to fit four corners of an object so that it matches the edges of another object. It is not quite morphing, but it allows you to put images inside TV monitors, picture frames, or the like.

There are so many things you can do with the 3D option that I helped write a one-day course just for that. Truly, this area calls for personal experimentation. Just taking some of the shapes and using them to warp images into interesting moving backgrounds requires parameters you must discover for yourself.

PAINT AND ANIMATTE

If you consider editing to be an interframe process—working between frames—then painting on a frame is intraframe. There is a full palette of familiar choices for anyone who has used third-party painting programs. Brushes can be changed and areas of the image can be blurred, color corrected, and generally affected like any standard paint program. Multiple layers of paint effects on the same frame are possible, but unlike the other, single-image paint programs, you can easily change the effects over time with keyframes.

You can also isolate parts of the frame, say the sky, and draw a matte shape around it to make it a deeper blue. The ability to draw on an object and create control points to adjust curves with bezier handles and move the edge

Figure 7.21 Creating Control Points on an Object

over time means that almost any part of an image can be manipulated separately from the whole. By creating points that change over time, any smooth, even motion in a shot can easily be followed with just a few keyframes. If the motion is jerky or unpredictable, you need more keyframes to adjust the control points. If the motion is truly difficult to follow or you are doing dozens of motion-following effects, you may want to use the tracking feature in Symphony that allows you to automatically track specific pixels over time. Since this is a complex and time consuming task, occasionally you may want to set up a separate graphics station for tracking and rotoscoping. This way you can split the tasks across multiple people if time gets tight. You can also set up your graphics station to render while you continue to edit on the Avid system. Adobe After Effects, discreet Combustion, Pinnacle Commotion, or Eyeon's Digital Fusion are good choices for this kind of work. DS Nitris is the best combination of graphics, paint, and 3D effects into a timeline-based editing program in the Avid product line. Paint and Animatte are fine for shorter jobs, but if you need to do lots of this type of work you may want to consider a DS Nitris finishing suite.

COLOR CORRECTION

There is much more control over color in the latest versions of Avid editing systems. It is clear that everyone needs a little color correction, even if the images are perfect, because they all need to match seamlessly from shot to shot. Of course most people's footage is far from perfect, so you may find that you can make major improvements to your project with just a few well-placed tweaks and the reuse of powerful templates. Chapter 12 covers the Symphony color correction tools in depths but for now let's touch on some of the most useful types of changes you will need from day to day.

Avid has many different kinds of color correction depending on the system you are using. The basic Color Effect has been improved to allow quick adjustment of contrast range and gamma using eyedroppers. Pick the darkest point of an image with the eyedropper and it changes to the blackest possible values. Choose the brightest part of the frame and that will become the brightest possible value for broadcast. An image that was dull because of a narrow contrast range can make use of the full range of levels available. You can also quickly take illegal levels and bring them to within broadcast specifications for luminance and black level using the adjustable clip levels.

The Spot Color Effect is the combination of the Color Effect with Animatte. It also includes NaturalMatch(TM), which is a powerful color-matching algorithm from Advanced Color Correction mode. This allows you to determine an area of the image, sample a color in that area, and change it in a way that many people consider secondary color correction.

Auto Contrast

For many reasons out of your control your images may not be using the full contrast range available. This might be because you are working with a one-light film transfer where the colorist has not taken the time to optimize each shot for exposure. It might be because you didn't have the time to light properly or because you were using an inexpensive camera that couldn't deal with changing light conditions well. Whatever the reason, your blacks and whites are dull, gray, or lacking richness. If you have a Symphony or DS Nitris system you can adjust the contrast range by hand using the histograms. If you are clever with eyedroppers you can do the same with luminance matching gadgets (see the earlier strategy with the Color Effect for something similar). Or you could just use the Auto Contrast button and move on to the next image!

In effect, you are taking the darkest part of the image and matching it to a reference black that is equal to the darkest possible black that is safe to broadcast or print to film. You are doing the same with the whites. In this process you evenly distribute the luminance values of the rest of the image to fit in between these two new brightness extremes. The blacks get blacker and the whites get whiter. This has an instant effect of making the images look more pleasing most of the time.

The power of the Auto Contrast button is that it will find the darkest part and the brightest part of the image automatically without your hunting with an eyedropper. You may find that you can get a better result by sampling another black area in the image so you may still have to use the luminance matching gadgets in the HSL controls.

Auto Black and Auto White

These will adjust only the one parameter that needs it the most. Your whites may already be blown out and will not improve by an Auto Contrast adjustment. In this case you may want to darken just the blacks. If an image is blown out (over-exposed) or has the black detail crushed (underexposed) you may have a hard time correcting these problems to bring back any detail in these areas.

Auto Balance

Most video cameras have a white balance adjustment that finds the white aspects of a picture and balances them so there is no "color cast" like a slight green or red shading over everything. The auto-white balance functions on some cameras take several seconds to look correct after the camera has been powered on so there may be sections of your video with an unpleasant color cast that doesn't match the rest of your shots from that day.

To counteract this, you may want to carry a white card with you and focus the camera completely on the card before shooting a new scene. Professional videographers have done this manually since the first days of video and still do. However, most likely you didn't shoot the material yourself and are still being asked to fix someone else's problem. The more money the producer has saved the more likely he or she is counting on you to fix it in post!

The Auto Balance button will take a look at the image and assume that there is a pure white, gray, or black without color saturation somewhere in the image. Auto Balance will remove the color saturation difference between perfect white or perfect gray from the entire image based on that automatic sampling. If the image shifts to an unacceptable color using Auto Balance, you will have to do the matching by hand. The Auto Balance button has a slightly different effect depending on whether you use it in the Curves tab (affecting RGB levels) or on the Hue Offsets tab (HSL levels). Big lighting changes like the teacup in Figure 7.22 will have slightly better results if you use Auto Contrast and then Auto Balance in the Hue Offset tab. In this case the result was slightly more neutral after the huge luminance change compared to the Curves controls, which left it slightly greener. For basic hue adjustments choose Auto Balance first in the Curves tab. Use the Remove Color Cast eyedropper icon to pinpoint the area you think should best represent a pure white, gray, or black. In the Hue Offset tab, you can choose Highlights Midtones and Shadows eyedropper to limit the change and in Curves you will be picking a specific RGB value.

You can set an Auto Correction to occur every time you drag the Color Correction effect to your timeline. This is a quick way to get large sections of your sequence to be corrected all at once. Go to the Project window and open the Correction settings. Go to the AutoCorrect tab and choose the kind of correction and the order you want the actions performed. By selecting multiple clips in the timeline through Shift-clicking with the segment arrows or lassoing in the effects mode you can apply this effect many times with a single double-click.

Figure 7.22 Teacup before Auto Correction

Figure 7.23 Teacup after Auto Correction

Figure 7.24 Teacup Split before and After

Here is a quick way to do this by hand if you don't have the latest versions with the Remove Color Cast function:

- Go to the Curves area of the Color Correction mode.
- Sample an area of the image with the eyedropper that is meant to be neutral, but has a color cast so that you have the same color on both color swatch areas (the before and after samples).
- Double-click on the after area (the one on the right). A color picker will appear courtesy of the operating system.
- Change the saturation value to 0. Then click OK and go back to the matching gadget.
- Choose R + G + B with Natural Match and click on Match Color.

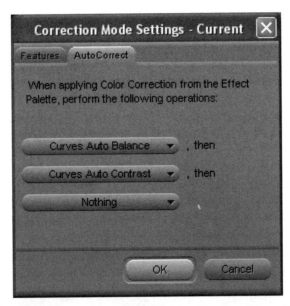

Figure 7.25 Correction User Settings for Auto Correction

You have now matched the exact same luminance value without any color saturation. So no matter if the sampled area was white, black, or gray, it should have no saturation in that area and the color cast should be removed for the entire image. By using Natural Match you have not changed the luminance value of the image if the two samples were slightly different.

Of course there are manual ways to tweak the color of your image—just make sure that your monitor has been properly color balanced as well. The video monitor should always be the guide unless the final playback will only ever be on a computer.

Once you have created a color effect that solves your most common problems, save the template in a bin. This is detailed in Chapter 12. You will find that you will have the same issues with the same camera or the same film transfer so often that you will save yourself hours by the judicious use of templates.

AVX

An expanded range of choices is available with the AVX (Avid Visual eXchange) plug-ins standard. This is an interchange format so that other third-party effects companies can modify their existing product line easily for use as an Avid plug-in. Make sure that the version of AVX effects you are using is also compatible with the Avid DS system if you want to take your sequence to the next step for high-quality, high-definition finishing. As computer processors get faster, there will be more you can do in realtime with AVX effects.

TITLES

Always try to work with titles at an uncompressed resolution in standard definition. Aliased edges and blockiness are eliminated when titles are uncompressed. If you are working on a Meridien system you have the ability to run the titles through the DSK. This is a special section of the hardware dedicated to being able to play uncompressed titles and graphics no matter what the resolution of the clip below. You also can run an uncompressed title on V2 and have unrendered real-time effects on V1 because you are not running three streams of video, just two streams and a PICT file through the DSK. With the host-based systems, your ability to play uncompressed titles is restricted only by the speed of your computer. You can mix compression types as much as you like with these systems.

Since you can copy and paste from a word processing program, instantly apply a custom style, and create title rolls on most systems, it is easy to use the Title Tool for large amounts of text. You can create a custom title template and map it to a function key. All styles are mapped automatically to the next unused function key and are enabled only when the Title Tool window is active. If you are creating titles with lots of font and size changes, you can highlight the text in the Title Tool and press the function key to apply the premade style.

You can open a title and edit it straight from the bin if you want to use it as a template for new titles. Ctl/Cmd double-click on the title in the bin to open the Title Tool. It will also give you the choice of promoting this title to Marquee on some recent PC-based Avid systems. After you make changes you can Save As to create a title that needs to be edited into the sequence.

The Title Tool has its own Safe Color setting. If you are picking colors from a color picker or trying to match a color in an image, it may automatically dull or change the color to make it broadcast safe. If you really need to match the color and take your chances later then turn off the Safe Color choice under the Object menu.

The Tool is trying to show you a title for position, spelling, composition, and other basic choices as quickly as possible. This is why it defaults to a lower quality draft mode during creation. If you want to see the finished quality of a title for approval purposes before you are finished, press Shift-Ctl-P or Shift-Cmd-P to turn on Preview mode.

The final tip for any title is to always check the kerning or leading of any title before saving it. This is the proportional spacing between letters and between lines. Basic fonts almost never have correct kerning on all words. You will have to kern the letters together or apart to make them even and aesthetically pleasing. The window marked Kern in the toolbar applies what a typographer would call Tracking to the entire text string. You can do this from the keyboard by using the arrow keys and the Alt/Option keys. Use the arrow keys to navigate to the font pair and then hold the modifier key to make the change. This will please both your inner and outer Art Director.

Marquee Title Tool

For truly complex manipulation of type, you will want to work with a program that can use vector-based graphics. Marquee now ships as a second choice for more sophisticated titles and animations on many models. When you choose Title Tool you will be given the choice between the old Title Tool and Marquee. If you want the choice to always be one or the other and you want to eliminate this pop-up option, choose Persist and your choice will be remembered. If you ever want to go back to being given a choice you can go to the Marquee Title setting in the Project window. There you can switch to the other choice or reverse the Persist choice. If you have been given titles that were created on the old title tool you can choose to promote them in Marquee and continue to add that extra level of polish and pizzazz. You can also continue to use a mix of old titles and Marquee titles in the same sequence. Once you have promoted an old title to the Marquee title you cannot go back; this may have implications if you want to send your finished sequence back to an older system for continuing offline work. There is a checkbox for saving a version of your original title before promoting that you should check "on". If you need to get to an older version of the title before you go back to a version that does not support Marquee, you can cut these titles back into the sequence.

Marquee, a true 3D type and graphics manipulation program, allows you to quickly create titles with textures, light sources, and extruded type. It will give you realtime rolls and fast render crawls on Meridien systems. You can manipulate each letter in a title on its own timeline and control all the movement with bezier curves. A static title is quick to create and plays back in realtime. An animated title will take longer to render, but it will preview in realtime using the Open GL board that ships with most systems. This render speed will increase over time, but if you find it too slow you can go into Marquee and turn off some of the quality settings under Render/options. Avoid making large, soft, drop shadows if render speed is a problem on animations. You can set up a Marquee animation to render and then go back to editing, but you may find the rest of your system has slowed down too much to do much serious work.

There is much depth to Marquee and many people only scratch the surface. If you have the time you should explore the scripts and perhaps write some of your own. The Marquee scripts can be written in Python programming language, so if you have a repetitive task you could write a custom function to handle it (or pay your favorite Python programmer to do it for you). You can also import images that are larger than standard frame size and zoom and pan on the image. This is great for simulating motion-controlled camera moves. Since Marquee has keyframing and bezier curves, you can get quite sophisticated motion on the images. And finally, Marquee can be used as a sophisticated multilayer 3D DVE if you are willing to spend the time to learn it.

CONCLUSION

The ability to layer, paint, and use plug-ins has given the Avid editor a whole range of tools and looks to create effects that look like they were made on much more expensive systems. Graphic looks continue to get more sophisticated and subtle, so taking the time to explore the Avid effects and the interoperability with third-party plug-ins and graphics programs will definitively pay off. Faster rendering and more realtime streams make more creative work possible in the same time frame. Faster CPUs promise that more work can be done without dedicated hardware. Networks will allow users to distribute work and share media and, like the DS Nitris systems, distribute rendering to unused or dedicated rendering systems. Editors and designers will always continue to experiment and push the technology to the limits, and with the tools now becoming available for nonlinear editing with the Avid, they have more choices than ever.

8

Preparing for Linear Online

Most of the preparation for a linear tape-based online happens at the end of a nonlinear project when you make the edit decision list (EDL). But there are things that must be done properly throughout the job in order for the online to go smoothly. The reality is that with all the twisted financing that goes on in this crazy industry, there are still reasons to buy a bunch of offline models and create an EDL when finished. However, the days of saving money by going to a linear suite are swiftly ending. Throughout this process keep one thing in mind: You are stripping valuable metadata from your sequence whenever you make an EDL. With the growing acceptance of AAF (Advanced Authoring Format) for transporting all of your sequence's valuable parameters, you may want to think hard before going this traditional, limited route to finishing your project.

This chapter discusses the most basic requirements to get a good EDL under the most common scenarios. Then it goes a bit deeper into the possibilities for increasing the speed of an online auto-assembly, the Holy Grail of all expensive online sessions, and almost as tough to find! But the good news is that many facilities want the list in a simple, bulletproof form so that they can do their own, more complicated variations.

How do you transfer a 24-layer effects sequence with nesting, color effects, audio equalization, and rubberbanding to a format that supports one layer of video, four tracks of audio, and hasn't changed much in almost 20 years? Answer: Not very well. Dedicated hardware is the rule in traditional online suites with a different piece of equipment and, many times, a different manufacturer for each function, like text, special effects, and deck control. The separate pieces of equipment were purchased years apart and there may be components that are 10 years old that still work fine for what they were designed to do. Compare the processor of a CMX 3600 with a high-end workstation Macintosh or PC. The workstations have consistently supported Moore's law and doubled the amount of processes on a single chip every 18 months. Yes, if you design the hardware to do just one thing, then you can maximize it so that it doesn't need as much RAM or processor speed. But just try to add new features! This is why there are so many third-party EDL management software choices. If the dedicated linear online systems cannot add new software features, then new third-party software does it and simplifies the EDL to a format the dedicated systems can handle.

Even with third-party EDL programs, there is still a massive mismatch in capabilities between what can be done by clever folks with the Avid and what can be represented by this limited format for a linear online. The only way to convert from one format to the other, preparing for that precious linear online time, is to dumb everything down. You need to understand the limitations of what can be conveyed in an EDL. Bring the Avid EDL to a point where the show can be built back up again, piece by piece, in the order necessary in the linear, tape-based world.

This is more of a challenge than it would appear at first, not because it is that difficult, but because of the limited experience of the Avid editor in the world of linear online. Unless you have actually worked as an editor both at an Avid system and at that particular linear online suite, you are not really in a good position to take advantage of all of that tape suite's advantages. We can discuss capabilities in generic terms, but that doesn't help you for the specifics of that edit suite. You need to know that for that suite, you needed to reserve the character generator a day in advance and you can forget using the DVE unless you book a night session.

This is where you must rely on two very important low-tech tools: clear communication and the ability to collaborate with experts. As you devolve the Avid list to something simple for the linear assembly, you must be very clear about exactly what must be done to every part of your sequence. What effect do you really want? How are you going to deal with all those channels of audio? This chapter will describe different ways to leave a paper trail that someone else can follow. Count on the expertise of the operations or scheduling people at the particular post-production house that has been chosen. Speak to the online editors if they are available. Depending on time available and the complexity of the program, you may even want to (gasp) plan a pre-postproduction meeting to discuss approaches with the editor! Imagine, warning them what to prepare for! Let's face it, there is no way to predict everything that can happen after you leave a project. It is really not fair for the offline editor to be blamed for changes that occur after the cut is locked. Sometimes changes ripple backward to affect a decision you made in good faith, with the information available weeks or months ago. Unfortunately, this happens all the time.

Coming from a post-production facility background, but having freelanced for a few years as well, I strongly advocate building a good relationship with a handful of production companies. Bring any work you can to these select production companies, large projects and small, and spend some time learning the facility's capabilities and quirks. If you create a sense of loyalty, they will look out for you. This happens in little ways, like working a little harder to fit you in for emergencies. They may come to know your work and style and be able to anticipate and fix problems for you before you even know they are there. If they can get you in and out on time and on budget, they can fit in more paying sessions, need fewer "make goods" for questionable mistakes, and lower their already unnecessarily high stress level.

If you can afford to sit in on a few online sessions, it will pay off for your customers in the long run. If they have a relationship with you, then it is in their best interest for you to continue to make informed decisions about effects that can be recreated easily, or faster ways to assemble the show in the final stage. Will it be faster to dub shots onto a selects reel? Is it better to have more tape decks available for a shorter time or fewer decks for a longer time? And really, what can you do at a lower rate per hour to make things go smoother? When you learn these things, you become more valuable to your employer and have more job security. You are more likely to stay on the A-list the next time your favorite client, director, or producer has a big job.

So it is in everyone's best interest that you understand the implications of every step of your offline on the final online. If you capture without timecode or without paying attention to tape names, you are directly hindering the next step. Don't blame "those Avid EDLs" when you have no regard (and, really, no respect) for the next link in the chain. It is your job to continue to inform the director or producer that choices they are forcing you to make will cost them money later. If they take just a little more time now, at the lower rates, things will go smoother at the crunch time.

What is the biggest mistake? Ignoring the importance of timecode and tape name. Everything must have a timecode reference—all sound effects and music, all graphics and animations, and all picture and sync sound. Let me say this again: If you are going to a linear online, every source must have timecode. It is tempting to import just that one cut from the CD music library or pop in just those three shots from the VHS dub. But you must be sure to go back and match by eye or ear with a timecoded original source before you make the EDL. This is a clear example of GIGO—garbage in (sources without timecode) and garbage out (a list that cannot really be assembled).

PREPARING SOUND

There is one exception to the timecode rule: when the audio captured into the Avid will actually be used as a source in the online (or the final mix for a film). It is becoming more common for the first edit to lay down a digital cut of the audio, low-resolution video, and then cut with original video sources the rest of the day. This assumes no changes in the online that will affect sync or length; of course, something you may have to mention repeatedly! In this case, you are working from the timecode of the digital cut, which should match the timecode of the master sequence. If the show starts at 1:00:00:00, then the sound source should, too. For transferring just the audio from the offline session, a timecode-controlled DAT or another digital tape format like the 8-channel DA-98 works quite well. You could go out of the Avid using the digital audio output format to a digital Betacam or another video format that supports the digital material you will be outputting. In a pinch, audio tracks 3 and 4 of a Betacam SP can be used since these audio tracks have a higher dynamic range and lower signal-to-noise ratio than tracks 1 and 2. This is good for effects and sound under, but should not be the main method for the most important sound in the project. Just make sure the Betacam SP decks in the online suite are models that can play tracks 3 and 4. Generally, this means a Sony editing deck with the letters BVW in front of the model number.

Did I mention you can't make changes?

USING THE OFFLINE CUT IN ONLINE

If you are using digital videotape to bring the Avid digital cut to the online suite for your audio, then you should also include the video as part of that first edit. Many online editors use a clever trick where they lay down both video and audio from the Avid to the master tape in the online suite. They put a big circle wipe with color bars in the center of the video. That way they can cut over the top of the old video all day and see if there is any discrepancy between the EDL and the digital cut. Any time they see a flash of color bars they know something is a frame or a field off and can make adjustments. It sure is better than squinting all day to tell the difference between one frame of 10:1 and the original!

The other way that the captured audio without timecode from the Avid can be used is when you make an OMFI export of the audio and the sequence for use on a digital audio workstation. But if the original sound was captured with a little distortion or mono instead of stereo, then the sound mixers have to recapture it on their workstation and match it back by ear. Not too much trouble if it is just a few cuts, but if the music comes from a five-minute cut and is used at every transition, then you are costing your client money. How do you think the sound guys will justify their higher bill?

When you are dubbing everything to timecoded sources, it is up to you (or your assistants, dubbers, etc., but really you) to make sure the quality is not degraded. Dub to higher formats—digital formats—and monitor the levels very closely. There is nothing worse than being forced to recapture a distorted media file and finding that the dubbed source is also distorted! Some facilities make sure that everyone who dubs a source puts his or her initials, date, and machine used on every tape. Being able to track a problem back to the source does a lot to ensure quality control.

DUBBING WITH TIMECODE

There are some basic requirements for good timecode when dubbing sources to be used in an online session. The first requirement is: Never dub timecode. That seems like an outrageously stupid assertion, but one that has confused people for many years. You should not dub timecode—you should always regenerate it. Timecode is a square wave signal, and straight dubbing tends to add noise to the signal. Eventually (sooner if the timecode wasn't great to start with), the signal becomes slightly rounded and suddenly, where that sharp edge in the signal delineated a specific number, it is now just a little too rounded to read accurately. You may have been dubbing timecode for years and sending your problems downstream or, luckily and more probably, your record deck has been left in the position for Regen. This is a switch under the front Control Panel of the deck that allows you to regenerate—resquare—the signal as it passes through the deck's electronics and lays it to tape.

It may be important to keep the timecode from the original source for accounting reasons, so the stock footage can accurately be paid for or because of the way the selects reel will be made. When making a selects reel, you can be dubbing many shots from separate reels onto one reel for the convenience of capturing or because you will lose access to the originals during postproduction. The shots on the selects reel may need to be traced back one more step to the original tapes before the online. Now you should set the internal/external timecode switch on the tape deck to ext for external source. Make sure that the source deck and the record deck have timecode cables connected between them so that you can jam-sync the timecode to the new tape. Give yourself plenty of preroll because there is a break in the timecode at every shot. You do not want the Avid or the edit controller to rewind over a break in the timecode when cueing up for a shot.

All edit controllers, whether they are part of the Avid software or part of an online suite, always expect a higher number to come later on the tape. The timecode is expected to increment continuously! If you put hour 10 before hour 5 on the tape, anyone who uses this tape will be confused. Edit controllers will search and fail to find preroll and edit points. Your name will be used in vain and no one will ever return your phone calls again.

If the beginning shots of the tape are drop-frame, then everything that is dubbed to that tape should be drop-frame, too. The Avid system associates a tape name with whether an entire tape is drop or nondrop. If you mix the types of timecode, you have to give the tape two names—very confusing and not recommended!

If you can lay down new timecode instead of taking it from the original source, you can keep the timecode nice and neat on the selects reel. If the ext/int switch is back in the int (internal) position, the record deck always generates its own timecode. Another benefit of leaving your deck in Regen is that any assemble edit (an edit that changes the timecode on the tape) picks up where the last timecode on the tape left off. During the few seconds preroll to make the assemble edit, the deck reads the timecode already on the tape. At the edit point, the deck neatly regenerates timecode without missing a frame. Either you should be purposefully presetting the timecode on the deck, jam-synching the code from the source tape, or regenerating it from the timecode that is already on the new tape.

Timecode is just another tool and can be used very effectively once you know how to preset it and actively use it. If you are making a selects reel, choose a timecode that is not used by the other field tapes. If the last field tape is tape 15 and uses timecode hour 15, then start your selects reel at hour 16. No matter how a tape is named in the EDL, you can trace it back to the right tape based on the hour of the timecode.

TAPE NAMES IN EDLS

Tape name is incredibly important to the Avid and, if you get almost nothing else right, at least this should be perfect. Every tape needs a unique name. A tape must be named while capturing or logging in the correct project. It is good practice to make the tape name match the timecode hour and never repeat a tape name. As mentioned before, but bears repeating, tape names should be short — five to seven characters maximum to allow a space for a "B" to be attached to the end when making dupe reels for CMX and Grass Valley Group (GVG) systems. This chapter discusses dupe reels later — you can't count on not needing them!

Assuming that all timecodes in the Avid refer back to real sources and all tape names are unique, let's look at the basics for preparing a list. Be sure that you have installed EDL Manager from the Avid installation CD-ROM. This piece of software creates EDLs from Avid sequences and is meant to be able to open bins and locate the sequences inside the bins to create EDLs.

EDL MANAGER

EDL Manager is Avid's EDL generation and list-importing software and is separate from the Avid software for several reasons. The first reason is so that it can be updated and improved upon with a development schedule separate from the editing systems. EDL Manager can then skip the long and complicated beta testing programs in favor of quality assurance just for this small piece of software. Being separate means that the offline editor can carry it to the online suite or the online editor can have a version running in the suite already. The offline editor can be spared some of the responsibility for making a good list if he or she brings a sequence bin to the online session with the (ideally) final version of the sequence.

The EDL Manager launches automatically when you choose EDL from the Output menu in Media Composer and Symphony. Xpress users will need to launch it independently. If it does not launch then perhaps either you have not installed it from the original installation CD-ROM or someone has cleverly hidden it somewhere on your system and now it cannot be found. Instead of troubleshooting, it is quicker to go back to the original CD-ROM and re-install the application in the proper location.

If both applications are running, you can use the simple arrow icon in EDL Manager. The sequence, which is in the Record window of the Avid editing software, will load into the EDL Manager with a single click. You can also use the arrow to go the other way and turn an EDL into a sequence, but since there is more art than science to that technique, we will discuss it in Chapter 10.

GETTING READY

There are so many variables in the linear online suite that you can never be too careful about preparing for any eventuality. There are many different kinds of edit controllers on the market today and, to some extent, they are not compatible. The

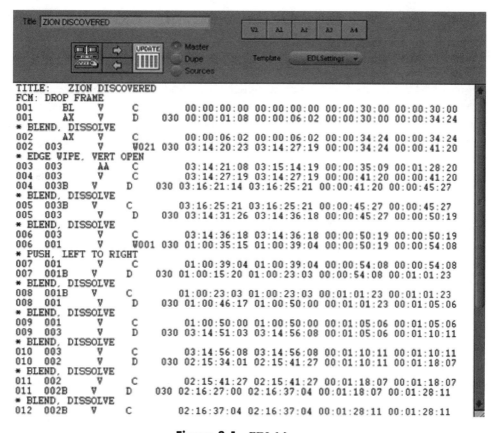

Figure 8.1 EDL Manager

main ones in use are CMX, Grass Valley Group (GVG), and Sony. There are also many other less-prevalent models like Ampex ACE and Abekas; however, within your market there may be an exception to this, and you may find that the most popular edit controller is something else. In Chicago, for instance, you will find many Axial controllers.

The first rule of EDLs is this: Find out which edit controller format the linear online facility needs. Notice I said what they need, not what they have. This is for a very important reason: They may have a CMX Omni, not a CMX 3600, but the lists are mostly compatible, so the facility may ask for a CMX 3600 list instead. Or they may have a bizarre 10-year-old Brand X model that went out of business seven years ago, but the darn machine just refuses to die. They may ask for a CMX 3600 as well because out of necessity they may have, or may have written, their own software to translate from CMX 3600. Most people consider the CMX 3600 the Latin root of all list formats.

FORMATTING FLOPPIES AND RT-11

You must format the floppy disks for CMX and Grass Valley Group edit controllers in an unusual way. CMX originally used a disk format that was useful with the DEC computers they used internally when they were developed. Because GVG developers came from CMX, Grass Valley Group also uses a version of that original CMX format. The format is called RT-11 and is problematic with both Macintosh and Windows because these operating systems are not designed to recognize it. Special commands must be written to make the computer ignore the fact that it can't read the RT-11 floppy disk you just popped in. The edit list software must take over for both reading and formatting these floppies. This is why if you just pop a CMX disk into the Macintosh, you get a scary message saying this disk is unrecognizable and asking if you want to format it. No! Eject, now! The Mac defaults to this evil question whenever it cannot recognize the formatting on a disk. You can work with RT-11 disks only when you are in the Avid, EDL Manager, or another piece of specialized software designed specifically for EDLs.

RT-11 was not written to make your life a living hell, but it sure doesn't help. Add to this the restriction that the disk used must be a double-density, double-sided disk. I want to say that again: double-density, double-sided. The only exception for CMX and GVG is the HDGVG option in EDL Manager, which will correctly format an HD disk. Look at all the floppies strewn around your computer (if you still use them). I'll bet none of them used for standard, everyday saving is double-density, double-sided. They probably all say HD, which stands for high-density. The disks you want should say DD on them. They are not interchangeable! This is an example of dedicated hardware being passed by very quickly by technological standards. Double-density disks hold only 800 KB of material. This could be a small novel, but it is not very many EDLs. The real world uses HD floppies because they hold 1.4 MB, almost twice the capacity of the DD disks. Even the HD disks are pretty pathetic in a world where gigabytes are becoming the standard measurement for storage capacity. If you are making the edit list, it is up to you to scrutinize every disk that is handed to you.

I strongly recommend that the DD disk should be unformatted or already formatted for RT-11 from the edit controller that will use it. The formatting of the disk for the EDL has caused many people the vast majority of their EDL woes. On older Macintosh systems you will need to "unformat" a floppy. First, quit all the programs you are running. Pull down the Erase Disk command under the Macintosh Special menu, as you would normally erase all the material on the floppy. Let the process run for a few seconds, but before it finishes you must force the computer to crash. The only way you may be able to interrupt the erase is to press the Restart button on the outside of the computer case, or on the Mac, press the Control-Command-Power keys all at the same time. This works to totally screw up the format on the floppy, but it sure is ugly. For all of the Windows versions of EDL Manager and later versions on Macintosh, this method is not necessary since the system will competently erase any original formatting and replace it with RT-11.

So now you have an unformatted double-density, double-sided floppy. If you pop it back into the Macintosh, the system tells you the floppy is now unrecognizable—this is one of the few times you want to see this message! Windows will give you a similar message. Now the floppy can be reformatted while you are running the EDL Manager or other EDL software. Is it any wonder that busy postproduction facilities would rather FedEx you a few of their own disks? Overnight mail with a proper set of disks is a good idea from their point of view, so give them the chance by letting them know you are coming far enough in advance. Sometimes there are differences that have developed over time between the disk drives at their facility and your brand-spanking new Apple floppy drive. Old drives get cranky and occasionally may not accept a floppy that has been formatted by another drive. The busy facility also has seen more than their share of HD disks with EDLs and offline editors who blame the Avid! If all else fails, to get your floppy to be accepted by their online edit controller, copy it to another computer that can read RT-11. This other computer potentially may have EDL Manager running. Turn the EDL into a text file using a word processing program, and copy it to a properly formatted disk.

SONY AND DOS

One way to avoid all of this is to work with a Sony list. Sony saves to a DOS-formatted disk that the Macintosh can easily handle. Early versions of the Macintosh needed one of two different extensions, either PC Exchange or DOS Mounter. These extensions need to be enabled in the Extensions folder to read or format a floppy for Windows. Choose Other during the formatting dialog on the Macintosh and choose the proper format for the size floppy you have. Early Sony edit controllers may read only DD disks, but the later ones read HD. If in doubt, get the DD and format 800K.

REAL-LIFE VARIATIONS

For one final twist, there may be weird combinations of formats for disks and formats for edit controllers. The later CMX controllers, like the OMNI, can accept DOS-formatted disks. The production company may have its own list optimizing software that runs on a DOS or Windows system, so they may want the list, no matter what it is, on a DOS floppy. There used to be edit controllers that worked from 8-inch floppies; most people have never even seen these before. Forget about compatibility. These unlucky people have devised a way to send information serially, through a cable from another computer that can handle the 3.5-inch floppies to their online system that can handle only 8-inch floppies. EDL Manager allows this, too, but it takes a 1–2–3 Go! procedure that you probably don't want anyone to see.

BE PREPARED

There appear to be many ways to get this whole EDL process wrong because there are so many ways to get it right. With all this potential for 9 A.M. thrash at the online facility, there are ways to cover yourself. First and foremost, you should ask the production facility what they want. If you are using the same production houses repeatedly, then you know their drill and can have procedures in place for each edit suite of each production company you use. There is a very useful feature on newer EDL Manager versions that allows you to make a template for your favorite edit suites.

EDL TEMPLATES

Regardless of whether you are an offline boutique that sends the lists to another facility or you are the facility and must make different EDL versions for each of your edit suites, you could use a little help. To make sure the lists are always correct and to make sure that you can turn this chore over to even the least experienced assistant, you can make EDL templates. These templates can actually be made by the post facility and e-mailed to you as an attachment. You would then place the template file in the EDL Manager Templates folder inside the Supporting Files folder of the EDL Manager folder. These templates are simple text files, so they can easily be copied to a floppy disk and loaded onto any system with the proper version of EDL Manager. As a facility, instead of faxing a complicated series of instructions, you can e-mail or overnight mail the small file on a floppy disk, which will guarantee that the offline editor gets it right.

You will also find that you may need to make multiple versions of every EDL for different reasons. You may need an audio-only EDL and an EDL for each of the video tracks used in the offline sequence. If you create a template for each of these often-used versions, you can make them all much faster. Just choose the Template pulldown menu for Video Suite 1 Audio Only or GVG Suite Video 2 and quickly save the new version. With a preformatted floppy and a template from the post facility, how can you go wrong? But don't bet everything on one floppy—use two, they're cheap! You may also want to save the EDL in several other formats. Always make exactly what the facility asks for and then make a CMX 3600 list. Pick your market's favorite format and make that your backup format. It is up to you whether you want to put it on DOS or RT-11. (If you choose both, it could make you look pretty slick.) Then you should burn the sequence bin to a CD-ROM and, if the project was small enough, include all the bins. You may find yourself actually making the list again from the sequence at the online facility's Avid or from your own copy of EDL Manager (which you brought with you). After this is done, trust nothing! Make two printouts and bring a copy of the project on VHS or some other format that can be played back in the online suite without a charge for another deck. An MPEG file on a CD-ROM could work, especially if you have your own laptop to play it back. You or your producer should control the playback of this copy in the online suite so you can follow along with the online assembly.

WHAT IS AN EDL?

Now that you are prepared to make the EDL, let's take a look at what an EDL really is and why it is so hard to make one from a nonlinear editing system. The only information the edit controller wants from the Avid is the tape name and timecode (sound familiar?). The edit controller cues up the source tape and inserts the shot onto a specific timecode on the master tape. So a basic event—a cut, for instance—needs four timecode numbers: an inpoint and an outpoint on the source tape, and corresponding in and outs on the master. It looks like this:

```
017 100 V C 02:00:24:19 02:00:25:14 01:00:22:02 01:00:22:27
```

When you need to make a dissolve with analog tape, most of the time you need a second source. If you must dub a shot onto a second tape, typically it is called a B roll. Thus, an A/B roll system is one that allows dissolves. A dissolve needs the inpoints and outpoints from the B roll in the event and looks like this:

```
012 100  V C     02:18:31:29 02:18:31:29 01:00:12:19 01:00:12:19
012 100B V B 030 02:23:26:15 02:23:28:26 01:00:12:19 01:00:15:00
BLEND_DISSOLVE
```

Any kind of fade up from black or from a graphic that has no timecode is an event that starts with black (BLK) or an auxiliary source (AX or AUX). It looks like this:

```
002 AX  V C     00:00:03:27 00:00:03:27 01:00:01:28 01:00:01:28
002 100 V D 030 01:43:44:28 01:43:50:29 01:00:01:28 01:00:07:29
BLEND_DISSOLVE
```

All edits that come from a nontimecoded source, like CD audio or a PICT file, come in as AUX. If you have 100 PICT files, this could be a problem.

TRANSLATING EFFECTS

Any time you create an effect like a wipe or a 3D warp, it must somehow be translated to this very limited, older format that is concerned with tapes and timecodes. There are standard SMPTE (Society of Motion Picture and Television Engineers) wipe codes, but some of the most popular effects switchers do not use them. EDL Manager has a list of the most popular switchers represented, and you can use these simple templates to represent your wipe effects. You can even take these templates for the switchers and modify them to your own switcher if you are feeling geeky. The wipe codes can usually be found in a manufacturer's technical manual, and these templates are just text files.

There is no way to create a generic 3D effect that any group of DVEs (digital video effects generators) made by different manufacturers can read. Many people don't have the capability to send DVE information from their edit controller to the DVE. These two machines easily could be from different manufacturers and have no communication between them except "Go Now." Different DVEs use different scales for computing 3D space, their page turns look incredibly different (if they can even do a page turn), and really, forget it, there is no standard for effects. There will not be such a standard for a long time to come. You have the timecodes for the two sources for the effect and you have the VHS. If you are really persnickety, you may even have written down some of the effect information as you made the effect. Adding a comment to the list like, "second keyframe is exactly 20 frames later" doesn't hurt.

This information is best delivered to the EDL itself using locators or the Add Comment feature in Media Composer and Symphony. You can use the text entered into locators as comments in the EDL. This is an option in EDL Manager's Options window under Show. To use the Add Comment feature you highlight a single shot in the sequence with the segment arrow, pull down the menu above the Record window, and use the Add Comment option. The comment shows up in the EDL as a little bit of extra text when you choose Show: Comments in EDL Manager. You can search by locator in the locator window or by comment in the sequence using Ctl-F/Cmd-F in the Avid timeline.

Even if you can represent exactly the way an effect was going to happen, you cannot foresee how the online edit suite will be wired. The effect may call for three levels of dissolves, but the switcher has only two levels it can do simultaneously. Your effect may require that the editor create multiple layers of video, which the editors should know how to do quickly and efficiently with their own configuration of equipment. If you try to be too specific about the exact procedure for creating an effect, you are denying the online editors the chance to do what they do best.

MULTIPLE LAYERS OF GRAPHICS AND VIDEO

The final insult in translating from nonlinear to linear is the way most EDLs represent multiple layers of video. They really don't. This means you need to reduce all multilayered sequences to a series of single video track EDLs. The nested tracks must be broken out of track one and put into separate EDLs for each track in the nest. This provides you with the ability to run more than two video sources simultaneously in the online edit suite. Don't count on it unless you have researched the specific suite you will be using.

The online editor loads the video tracks as separate sequences into the edit controller and combines them. You can help the online editor figure out which tracks go where by the way the separate lists are numbered. Each edit is called an *event* and has a number. If the first layer of events ends at event 110, then all of the events for the second layer of video can start with event 200. The third layer can start at event 300, the fourth layer at event 400, and so on. The editor knows that when two events overlap in the EDL and if one has event number 50 and the other is event 210, then 210 is part of a multilayered effect. Any multilayered sequence should be discussed carefully with the individual online editor. Make sure they can even get close to what you are doing on the Avid. You may find they say, "Sure, we can do that—in the compositing suite!" Listen for the cash register sound when they want to put you into a special compositing suite. There may be no other choice but to go this route. On the other hand, a slight redesign and some extra goofing around may save you thousands. A freelancer using Adobe After Effects or another desktop compositing program may be the right answer if you have the time to render. There is a tradeoff between waiting for a short render in the expensive compositing suite and waiting until Monday for After Effects. With the faster CPUs the render times on the desktop are becoming more competitive. It may make sense for you to get a cheaper online suite and arrive with all the effects already made and output to tape so you just slug them in. Did I mention that you can't make changes in the online suite?

SIMPLIFY THE EDL

If you try to make an EDL from a complex sequence with layers and nesting and you just load it all into the EDL Manager and press the purée button (or Update), what comes out will not be very useful. You may get some error message as the system warns you that what you are trying to do is not very nice. This is usually some kind of a parsing error or a message that an area of the sequence is too complex to be represented. Make a copy of your sequence and start to simplify.

First, get rid of the nesting. Figure out what are the most important sources for that effect. What timecodes do you need? Keep just that source. If you absolutely must have the timecodes from multiple sources for that effect, you should not have nested it in the first place. Yeah, right, that assumes you knew you were going to an EDL! If you are going to an EDL, do not nest. Chapter 7 describes how to unnest. Enter the extra timecode and source information by hand as a Comment or a locator. In EDL Manager, choose which video track to use, starting with V1.

If you know you are going to an EDL, don't use imported graphics. An online suite may be able to use PICT or TIFF files through a sophisticated character generator, but if you know you're going to an online room, all graphics should be fed to you on tape. The timecode numbers of the fill, matte, or backgrounds can go cleanly into your EDL with few questions asked. Also, if you use imported graphics, it means additional work for the graphics department because they must output the electronic version for you, then they have to make a fill and matte and lay it to tape for the online. If you have subtle variations in logos or graphics, the online editors can put in the wrong graphic since now you expect them to figure out which graphic on the graphics tape corresponds to your nontimecoded source. Even if you use imported graphics to simplify your edit or because you didn't know you were going to an EDL, I strongly recommend recutting those sections with the same graphics tape that is used for the online.

SOUND LEVELS

How did you change your sound levels when you were mixing? If you used rubberbanding, there is no problem. Those levels are not in the EDL, but the sound edits do not confuse the online editor either. If you used the older method of creating Add Edits, changing the levels and adding dissolves to smooth out the level change, then you have some work to do. These Add Edits are real edits and the EDL Manager is smart enough to ignore them, unless you have added a dissolve. The EDL Manager puts an edit in the EDL and tells the online editor to dissolve to a B roll, a dub of the exact same audio source, just to change the audio level. This is confusing and annoying. You need to remove the dissolves to any audio transition that is just for a level change. This is actually easy because you can go into the effect mode and lasso or Shift-click them and delete them all at once. This is a good reason to use rubberbanding. You can also use a feature in EDL Manager called Audio dissolves as cuts.

For the high-end jobs, the sound usually is sent out for audio sweetening and is laid into the online session in one big edit. For lower cost jobs, a digital cut is laid to tape (sometimes split tracks, sometimes stereo split) along with the low-resolution video. Then the online editor cuts his or her video on top of that, covering it. For these EDLs, disable the audio tracks, creating a nice, clean video-only EDL. To be safe, make an extra EDL with audio and video, just in case they need to find the video for a sound bite that has been covered up entirely. It will also help if there is a problem with the sound from the Avid in a certain spot.

SETTINGS

There still seem to be many choices in the EDL Manager that haven't been described yet, so don't get nervous. Most of these functions are for streamlining an auto-assemble and other more obscure situations. We will get to them at the end of the chapter, but for now let's go with the simple answers.

SORT MODES

A sort mode is a way of ordering the events in an EDL so that they can be edited out of linear order. Linear order means that you load a tape into the deck, adjust the levels from the color bars at the beginning of the tape, fast forward to the spot on the tape where the shot is, and perform the edit. Then you load the tape for the next shot in the program and do the same thing. You do this whether the source tape is 20 minutes long or 2 hours long. There are a lot of

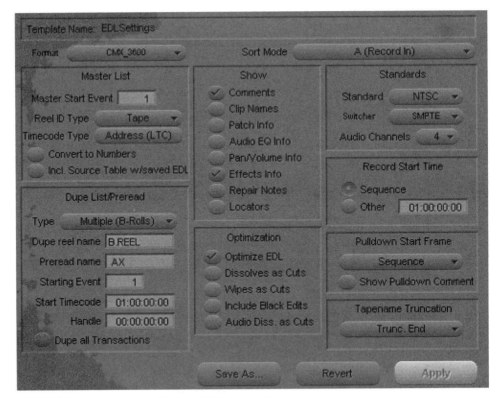

Figure 8.2   EDL Manager Options

repeated actions and wasted time shuttling the source tapes back and forth. Unfortunately, a linear order is the easiest to follow and the safest way to assemble a show and be able to compensate for a mistake or make changes. You know immediately if a shot is missing or a sound bite is wrong because it does not make sense as you put it in place.

The time savings are significant if you can assemble a show out of linear order. It means you need to load a source tape only once, do all that color bars setup, and skip through the show, dropping shots wherever they are needed from that source reel. Then put the tape away. You might have to bring it back for effects later, but if your show is just cuts and dissolves, you have made serious savings in time and money. But imagine making a change that makes the show just a few seconds longer. If you have been inserting shots all over the master tape out of order, this is pretty serious. How can you ripple that change and push down all the edits after it in the program if there are already 100 shots laid down after the change? You can't really; if you are working with digital tape, you can clone the master and make an exact digital copy. If it is an hour-long program and you are halfway done, then kiss goodbye a good 45 minutes of setup and dubbing time. Did I mention that you can't make changes in online?

This kind of penalty for being wrong or making changes at the last minute tends to frighten even very brave online editors. For this reason, they want to be the person who decides how to sort the list. When asked, they inevitably tell you to use the A sort mode because that is the linear mode and they sort it themselves in their edit controller to make it a B, C, D, E, or S mode.

Let's do a quick recap of what the different modes are so you can discuss them intelligently at cocktail parties. B mode is for long-format shows where the source tapes are shorter than the master tape. The source tape fast-forwards and rewinds and finds shots that go in a linear order on the master. This is easier to follow because the master moves forward while a specific source tape is being used. As soon as a new tape is loaded, the master rewinds to the first place the new tape is used. It then starts to move forward again until all the shots from that source are inserted on the master.

A C-mode list is for when the source tapes are longer than the master tape. You may have a 90-minute film transfer for a 30-second commercial. (I hesitate to call them "spots" anymore since in the UK that means pimples!) Here, the source tape always moves forward in a linear way and the master tape flails around rewinding and fast-forwarding to assemble out of linear order. Potentially this is faster than B mode but harder to figure out what is going on. It is very disconcerting for clients who should just look away if they are feeling queasy.

D and E modes are like B and C modes except that all the dissolves are saved for the end. This works only if you have time base correctors that have memory for individual tape settings. More common these days is a serial digital interface (SDI) signal from a digital deck. Supposedly, the SDI signal does not change from deck to deck like an analog signal and so, theoretically, never needs adjusting. The editor must match a frame perfectly from something that he or she adjusted this morning or yesterday so that you don't see where the pickup point is to start the dissolve. D and E modes get used more in PAL countries where they don't worry about analog hue adjustment, unless it's terribly wrong to start with!

S mode is useful for getting a list that will be used for a retransfer from film. If you have been cutting with a one light or a best light, inexpensive film transfer without shot-by-shot color correction, then you want to go back and retransfer the shots that were actually used. S mode is the order that the shots came on the original film reels and allows the colorist to identify quickly only the used shots for the final color correction.

DUPE REELS

After choosing the format and the sort mode, it is time to think about dupe reels. What are you going to do when you need to dissolve or wipe? If you haven't experienced the difficulty of dissolving between two shots on the same tape, then you are about to. With a traditional analog tape suite, you must physically copy the second shot onto a dupe reel so that you can roll two tapes at the same time. It seems kind of old-fashioned and quaint now, but a dupe reel is still the necessary evil in many suites across the world. The biggest change to this procedure was *preread*. Someone figured out that since the digital format tape was just outputting digital information, short amounts could easily be held in a buffer, or temporary memory, in the record tape deck. During the preroll, the master record machine loads the digital images of the shot already on the master tape into the memory and plays it back as a source deck. With the last shot held in memory, the master tape switches from being a source and records the dissolve with the new incoming shot.

Some editors also use a kind of nondestructive preread called *auto-caching*. Auto-caching holds the end frames of a shot for dissolves and various layers of multilayered work in a cache that uses a digital disk recorder (DDR) or a digital tape. The edit controller figures out what needs to be cached for later use in a dissolve or effect. Then it automatically lays the image on the DDR or digital tape. When the tape with the other half of the dissolve is put into a deck, the controller performs the dissolve between the tape and the auto-cached piece of video. This works better than preread because prereads are destructive, meaning you are committed to that edit permanently because the last image is recorded over itself when the recorder becomes a source. You get only one chance to do a preread correctly, and then you must reedit the previous shot. Auto-caching uses a preread EDL to create the auto-caches that can be executed and reexecuted if needed.

Most of the choices that deal with how a tape is dissolved are refinements on these basic requirements for either a dupe reel or preread. Under certain conditions, you can significantly reduce the amount of tape loading and shuttling, but most of the time you choose the simplest technique. The online editor has a very specific requirement and again, as with sort modes, can modify the simplest format into something he or she can use best.

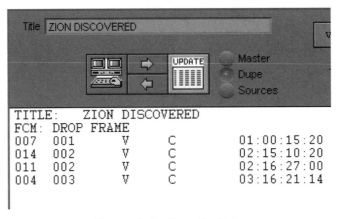

Figure 8.3   Dupe Reel List

The simplest, best answer for most circumstances is the Multiple B Rolls option. It is also the least clever, but with more sophisticated editing systems out there, editors can modify this list to speed up the assembly. The Multiple B Rolls option assumes that there is a separate B roll or dub of every tape in the show. This is probably not true unless you have a very small amount of material or have made simultaneous duplicates during a film transfer. It is more likely that the online editor looks at the list, sees that a dub is required, and hunts down the original tape. The timecodes on the B roll are the same as on the original tape, and the name of the B roll is a variation of the original tape name. Finding the shot is easy because it has a B after the tape name or the tape number is incremented by 500, depending on the type of list format. The editor can then just dub that shot when it is needed or change it over to preread in the edit controller.

Making a Special Dupe Reel

The other B-roll choices allow the editor to prepare before the online session by taking all the shots that need to be dubbed and putting them onto one reel. This keeps the tape loading down to a minimum and allows you to move faster with fewer decks configured (and charged for).

The most dangerous dupe reel choice of all is the One, New Timecodes option, although it is extremely popular in Europe. This requires that you make two lists: one for the master tape and the other for the tape where all the B shots will be dubbed. This new dupe reel has perfect new timecode because you are ignoring the original timecodes of the source tapes and starting all over again. You take each tape and dub one after the other onto this new reel, which starts with timecode that begins at 1:00:00:00 (although you can change this). This means that the timecode link between the original source tape and the dupe reel will be completely broken as soon as the dupe reel is made. The dupe reel name, which defaults to B.REEL, shows up in the list every time a shot needs to be dubbed for an effect. You cannot look at the EDL dupe reel name and figure out where the shots came from originally. Many people choose One, New Timecodes and then do not make the dupe reel list—a big mistake. When the online editor turns around and asks, "OK, where is this B.REEL?" you really should have another list to hand over (maybe even the premade reel). Unless you make the B.REEL yourself, that reel needs to be assembled at the beginning of the session.

Because it is assembled before the editing really starts, changes to the B.REEL are also very difficult. You *must* make a second list for the dupe reel! If you must make a dupe reel, then One, Jam Sync is better. If you forget to make the list for the new dupe reel, at least the original timecodes are in the list and, if the timecode hour matches the tape name, finding the original is a snap.

Another dangerous choice for the unsure is None. This gives you a list that most edit controllers cannot read, which at first seems like a rather bad choice. Not so if you are going to another NLE or an older compositing system. These systems do not need to dub a tape in order to dissolve because, once captured, individual shots are divorced from the physical restrictions of tape. Loading the list into the NLE system forces it to see the tape 001 and another tape named 001B. Is this the same tape? There is no way to know for sure, so the NLE system makes the safe but incorrect assumption that 001 and 001B are two distinct tapes. When you are capturing this material, the system asks for tape 001 and then proceeds through this tape to the end and grabs all the shots that are needed for the sequence. Then the NLE system asks for tape 001B, rewinds, and does the same thing again with the same tape! This is not efficient and, at those room rates, you need to be efficient. The None choice is the best for this situation since it shows just one tape, 001, and the NLE system requests that tape only once.

PRINTING THE LIST

When you print out the list, you should make two versions. Once you are ready to print the EDL, think about who will be using the printout. The people supervising the online edit certainly need it in front of them. They may not be the most list-savvy people in the production, so there are ways to make the list more readable. Include comments in the list when you are about to print. Use comments like "clip name" and the source table, which is a list of all the sources needed for the online assembly. Then print a stripped-down version for the online editor. Again, ask if there is any special information the editor wants, like audio patch information or repair notes. I know some offline editors are reluctant to have repair notes show up in their list because it looks like they did something wrong. Repair notes are important because it may just be a case of the effect being too complex and not being able to represent it correctly. If you have any repair notes, then the online editor should see them.

THE SOURCE TABLE

The important function of the source table is when occasionally the EDL Manager finds a tape with a name that is too long and must shorten it. Sometimes there may be a duplicate tape name from another project. There may also be two separate versions of the same project that have been combined. The EDL Manager has to change that tape name in the list. The source table tells you that since you had a tape 001 from project X and a tape 001 from project Y, that with a Sony format, EDL Manager changed the second tape name to 999. A CMX list changes a second tape 001 to 253. According to the system, these are two different tapes! This is why the original logging is so important. If you don't have that information, then you could get very lost. Never attach a source table to the

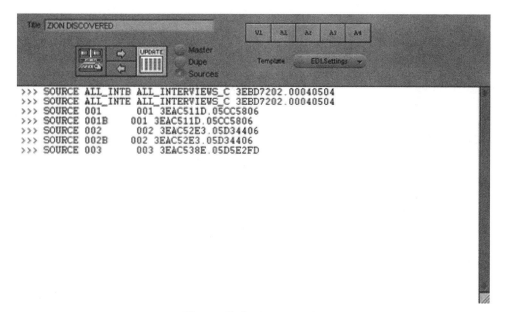

Figure 8.4 Source Table

electronic version EDL on the disk because it is really just a very long comment added to the last event. Many systems have a limit as to how many comments they can read per event, and this will cause an error when reading the list if you have a lot of tape sources. Make the electronic version without the source table, but always print it out separately to take with you.

You can truncate the names of your tapes in a very specific way. This is important if your tapes have been given names that do not match the restrictions imposed by the EDL format. Most formats impose a character length limit of between six and eight characters. You may find that the way the system truncates your tape name makes very little sense. You now have control to preserve the end of the tape name if that is where the most important digits of the tape library code exist. This is especially useful for tracking back tapes to a stock library.

PAL LISTS

If you have a PAL project, be aware that the very first time you open EDL Manager, it may default to NTSC. You can change this to PAL under the Standards Setting and it will stay that way. PAL lists become complicated when you are working in film and want to go back to the negative because of the two different methods of dealing with the difference between 24 fps and 25 fps. Sometimes you slow down the audio and sometimes you include extra frames in the video (that pesky 25th frame). The PAL Method 1 is for film projects in which video and audio are transferred together at 25 fps. PAL Method 2 is for when video and audio are captured separately and synched in the Avid. If you have a video project without any matching back to film, then choose plain vanilla video PAL.

MATCHBACK AND 24 FPS EDLS

Matchback is a method preferred by low-budget films or film projects that are primarily for video distribution, but will be projected sometime in the future. This is an option that will output a 30 or 25 fps project as a 24 fps cut list. Film transfer to video adds extra frames. There is no way for the video-based project to keep track of the A frame, so every outpoint in a matchback project can be accurate only to plus or minus a video frame in NTSC. The matchback is much more accurate in PAL. For many people, depending on the material, this is quite acceptable compared to the cost of conforming a negative first and then transferring to video or using a Film Composer. They get a video project at a high, two-field resolution that they can shop around or distribute, and when the funding begins to flow in, they can get the negative cut.

There is an easy way to make a 24 fps EDL for cutting negative or to import into a 24p finishing system like Symphony or DS. Highlight the Start column in the bin with all the sources used in the final sequence. Use Ctl/Cmd-D to duplicate the column. You will get a pop-up dialog asking where you want to copy this information. Choose TC 24 and the 30 or 25 fps start times will be converted to 24 fps. You can then use the 24 fps choice in EDL Manager to make your list. This is one of the least known tricks for taking any film project to a 24 fps finish without using the Avid film options.

Using Symphony Universal at 24P allows the creation of both a high quality output when working at 24 fps and a perfect negative cut list using Film-Scribe, a cut list utility. This will allow you to broadcast the finished project today and cut the negative when theater distribution is required. Using the Universal Editing option on Media Composer, you can output at 14:1 progressive (which looks pretty darn good for offline) and output the perfect list for cutting negative.

CONCLUSION

It should be clear by now that creating EDLs is rather complicated. It is also clear to the folks who receive these EDLs from Avid systems that many people get it wrong. The amount of 9 A.M. thrash that is completely avoided by asking people to bring only a copy of their sequence bin more than pays for the EDL

Manager software, so every post facility should own a copy and have it quickly accessible. Of course, the translation to EDLs becomes less desirable every year as Avid solutions to finish using AAF are used by more facilities.

Here is a checklist of steps that should be taken before the linear online assembly:

- Simplify a copy of the sequence before you make the list. Do this based on feedback from the online editor.
- Format the floppy for the needs of the production house (either DOS or RT-11 or both).
- Use the EDL template sent by the post facility or their faxed instructions for formats and options.
- Make any extra versions that might be important (audio only, video only, V1 separately, etc.).
- Use the same template to make a CMX 3600 version.
- Print the list twice. Make one for the editor and one for the producer. Include the source table in the printed versions only.
- Make a VHS copy of the sequence that you can control yourself in the suite.
- Don't forget to bring the source tapes!

9

Overview for Finishing

One of the most exciting aspects of the digital nonlinear technology is the promise of finishing a project completely digitally. Many look to the very near future of never actually going to tape from beginning to end of their production. Some have even gone as far as to claim that the distinction between offline and online has disappeared. I disagree with this pronouncement and feel that there will be a finishing stage to every project for the foreseeable future. The project that originates on DV25 and finishes on DV25 may be started and finished on the same system, but the final polish is still valuable and takes time to do right. More inexpensive systems will be able to handle uncompressed standard definition and the cost of media storage continues to plummet, but currently it is rare for someone to work on a project entirely at a high-definition resolution. It still makes sense to assemble a project at low resolution for a high-definition finish. It also still makes sense to have the storytelling and the technical polish done at different stages in the timeline of the project, but not necessarily by different people.

The postproduction ideal is the ability to input everything into one system and free yourself from having to go to any other facility to output a high-quality, professionally finished product. As we discussed earlier, this means mastering new skills and taking on the responsibilities that other specialists have taken care of in the past. Specifically, it means mastering details like learning to set your own video and audio levels. It also means keeping the entire workflow in mind at even the earliest stages and scheduling enough time for all the details at the end of the project.

This chapter deals with techniques for bringing your project over with all the effects, graphics, and audio from the offline stage to the finishing stage at uncompressed standard definition or one of the many types of high definition. If you have been capturing at high resolution throughout the project, you can output to tape and be done with it. All your levels should get close scrutiny at this stage because you are the last person to run the project through a quality check. This is a serious responsibility, and you should take nothing for granted. Symphony gives you an automated method to limit all your levels using Safe-Color, and DS has a legalizer effect, which we will cover later in this chapter.

If you have been working at low resolution throughout the project, then there are quite a few more steps. In the simplest project, you recapture the shots in the sequence, add in the high-resolution graphics, render the effects, and digital cut to tape. If your sequence is that easy and straightforward, this process will be simple and efficient. If you have a long, complicated sequence or some not-so-simple demands from the client, then you need the extra power and flexibility that the techniques described in this chapter give you. Some techniques are basic and intuitive and some are pretty complex. I invented some of the more complex techniques in response to real-world complications. These techniques have previously been published on the Internet, given out at seminars, and faxed by Customer Support. One of the philosophies of finishing on an Avid system is that simple is good, but flexible is better.

THE MERIDIEN VIDEO SUBSYSTEM

Starting from the ground up, Avid created its own video board to edit with uncompressed YUV digital video. Avid also made it much easier to get signals in and out of the system by using most of the industry standard input and output signal types and putting them on a separate I/O subsystem. You can input and output composite, s-video, component, and serial digital all from the same rack-mounted box. Audio can be tied to a two-channel audio card (Xpress) or the standard 8:8:8 channels (input, monitor, and output) we have come to expect from the Media Composer and Symphony.

Because all this hardware is in a separate box, it is away from the CPU and thus better shielded from any electronic noise that might be generated by the computer's internal systems. It also means that a single connector to the host computer is all that is required for audio and video capturing and video compression. This saves valuable slots and allows you to continue to add network cards, Fiber Channel, SCSI, and DVE with perhaps a slot left over depending on CPU choice. The Macintosh systems must occasionally still use a PCI Extender since the high-end workstations contain only three slots. Important Avid boards are attached through this customized, rack-mounted unit.

The Meridien resolutions are named by their compression level, such as 1:1 for uncompressed and 2:1 for a 50 percent compression. There is a low-quality two-field resolution called 20:1 that is mixable with the 3:1 and 2:1 and will quickly make you forget AVR 12. The single-field resolutions are marked with an "s"—as in 15:1s—to make the difference clear. As with the ABVB, Avid's previous video board, there is mixing of single-field with single-field, two-field with two-field, but no mixing at all with uncompressed. There are several multicam or "m" resolutions that allow you to play up to nine simultaneous video streams and switch between them while playing.

Adrenaline and Nitris Systems

The latest configurations from Avid handle a wide range of compression types. You will be able to mix uncompressed graphics and animations with DV25 footage. This means that you will be able to ingest or capture footage in its native format through firewire, SDI, SDTI, or standard analog connections. There will still be the need to convert analog to digital and vice versa for some time to come and quality in this conversion makes a difference. Much backward compatibility is built into the new Avid systems; they will play JFIF and AVR material together in the same sequence with newer formats. This is so new projects and material fit seamlessly into existing production companies' infrastructure and mix with legacy media.

The Avid DS Nitris handles all the major types and frame rates of high definition video. It can also handle standard definition in a wide range of existing compressions and a custom compression board that can continue to be programmed to stay current with emerging formats. There are high-quality resizers, codec compressors, and color correction algorithms programmed into the custom Avid hardware. The audio inputs, outputs, and sample rate converting are as high quality as anything in ProTools 96 kHz systems. The Nitris video and audio subsystems are flagship engineering without functional or quality equal in professional postproduction.

NEW COMPRESSION SCHEMES

Two very interesting aspects to this new hardware don't grab the kind of attention they deserve. The first is that the compressed video resolutions are stunning. It will take a very sharp eye pressed up against the screen to tell the difference between the 2:1 compression and the uncompressed resolution, especially coming from a very clean signal like SDI. The new Adrenaline and Nitris DNA hardware have better filtering, oversampling, and decoding, so even noisy signals compress better. Noise in the signal tends to take up valuable bandwidth with compressed images, but if you are working with a film transfer to Digital Betacam or a downconvert from high definition and can capture using the SDI input, you may find yourself using uncompressed only for your most finicky customers. Why then is uncompressed such a holy grail? And why do clients insist that they cannot accept a compressed signal—only Digital Betacam or HDCAM? Well, first, uncompressed will never introduce compression artifacts under any circumstances so you will never have that rare but awkward moment when suddenly a complex scene doesn't look as good as it should. Realistically, uncompressed images are required for many network deliverables as a format for archiving. No one wants to save a compressed version for future release or syndication because of all the upconverting that will occur when the world slowly shifts over to HDTV for broadcast.

The other concern with compressed images is something called *cascading*— the further compression of the signal for transmission purposes. Broadcasters are concerned that if they compress something 2:1 and then compress it again 12:1 for satellite transmission, it will not look as good as an uncompressed signal compressed 12:1. In my humble opinion, the best answer would be not to compress it 12:1 if they really want a high-quality image, but the economics of satellite time put the burden of quality on the production company, not the network. Findings show that after the first compression of a signal, further compression cascading has little effect (Cornog, Katie. "Factors in Preserving Video Quality in Post-Production when Cascading Compressed Video Systems." *SMPTE Journal*, Vol. 106, Number 2, 1997). Ongoing studies have shown that further compression steps really degrade a signal only when there are major changes made to the image. For instance, to emulate a worst-case scenario for codec testing, Avid will take an image and move it horizontally by one pixel for every new compression stage. This forces the codec to recompress the entire image. Without this artificial torture test, it is very hard to see differences in multiple compression generations using high quality codecs.

Archival concerns are legitimate, so there is definitely an important market for uncompressed video; however, if you want your production to be completely uncompressed, do not use Digital Betacam or HDCAM. Both formats use a high-quality compression scheme, but are compromised to utilize the half-inch tape format. Digital Betacam comes out at a little over 2:1 compression with color sampling at 4:2:2. HDCAM is approximately a 7:1 compression with 3:1:1 color sampling. These formats are widely used by professionals who are usually quite happy with the results, so this example just goes to show that there are still a lot of misunderstanding and ignorance about the value and dangers of using compressed video.

PICKING THE RIGHT RESOLUTION

Look at the total amount of material that you can realistically put onto the storage available to your system. At this point in the technology the limitations have much to do with the throughput of the computer PCI buses and whether you are using SCSI or fiber channel. SCSI is still faster than fiber if it is striped into large clusters with RAID protection, but this may change with 10-gigabit fiber channel switches in the near future. The types of drive configurations are changing pretty fast and, with the widespread acceptance of fiber channel, you will see some pretty amazing drive farms, perhaps with "self-healing" capabilities to continue past drive failures without the user even noticing.

For offline, use the lowest resolution you can stand without squinting; think of the low resolution as a new form of impressionism! If you are confident

in the quality of the footage—for example, a film transfer from an excellent
director of photography—all you really need to see is boom shadows, lip
synch, and distracting backgrounds. The client may determine the quality, however, and leave you no choice but to figure out how to get all your media on
your drive at a higher resolution. Some people, especially those not used to
working with compressed images, get very nervous when viewing low resolutions and just don't want to put up with the lower quality. They must have the
better-looking image throughout the project. The widespread adoption of DV25
has made it the current resolution of choice for offlining; however, some people
have too much material (more than hundreds of gigabytes) or want to offline on
a laptop and do not want to hang large external drives from the firewire connection when they are editing their magnum opus in an airport. We may see the
automatic creation of low-quality proxies at the time of shooting and these may
take the place of any offline resolution, but that is still emerging. If you were
using the Meridien board, you would want to offline your timecoded material
at the 20:1, two-field resolution to mix with the 3:1 or 2:1 for the graphics. You
cannot mix uncompressed resolution with any other resolutions; however, if a
graphic is imported and playing in the Meridien DSK it is considered uncompressed and can still be mixed with other resolutions.

Many film projects are using the multicam resolutions because they take
up less space than similar "s" resolutions. Only on ABVB systems, if you are
using the "m" resolutions, make sure to turn off or disable your 3D effects. Do
this by quitting the Avid and relaunching while holding down the F and X keys.
If you are going to be using "m" resolutions consistently over a long period,
then you should disable the 3D in the Console. Go to the Console, type "dis-
able3D" and press Return. After you relaunch the application, your 3D effects
will be turned off until you go back to the Console and type "enable3D" and
relaunch the application. Your sound will always be CD quality or higher no
matter what resolution you choose, so there is no need for recapturing there.

Nitris and Adrenaline make it possible to mix and match many compression types in the same sequence. The Adrenaline systems allow you to do this in
realtime, but for some compressions the DS Nitris system requires you to process
the media into a cache before you can play it. This makes the picking of resolution simpler since you will normally pick a native resolution from a format type
(like DV25, DV50, or IMX) and mix it with uncompressed titles and graphics.

AVID CONFORM

Many nonlinear systems claim to be able to conform Avid sequences, but none
does it quite as well as Avid. The integrated use of OMFI and AAF information in
the sequence allows Avid to open a project from an offline system inside a Symphony or a DS 6.x and later. All the keyframes are present, along with other

detailed, important information required for the most efficient recreation of the creative work done by the offline editor. The more specific and painstaking the offline editor was, the more precise to that vision the final version must be. Avid Conform will be covered in much more detail in Chapter 12.

NECESSARY EQUIPMENT

Your Avid suite is now an online suite if you plan to finish there. By purchasing the highest resolutions and the fastest drives, you have not completed this transformation! Online suites have some important components that must be present. There must be a high-quality third monitor to view reference video. This must be an engineering-quality monitor. You need one that can be adjusted through standard adjustment procedures like using the blue bars setting. The monitor must hold that calibration in a stable fashion over time. This generally means more money for the monitor, but if you ever get into a dispute about color or brightness, this monitor is your absolute. It is the end of an argument and the last word on "what it really looks like." Don't put your faith in cheap equipment — this finished image is your reputation!

Some people like to have a consumer-type monitor in the room as a low-end reference to "see what it will look like at home." Be careful of this since it is very hard in NTSC ever to get certain colors to look exactly the same on both monitors. When your client insists that the yellow on the low-end monitor must match your high-end monitor, you have a frustrating no-win situation. This is why many online tape suites have only one color monitor and everything else is black and white.

There are very precise color-measuring devices that allow you to adjust monitors to match more closely. After such an adjustment, a video monitor should be left on all the time to minimize the drifting that occurs with warming up and cooling down. If you need to adjust a color monitor, always wait until it has been on for several minutes, longer for older monitors.

It is also nearly impossible to get the computer monitors to match up to the video monitors. They are two very different types of monitors and, even though you can adjust the computer monitors, you cannot adjust hue through hardware adjustments. Color-matching extensions give you more control than the front panel buttons. Many professional graphics programs include these gamma-adjusting extensions, but you will never get the monitors to match completely. You can also use an automated device that can run test patterns and self-adjust. Using one of these calibration devices (for example, from Monaco Systems) can help you match the computer monitors with your video monitors. Do not let the client pick a final color on the computer monitor. They must have final approval on the well-adjusted, carefully lit, high-end video monitor. And yes, it will look different at home.

You will need external waveform and vectorscopes as independent references for all signals. Use them to monitor input, output, and dubs. If you can set up a patch bay or router, everything should go through these scopes. You cannot consider yourself an online room until you have waveform and vectorscopes.

More and more people are deciding that having a color corrector in the suite gives them the extra protection and ability to deal with sources that have difficult color problems. It also gives you an extra tool to make sure that sources shot in very different locations really match up when they are next to each other. Since there is no hue adjustment for PAL, component NTSC, and no adjustment at all for serial digital input, the color corrector gives you the level of control you occasionally need. It can also give you some illegal levels if you are not careful! If that black looks really rich and the shadows have that deep dramatic look you really want, check that the black level never goes too low. The Advanced Color Correction in Symphony and DS Nitris will change the way many online facilities do color correction. We will discuss that later in detail.

You need studio-quality speakers, and they must be mounted far enough apart so that you can listen critically for stereo separation. If you can afford it, get a good compressor/limiter. This takes your audio levels and gently compresses the loudest parts of the sound so that they don't distort. This means you can have your overall sound louder and not worry about the occasional spike in sound level. You can use it during both the capturing and the output or you can use an AudioSuite plug-in before you output. However, if you are sending your audio to a sound designer to finish, don't compress it. Leave the sound alone except to adjust levels and make sure nothing is distorted.

CHOOSING THE INPUT

It may seem obvious that you always want to choose the highest quality input possible, but it is not always obvious exactly how to go about making the right choice. Let's start with the worst quality choice possible: composite. This will be your only choice for inexpensive VHS and three-quarter inch U-matic decks. Composite video has the color information and the luminance information all crammed into one antique signal. This signal has changed little since the early days of television. The equipment that generates, carries, amplifies, and broadcasts the composite signal has improved significantly, but not the signal itself. The PAL composite signal is better than the NTSC composite signal because it has more scan lines and the color is more stable, but both can still be improved.

The main problem with composite is that it must be decoded to separate the color information from the luminance. The decoding process is hit or miss unless you have dedicated hardware doing the separating of the signal:

a decoder. Separating luminance information from chroma in the composite signal is as difficult as if I told you to tell me the exact two numbers I used to get the number 12. And the two numbers change all the time. You probably would get it wrong occasionally, maybe even most of the time. That is how a decoder must figure out where the luminance stops and the chrominance begins in a signal where they are combined. Good decoders actually change how they decode over time and have expensive, adaptive comb filters. Frankly, good decoders do a better job of decoding than simpler luma/chroma separation methods. If you are relying on composite sources for online quality, it is well worth your money to invest in a composite-to-component decoder. Buy only the types of converters that have a minimal delay since some models use a frame buffer and put you out of sync by one frame.

S-Video, sometimes called Y/C, is a better quality signal than composite because it separates the luminance (Y) and the chrominance (C) information from the beginning. But the chroma is just like composite chroma and has all the same problems with crawling and bleeding. You have more bandwidth for the detail in the luminance part of the signal, and the chroma does not have to be separated, but it is still not as good as real Betacam component. Y/C will be your choice for higher end VHS, S-VHS, Hi-8, and other "prosumer" analog sources.

Then there is true analog Betacam component, more technically known as Y, R-Y, B-Y. The Y stands for the luminance and holds all the detail of the image. The R-Y is the red component of the image with all the luminance information eliminated, and B-Y is the same for blue. Together, they can create all the colors in the YRB color space, or what you normally can reproduce with Betacam SP images. This choice would be for Betacam SP and other professional analog formats.

When you get into the digital formats like D1, D5, and Digital Betacam, the best way to go into Avid is SDI. There was a special ABVB top card that could be ordered to take SDI input at a significantly better quality than analog component. Most good quality DV25 decks have SDI I/O, which can be used if you have an Avid system that doesn't have firewire input. Serial digital is also a good way to input DVC Pro although you may be able to use SDTI inputs for faster than realtime input and output.

SDTI-CP is a variation on SDTI from Sony that is designed for faster than realtime I/O of their IMX format. Serial digital input means that you don't have to convert from digital to analog and then back to digital again when working with digital tape sources. The digital information is copied over to the Avid just like copying digital information between two hard drives. Some formats need to be decompressed when sent over an SDI signal. This is true for DV25 when not using firewire. The signal must be decompressed to an uncompressed format and then recompressed in the Avid. You are not adding any quality during this process, but because there is no change in the image between the time it was decompressed and then recompressed, the difference is negligible. Firewire is simple, convenient, and handles the entire compressed signal with audio, but the difference in quality with an SDI input is not worth worrying about (remember DV25 is compressed 5:1 to start with!). In fact, any compressed format that needs to be decompressed to go over the SDI input will not have visible quality changes. The main reason to use the native input types like SDTI and SDTI-CP is for the faster than realtime transfers, and you will need to use the few special decks that support that transfer speed.

One of the more interesting methods of ingesting media into an Avid system is Sony's e-vtr. This deck is basically an IMX deck that can handle multiple formats and send them all down an Ethernet connection. Since the deck itself has an IP address you can control it via the Internet from anyplace in the world with a network connection. You will be able to send IMX format signal (a 50 megabit I-frame-only MPEG-2 signal) over the fastest Ethernet connection you can afford because you literally are copying digital media across a network.

Occasionally, you will need to work with very noisy signals because of either low light conditions or a very grainy film transfer. Keep in mind that the SDI retains all of this noise as well as all of the picture quality, which directly affects compression. It is much harder to compress a noisy signal, and you may find that a noisy signal through a straight SDI connection looks softer than component input when compressed. In this case, you should consider using the component inputs to the video subsystem. Ironically, by reducing the amount of noise and the amount of complexity in the image, you may make your compressed image appear sharper. This is a form of prefiltering before compression that many compression applications use before compressing for DVD.

SETTING LEVELS

In a perfect world, no tape should be captured without being set up to broadcast specifications. This means looking at the tape carefully, through both the Avid internal waveform and vectorscopes and external scopes. Both types of scopes serve important purposes and both are necessary. Don't set up an online suite without scopes! Use them when you capture and when you are adjusting color effects, imported graphics, and dubbing. Getting a show rejected by broadcast engineers or a disc facility because the levels are outside standard limitations is costly and embarrassing.

Video levels on Avid systems go from 0 IRE (not 7.5 IRE) to 108.4 IRE in NTSC and 0 volts to 1.063 volts in PAL. This range is above and below the levels for legal broadcast for a very good reason. The ends of this range are called *footroom* and *headroom*. Footroom is important for graphics using superblack (levels below normal black). Headroom helps to keep the signal from being clipped off if it is just a little outside of broadcast specifications. Preserving headroom and

Figure 9.1   SMPTE Color Bars

footroom makes a big difference when color grading and you need to grab those extra levels to change the gain of the image. Every end use has its own set of standards. Public broadcasting in the U.S. has slightly more conservative standards than other broadcast outlets. Laser disk manufacturers have different specifications and may require different kinds and amounts of color bars and tone on the master tape. Find out what your requirements are and use the scopes throughout the project to conform to those specifications.

Many people memorize which scan line to use for each of the level adjustments using the Avid video tool, but this method will fail you the first time you are faced with different kinds of color bars. Instead, understand why you are looking at a specific part of the screen and what information the color bars are showing you there. In order to see the perfect color bar pattern, make sure the Capture Tool is open; then open the waveform or vectorscope and hold down the Option (Mac) or Shift (Windows) keys (see Figure 9.2).

There are several different kinds of color bars depending on where the tape originated. The most common color bars for NTSC purposes are SMPTE bars, which have several different sections or fields for different adjustments. The top section is used to adjust color on the vectorscope. The very bottom section is used to adjust brightness since it contains 100 IRE white. It is also used to adjust black level, or setup, because it contains three kinds of black, 3.5 IRE, 7.5 IRE,

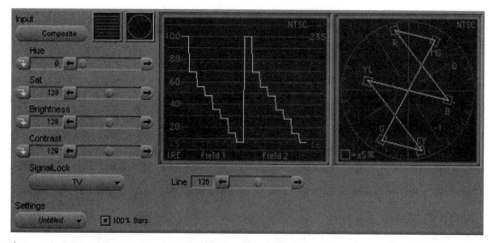

Figure 9.2 Holding Down the Option (Mac) or Shift (Windows) Key with Capture Window Open Shows 75 Percent Bars

and 11.5 IRE. The middle value should be on the dotted 7.5 line. The other two values of black in NTSC are called *pluges* (Picture Line Up Generating Equipment) and can be used to adjust your video monitor brightness. When you adjust brightness level on the monitor, you should see no difference between the pluge at 3.5 IRE (the one on the left of the group) and the setup at 7.5 IRE (the middle bar). PAL productions don't usually use color bars with a pluge. You can set the monitor to a cross pulse or underscan setting and compare the black of the now visible blanking signal to the black in the color bars. When you are adjusting the video input levels for gain (brightness) and setup, you must move

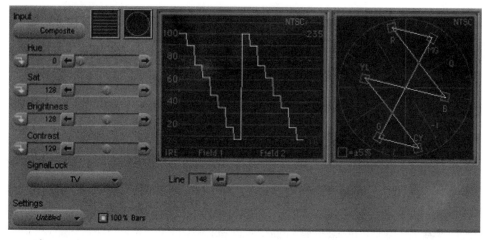

Figure 9.3 The Same Keys Show 100 Percent Bars When that Option Is Enabled

the scan line being monitored to the bottom of the frame until it looks like this on the waveform (see Figure 9.4).

The second most common type of color bars are full field bars, which have straight bars from top to bottom. These are simpler to use, but they tell you less. You can put the scan line just about anywhere in the image (except the very top and bottom few lines) to get a proper reading for brightness and color. The important thing to determine about full field bars is whether the white bar is 100 percent or 75 percent white and whether the chroma saturation is 100 percent or 75 percent. Preferably, you should use the full field bars that have 100 percent white and 75 percent chroma saturation. These are commonly referred to, rather vaguely, as 75 percent bars and more specifically as 100/75 bars or 100/0/75/0 bars. If the white bar is at 100 percent, then set the white level to 100 IRE (NTSC) or 1 volt (PAL). If the white bar is 75 percent white, then in NTSC set the white level at the dotted line at 77 IRE. The math is

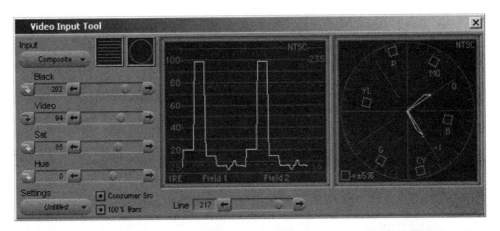

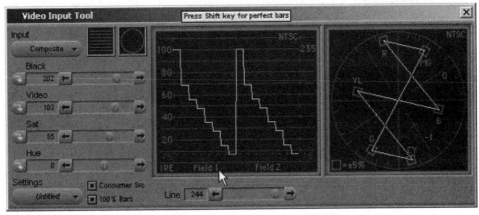

Figure 9.4 Perfect NTSC Gain and Black Adjustment with the Line at the Bottom of the Image

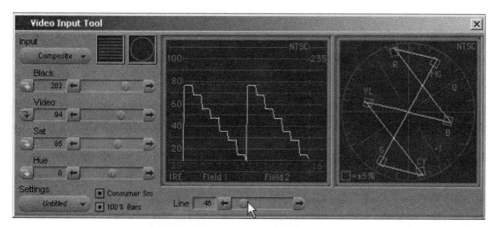

Figure 9.5   Perfect NTSC Hue and Saturation with the Line in the Middle of the Image

strange in NTSC because of the 7.5 IRE black level, so 75 percent becomes 77 IRE. PAL has no setup, so 75 percent white should be at .75 volts.

There is yet a further complication and variation with the common 100/75 bars, which is the 100/100 bars. 100/100 color bars are not used often in NTSC because they cause composite video levels to rise to 131 IRE, and anything

Figure 9.6   Full Field Bars

over 120 IRE would be clipped by the transmitter if it wasn't rejected immediately by the broadcaster. However, these 100 percent saturation color bars are quite popular in parts of the world that use PAL. This causes a problem with ABVB Media Composer Video Tool Settings, but it is handled with an optional checkbox on the Meridien systems and later. Most waveform and vectorscopes that can deal with 100 percent saturation actually allow you to change the way the scale is drawn to accommodate the higher level of chroma, which is what this new 100 percent checkbox does. I would strongly recommend against capturing 131 IRE through the composite input of the ABVB.

In NTSC there is hue adjustment with a composite signal. You can move the color wheel around until the points of the vector signal line up in the little boxes of the vectorscope. With component input, you can only have more of B-Y (blue minus luminance) or R-Y (red minus luminance). More or less of these colors puts the vector points in the boxes. If the points don't fit in the boxes, you need to use a component color corrector during the capturing or apply a color correction to the clip after it has been captured. You can just set everything to the factory-set default levels of the video boards and let the Symphony or DS Nitris online editor handle the problems! Actually, if you clip the signal by letting it get too bright or crush the signal by letting it get too dark, any color correction afterward will only mask the damage. Get the levels right the first time, and any further manipulation will look better.

A serial digital signal has no adjustments for video levels within the Avid Video Tool. You can adjust the levels on the tape deck TBC (the timebase corrector that has all the video adjustments) or use a color correction after the image has been captured.

WHERE TO ADJUST LEVELS

There are several schools of thought about the best place to set input levels: the deck TBC or the Avid. Frankly, according to the engineers who designed the Avid video boards, they are both good, with certain caveats. Let's look at both. You can decide which one is better for the way you work.

Output

Both methods require first setting the output levels. If you do not set the output levels before you set the input levels, you may be masking a problem. In audio, if you turn down the volume of a signal that is already distorted, you get quiet distortion. In video, you get a darker image that has already lost all its detail. So make sure that the output levels are not affecting the signal at all. There is a very easy way to do this, but it requires using external scopes and a built-in test signal. If you (foolishly) do not own external scopes, make sure the levels in the

Output Tool are in the green, preset position. Otherwise, use the toggle arrows at the bottom of the Output Tool (More Choices in early versions) and choose one of the color bar patterns under Test Patterns. These are uncompressed PICT files that are designed to show digital video levels (ITU-R.bt.601) and should register as the most beautiful and perfect pattern you have ever seen on your waveform and vectorscopes.

When you are outputting this perfect signal and looking at the external scopes, occasionally you may see a slight discrepancy. This may be because of other devices that are in the signal path between the video board output and the scope. You can eliminate this guesswork by putting the scopes immediately at the output of the video board. If this looks good, then you have a problem down the line, perhaps with a distribution amplifier (DA). A DA is meant to keep the signal from fading over long lengths of cable or to allow you to split the signal into many outputs for routers or multiple decks. The DA must be adjusted as well!

If the signal is still a little off and you have eliminated the obvious, then you may have to adjust the output levels. If you do this, save them as settings, put them in the Site Settings, and call them Default. That way, any new project will have access to these settings and can use them instead of the factory presets.

Input

The first method makes the most sense from the point of view of editors who have been setting all external devices to unity. In other words, before they send a signal to the DVE (digital video effects device), they adjust the input levels of the DVE so that all signals coming in look exactly the same coming out. That is called *unity*. What goes in is exactly what comes out. This method is also good for editors used to working with a complicated video pathway through a large online suite, routing system, and machine room.

Now feed the Avid a color bar signal, preferably from a color bar generator, but any tape that is known to have good color bars will do. You may want to make a tape of just color bars and 1 kHz audio tone and keep it around as your reference tape. The Avid outputs perfect color bars and tone from the Output Tool and the Audio Tool, so you can make your own reference tape by recording five or ten minutes of this output. Put the reference tape in a playback deck, set the output of the deck so that it looks perfect on the external scopes, and then feed the signal to the Avid. When you feed the bars to the input of the video board and set the levels so they are perfect on Avid's internal scopes, they should look perfect on the external scopes on the output of the Avid, too. Save these tape settings and call them Default. Save these settings somewhere safe and drag them to your Site Settings. You now have standard default settings for both input and output. This is what should be used with all tapes when capturing.

Now adjust the output levels of the video deck using the time base corrector (TBC) to adjust for the differences in levels between tapes. You will never touch the settings on the Avid. This way you know the Avid is not introducing any noise or affecting the signal level in any way.

There is one big drawback to this method: There is no way to save the settings on the deck, but there is a place to save the settings on the Avid. Some TBCs have a memory setting, but it is up to you to recall them by hand when you load each tape. What if you want to use the same tape two weeks later and you want the shot to match exactly the same shot captured at the beginning of the project? You had better hope that you set up everything perfectly to bars and then did not make any further adjustments because bars at the head of the tape are the only reference you have. Chances are, though, that you made minor adjustments after you set up the deck to bars because the picture didn't look just right. Unless you can reproduce that little level tweak by eye, you will have mismatching shots. It is much better to use the ability to save those levels and call them up automatically, and you can if you use a different method.

If you have not blown out or crushed a video signal while inputting, you can correct for any little adjustment in the Avid Video Tool without adding any extra noise to the signal. This is because the signal is digital, and as long as it stays within a particular range (not distorted), it can be moved around without any further degradation. In this scenario, you want the deck TBC to be always at the preset level. Throw all the video level switches on the deck into preset and then make the adjustments with the Avid Video Tool. Save the level settings based on tape name. Then, when you batch capture, the settings for the tape are recalled automatically. If you have 100 tapes, this is a lifesaver. You set the levels once during the low-resolution offline stage. When you batch capture to a higher resolution, and are in a little more of a hurry, the levels are right on without any further intervention by the capturer. This method works only if both the offline and the online are done on the same video board.

If you are adjusting the deck, look into a remote control panel for the deck's TBC controls so you can adjust quickly on the fly. You really can't count on the settings for the color bars being the absolute perfect setting for everything on the tape. So even if you are working from the Avid setting, you may still want to tweak the deck TBC occasionally. The problem with taking a deck's TBC out of preset is that it jumps to wherever the knob was set the last time you used it. Then you quickly have to tweak the levels to get them back close to the preset setting. This makes it hard to do just a little adjustment, for instance, while the deck is prerolling and you see something that is just a little off. A good method, although it is an extra step, is to put the deck in preset, set up the Avid Video Tool Input levels, and save them. Then, while watching the external scopes, take the deck out of preset and manually adjust the levels on the TBC so that they look exactly like they are in preset. You are using the power of the reference point—preset—to get everything right, then allowing yourself to make minor adjustments based on those original, perfect levels.

VERTICAL SHIFTING AND VITC

Sometimes the entire image can shift down vertically and show you a black bar at the top of the frame and flashing little white lines. What you are seeing is VITC (vertical interval timecode). This is a second type of timecode that is recorded as part of the video. VITC should be on a video scan line just above the visible image. You will see this problem quickly if you have set your video monitor on underscan. The source and record monitors in the Avid show you this after you have captured, but ideally, you want to catch this problem right away because it is so easy to fix.

This alternate type of timecode, VITC, is useful when you are moving the tape very slowly or are stopped completely and the main form of timecode, LTC (longitudinal timecode), doesn't work as accurately because it is an audio signal. Instead of guessing for the right timecode number at slow speeds, interpolating from the last well-read bit of timecode, a deck controller looks for the VITC. VITC also allows you to record extra pieces of information like tape name or film information that can be read by some timecode readers and even burned in to the image while capturing.

VITC is helpful until someone sets his or her record deck wrong and the LTC and VITC have different numbers. You see this problem if you are reading the correct number when the tape is playing and a ridiculous number when the tape is in pause. In that case, you want to go under the front panel of the video deck and change the switch to force it to read LTC Only. Imagine what this would do to your logs if you didn't catch it!

Back to the vertical-shifting problem: What is happening is that the video deck is being put into Edit mode in order for Avid to control it. When a deck goes into Edit mode, it changes where it looks for a reference signal. The video deck changes from using the blackburst it is usually being fed from the blackburst generator for reference and instead, the deck looks to the incoming video signal. Many people are not feeding the deck a stable video signal to the video inputs when they are capturing. If there is no stable signal, it shifts the playback of the picture down several scan lines and reveals the otherwise hidden VITC. There are three ways to prevent this situation: make sure the deck does not really go into Edit mode, always provide a stable video input to the deck (like a blackburst generator), or use a playback-only video deck.

The high-end series of record decks have a record inhibit switch that keeps the red light above the Record button on the deck always on, preventing the deck from going into Edit mode and introducing the vertical shift. This is an easy and fast answer until you want to record a digital cut. If you don't turn the record inhibit switch back off, Avid gives you an error as it attempts to cue up and make the edit. It may even ask you if your deck is in Record mode! The next method to prevent Record mode takes a bit more preparation: Make sure that every tape you put into the deck has its record inhibit tab in the inhibit position. Betacam SP tapes and higher formats have a little red switch or a screw on the bottom of the tape cassette. Three-quarter-inch tapes require a little red plastic tab to be filling a hole in the bottom of the cassette. If you don't have the plastic tab for the three quarter-inch tapes to fill the hole, then, as with most things in life, gaffer's tape does the trick. Open-reel formats, like 1-inch, have no protection on the tape itself, so you need to rely on the record inhibit switch on the deck. Confirming that the record inhibit tab is in the inhibit position should be done by every tape operator, every assistant, and every logger in the world as standard operating procedure. Accidentally whacking a deck with a quick elbow in the wrong place and an unprotected tape in a record deck will create a big ugly hole in your master source tape. Try explaining that to the producer! Once you have inhibited the video deck from being able to make an edit on the source tape, the image correctly repositions itself vertically. If this does not work, then you have a more serious blackburst problem and you need to evaluate that all pieces of equipment in the room are getting the same blackburst reference from the same blackburst generator.

This is an exhaustive look at setting levels, but they must be right. This is the biggest obstacle for many editors who have never been required before to care. Look at levels at every stage of the finishing process. What does that imported graphic really look like? I'll bet the print graphics designer didn't take into account broadcast chroma levels when he or she designed the logo. Train your eye to see color on your carefully adjusted monitor and scan for reds, cyans, and yellows that look too hot and double-check them through the scopes. You will learn what to look for, and you will feel much more comfortable with the whole process.

AUDIO LEVELS

Input and output audio levels must be analyzed before you do anything with audio. Generate a 1 kHz audio tone at 0 dB analog and input it to the Avid. There really is no industry-wide agreement on reference level for the digital level for audio reference tone, and it may vary from −20 dB to −12 dB. Used in video production, −20 dB is a new emerging standard audio. Digital video decks are calibrated straight from the factory at this level. Many engineers in the past have immediately adjusted everything that came from Avid to the −20 dB standard.

The digital reference level is adjustable in the Audio Tool under the right-hand pulldown menu (PH), Set Reference Level, and the scale on the Audio Tool will change. These adjustments may also be right in your timeline if you have the audio meter displayed there. This is also now a Site Setting, so it will stay set to −20, −18 or −14 dB. If you have an older system and need to be compatible with a newer system I would strongly suggest that you recalibrate your Digidesign hardware to −20 by using the methods we will outline in the next section.

Play back the videotape with the 1 kHz tone at the proper level by adjusting the audio playback controls. If you do not have these on your deck, you need to double-check that anything recorded at the reference level of 0 dB analog also plays back at that level. Otherwise, your deck needs adjusting first. Then you must decide if you are going to use a mixer during the input stage. If you have many different audio sources like DAT, cassette, CD, or microphone, it might be better to run everything through the mixer. This also gives you a place to equalize or affect the audio one shot or one tape at a time. Now you must calibrate the levels through the mixer. Make sure that the 0 dB tone from tape is at 0 dB in the mixer when both the individual faders and the master faders are in the detent or notched position. Many mixers have trim pots for each input—little knobs in the back where each source can be adjusted individually as they come into the mixer. Be sure to pan odd-numbered channels to the left and even-numbered channels to the right. Finally, look at the Audio Tool or Timeline Audio Meter. Is it at 0 dB, too? If not, you have three choices depending on the hardware of your system. Whatever you do, do not touch the slider in the Audio Tool with the large speaker icon (if you have one). This adjusts the output level, not the input level. This should always, always be at 0 unless you are specifically adjusting audio output that cannot be raised by any other means.

Eight-channel input is standard on many systems. This is a little trickier since it involves adjusting the levels by tweaking trim pots on the hardware that came with this system. These input and output levels may need some adjustment to be precise. The benefit to adjusting the hardware is that once the hardware levels are set, no one should touch them again. The audio I/O hardware requires a small screwdriver, sometimes called a *tweaker* or *greenie* because it is used only for making these little tweaks and is usually green plastic. The larger Avid screwdrivers used for assembling the system are just a little too big and may strip the screw adjustment if you are not extremely careful. Input a 1 kHz reference tone and carefully tweak the input levels until there are no yellow LEDs lit in the Audio Tool or Timeline Audio Meter. For a quick look at where the LEDs should be, call up the reference tone generator in the pulldown menu that is part of the metering tool and observe where the perfect tone should be. Then, with the levels adjusted perfectly, capture a minute or two of this tone into the system.

Play back the captured tone from the Avid and see if it plays exactly the same way that it was captured. It should. Now stop that playback and call up the reference tone again from the Audio Tool or Timeline Audio Meter. Where does your audio go out of the Avid? If it goes through any distribution amplifiers, then these amplifiers should be checked out for unity. Then, does the signal go to the mixer again or directly to the deck? My vote is for it to go to the deck, but many people like to ride the levels one more time as the show goes to tape by putting it back through the mixer. This is fine, but it adds another level of complexity. It also means that there will never be two identical digital cuts of the same sequence unless the levels in the mixer are not touched. My approach requires getting it right in the Avid and then outputting as simply as possible, direct to the deck, to avoid those inevitable 2 A.M. mistakes.

If the output of the tone goes through the mixer, then put all the mixer faders to their notched position again and look at the level. Is it 0 dB? If not, adjust the trim pots for Avid inputs to the mixer. If you are not going through the mixer, then look directly at the deck. Set the deck record levels so that they record the tone at a perfect 0 dB (or the proper digital equivalent if you have a digital deck). Here, if you have the Digidesign 8:8:8 and are going directly to the deck, you can cheat a little if you want. If you trust the deck preset recording level, tweak the output levels of the audio hardware to be at 0 dB when the input setting of the deck is in preset.

The third method can be used when you have no mixer and you need to adjust the monitoring levels of the audio while capturing. You should go to the Audio Project Settings and choose input. Click on Passthrough Mix Tools and adjust the volume levels of your mix. You will not be adjusting the level of the input, just the monitor.

Let's go back to our captured tone already on the Avid and play back tracks 1 and 2 only. Look at the levels on the record deck and note where they are; they should be 0 dB. Go to the Audio Mix window and pan both tracks 1 and 2 to the right. How much does the level rise? About 6 dB. Now turn off the

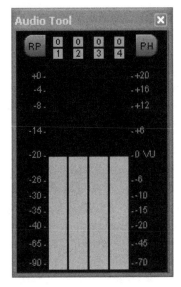

Figure 9.7   Audio Tone Calibrated to –20 dB

monitor for channel 2. There is a drop of 6 dB on the record deck levels. Keep this in mind as you mix your tracks. As you add audio tracks to the mix and then mix them down, the overall level gets louder; this is basic physics. If the stereo audio comes from a digital source like a CD and the mono sound comes from an analog source, you may have to drop the digital source level because it will have more headroom. There is a greater dynamic range on the CD audio than an analog source is capable of reproducing. If you have a sound compressor or want to use the AudioSuite compression plug-in, you may want to compress the CD music so it is more even in level and does not get too soft or too loud in comparison to other analog sources. Some CD library music is already compressed for this reason.

Changing the Calibration from –14 dB to –20 dB

Many facilities are already changing from –14 dB to –20dB so that they will be better calibrated with their digital video decks. You will increase the headroom, but lower the signal-to-noise ratio from the published Avid specifications by plus or minus 6 dB. Most people will never hear the difference, and you will be rewarded by having fewer problems interfacing with these digital decks.

Before you begin changing the calibration of the hardware, you must make sure your software is also set to the right calibration. You can go to the Audio Project Setting and change the calibration from –14 dB to either –18 dB or –20 dB (some countries prefer –18 dB as their standard). The Audio Project Setting is a Site Setting and will automatically be used for every new project created on this system from now on. After your software is set correctly, calibrate the hardware:

- Input a 1 kHz tone, +4 dBu @ 0 VU to audio channels of the audio hardware.
- Open the Audio Tool and click the In/Out toggle buttons for all the channels to display I for input. The tone will appear as green level units.
- Click on the PH (Peak Hold) pop-up menu in the Audio Tool and choose Calibrate. This changes the scale to make it easier to be precise.
- Carefully insert a small screwdriver into the input trim pots on the outside of the audio I/O hardware. Adjust the level until the green level indicators of the Audio Tool show 0 VU.
- Do this for each channel until all eight are calibrated.

You must calibrate the input channels first, before calibrating the output channels. This is because you will actually patch the outputs back into the system to check their level and must rely on the input settings to be correct.

- Take output channels 1 and 2 and connect the audio cables to input channels 7 and 8.

- Click the same I/O toggle icons to display I for the channels 7 and 8 since we are using them for input right now. Toggle the I/O icons to display O for channels 1 and 2, which are now being calibrated for the output.
- Choose Calibrate from the PH pop-up menu.
- Under the PH menu, choose Play Calibration Tone.
- While watching the green level indicators in the Audio Tool for channels 7 and 8, adjust the trim pots on the Digidesign box output channels (1 and 2) and bring it to 0 VU.

Perform this calibration for all the other channels. When you are done, leave them alone!

TIMECODE ON THE MASTER TAPE

When you are ready for a digital cut, you need to consider how the timecode on the master tape has been set. The default timecode for sequences is 1:00:00:00, so you should either black the tapes so that you have enough space for bars and tone and start the show at this sequence time or change the start time on the sequence. If you start blacking your tapes at 58:00:00 or 58:30:00, then you have enough room for a minute of color bars and tone starting at 58:45:00 and then 15 seconds of black or countdown. (Actually, you have 13 seconds since there should never be anything in the last two seconds before the program starts.)

Some countries use the 10:00:00:00 as the start time for their program on the master tape. To black tape in these countries, start at 9:58:00:00 or 9:58:30:00. If you are using the 10-hour start time, then you must change the Sequence Time (under General Settings) to start at 10 hours as well and save that in your Site Settings. This will affect only future sequences—any existing sequences will need their start time adjusted individually. If you do this correctly, then you can use the Sequence Time option during the digital cut. Sequence Time is preferable because there is a permanent relationship between the master tape and the sequence. This timecode relationship can be relied upon, even years later, if something has to change. You may also be able to edit into the master to make a small change without recapturing the entire sequence.

For editors who are working with tapes blacked and timecoded with no regard for the Avid settings, you will have to change the sequence time of all the sequences to match the tape. Many editors leave their record decks in the preset timecode mode and whenever they do a crash record to start blacking the tape, the timecode starts at 00:00:00:00. By the time the bars, tone, and countdown are finished, the starting time of the sequence is around 00:01:30:00 or 00:02:00:00. You can change the starting time of all future sequences by also going to the General Settings and then saving the change as a Site Setting.

General Settings are where you change the default sequence time from non-drop-frame to drop-frame on NTSC projects. Highlight the number already in the setting and type a semicolon. The type of timecode on the master tape, drop-frame or non-drop-frame, must match the type of timecode of the sequence or you will get an error. If you are putting multiple sequences on the same master tape, you may not want to take the time to change the start timecode on each sequence. In that case, there are the other two choices in the Digital Cut window: Record Deck Time and Mark In Time. Record Deck Time is wherever the tape happens to be cued up at the moment; the tape backs up and prerolls to insert edit at that point. Mark In Time allows you to type timecodes in the deck controller at, say, regular intervals of two minutes if you are laying down many commercials in a row.

ASSEMBLE EDITS

Assemble editing has saved many a session from stopping just to black and encode a tape for a digital cut. Using an assemble edit also saves many hours of tape deck record head wear and tear.

Why would you use assemble editing? If insert editing was the only choice, you could not perform a Digital Cut until your one-hour show had a prepared black and encoded one-hour tape (timecode laid down over the entire tape while recording black). Maybe the plans change at the last minute and now there are two versions of the program, one with titles and one without. Do you have two blacked tapes? If not, don't black another tape before you begin— perform an assemble edit for the digital cut.

An assemble edit lays down new timecode during the edit. That timecode is determined by the settings on the record deck. There must be enough time-code on the tape for the preroll—a minimum of three to five seconds that you can set in the Deck Settings. But since the worst quality of any videotape is at the very beginning and the very end, give yourself a healthy 30 seconds if you can afford it. All professional videotapes actually contain well over two minutes more than the length on the outside box would imply. After blacking just that 30 seconds at the beginning of the tape you can perform an assemble edit and the digital cut lays down the new timecode during the playback of the program.

If you set the deck correctly to start the tape blacking at 58:00:00 or 58:30:00, then during the preroll for the assemble edit, the deck will read that code and regenerate it. The deck setting to start blacking the tape with correct timecode should be Preset so you can set the correct starting timecode. Then the switch must be changed to Regen before the assemble edit is performed. If you do not change the switch from Preset to Regen before the assemble edit digital cut, you will create timecode 00:00:00:00 at the edit point. Try to do another edit at that point and see what happens when you search backward from 0.

The deck thinks it must rewind the 23 hours and 2 minutes to find the number for the inpoint, 58:00:00.

If there is no choice for Insert or Assemble in your Digital Cut window then it must be turned on. Go to the Deck Preferences Setting and check the choice for Enable Assemble Editing. Now you have a pulldown menu in the Digital Cut window with two choices—Insert and Assemble.

What happens if you do an Assemble Edit instead of an Insert Edit in the middle of the program? This is the most serious mistake you can make with a master tape and it is the reason why turning on the Assemble Edit function is not only just a little hidden, but it is also displayed in emergency red when activated. When the Assemble Edit is over, there are several seconds afterward where the control track of the tape is unstable as the recording comes to a stop. This can cause any image on that part of the tape to be unstable and unusable; this cannot be fixed except by continuing the Assemble Edit to the end of the entire sequence. You cannot perform a clean Insert Edit over a point in the tape where the control track is unstable like at the end of an Assemble Edit. You will always have a glitch. Use Assemble Edit mode only if you are planning to lay the sequence down all the way to the end and there is nothing else on the rest of the tape. If you plan to add other items to the tape, use Add Black at Tail to continue the timecode after the end of the edit.

OUTPUTTING TO NONTIMECODED TAPE

If you are outputting to a nontimecoded master tape, you cannot control the record deck from the computer. This method is strictly for a quick crash record or going to VHS for a preview copy. Use the countdown built into the software to make the start of the presentation look a little cleaner and more professional. This countdown is purely for letting the tape come up to speed; it is not frame accurate and is slightly different each time you use it. If you want a frame-accurate countdown, make one and edit it onto the front of the sequence. Spend a few minutes and change the PICT file for this countdown if you have a company logo. Any correctly sized PICT file for your video board and format will do, and it adds a nice touch.

COLOR BARS AND TONE ON THE MASTER

Where should you get the color bars and tone to put on the head of your master tape? One thing is for sure: Bars and tone should be representative of the levels in the show that follows! Just sticking a pretty picture with lots of colors at the beginning of your show doesn't make the show look good.

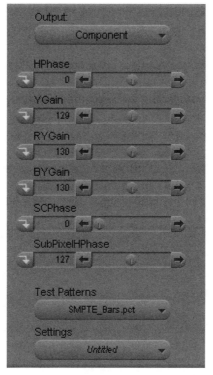

Figure 9.8 Output Tool Set to Show SMPTE Color Bars

If you set a deck up by using these bars, will the show be legal, broadcastable, and listenable?

A few choices allow you to decide what works best for you. You can output both a perfect color bar signal and perfect 1 kHz tone directly from the computer as discussed before when setting all the levels in the edit suite. If you perform an Insert Edit using the edit controls on the front of your deck, you can lay both bars and tone to the head of the tape simultaneously. If you are confident that your video and audio that follow are just as perfect, this works out fine, with one exception: Make sure that there is no slight jump in black levels between the black on the master tape and the beginning of the fade-up at the start of the sequence. The little jump in black levels is visible to some people when they have the black levels turned up a little too high on their monitors at home. You could add two seconds of black filler to the start of your sequence and then change the start time of the sequence to 59:58:00. You can play back black filler from the Avid and use the front panel of the editing deck to insert edit onto the tape after the end of the color bars. Or you could splice all the bars, tone, and countdown to the beginning of the sequence and change the start time of the sequence to be 58:45:00.

IMPORTING COLOR BARS

The best way to get absolutely indisputable color bars and tone into the Avid is to import them. Avid systems ship with several kinds of color bars that are designed specifically for the Output Tool. If you call up color bars as the choice for a reference signal directly out of the Output Tool, you can adjust your Output Settings to be perfect. The files for SMPTE bars, full field 75-percent and 100-percent bars, are in the Test Patterns folder in the Supporting Files folder. These color bar PICT files are created at ITU-R.bt.601 values and are meant primarily for direct output only, not importing into your sequence, but it can be done.

If you import these files, you need to make sure that the import settings are set to 601 file color levels and aspect ratio, pixel aspect: 601 (CCIR on some systems). Do not import them at RGB file color levels! If you don't do this right, you will have bars that are dangerously wrong! The white levels will be around 80 IRE and the blacks will be around 14 IRE. You will get your program rejected for broadcast if you import these test pattern bars at RGB file color levels. I have heard too many stories of the dubs looking wrong or the show being rejected, and the editor always claims, "These were the bars from the Avid!" Yes, but they were used wrong. Let me repeat: If you import the color bars from the Supporting Files folder, you must import with 601 file color levels and use the 601 aspect ratio.

There used to be another graphic that shipped with the software in the Goodies folder called Color Bars for Import, and it may still be floating around your facility. This graphic must be imported with RGB levels. Using RGB levels means that this graphic is for versions of the software and hardware that do not give you the choice of 601 values during import. There is nothing wrong with this graphic except that it is designed to be used in NuVista+ systems that cannot use any black level below 7.5 IRE. As discussed earlier, the split field SMPTE color bars have a pluge that descends to 3.5 IRE. This lowest pluge is not included in these bars. This doesn't affect the usefulness for setting up a signal; it means you have one less tool for adjusting the brightness of a monitor. If you have a NuVista+ system or your software version cannot import at 601 levels, this is the only graphic you should import for correct video levels.

For importing color bar graphics go to the Avid program folder and look in Supporting Files and then Test Patterns. Choose NTSC or PAL as required. If you want to confirm that the levels will be correct for broadcast then import the graphic using 601 color levels, cut it into the sequence, and go into Color Correction mode. Use one of the eyedroppers in the color swatch matching interface to sample the whites and the blacks of the test pattern. With SMPTE bars the blacks should be 16, 16, 16, and in the bottom pluge area they should be 7, 7, 7 then 16, 16, 16 and 25, 25, 25 for the three different pluges. The brightest white should be 235, 235, 235 on the bottom strip for 100 IRE and the white bar should be 180, 180, 180 for 77 IRE. With PAL standard color bars the white bar should be 235, 235, 235 and the black bar should be 16, 16, 16. With the PAL 100% Color Bars these levels will be the same, but the color levels of the other bars will be significantly (25%) higher.

To summarize:

- Import standard color bars from the Test Patterns folder with 601 Import color levels. These graphics ship with every system and are preferred.
- Import Color Bars for Import graphic at RGB Import color level. This graphic is rare.

If you are confused or suspicious about what color bar levels you have, check the bars at the head of the sequence, either through an external waveform or a vectorscope. If the color bars generated by the Output Tool look perfect, but the color bars at the head of your sequence do not, then the bars in the sequence must be replaced.

Audio tone must also represent the level of the audio to follow. Recent versions give you the ability to create a perfect tone media file in the Audio Tool. This can then be edited onto the front of the sequence with the imported color bars. All of this preparation can be done as a last step before performing the Digital Cut. Personally, I hate editing with a sequence that has all of this reference information at the head, so I put it off until the very last stage. When I use the Home button, I want the sequence to go to the first frame of the show, not color bars.

Another school of thought says that the color bars must be captured through the video board if they are going to match the rest of the video that has also been captured through the video board. I think it is much more important that the levels all be the same throughout the sequence and run through the same scopes. This argument does not take into consideration that a substantial portion of your sequence may have been imported and not captured at all! But just for convenience, you may find that carefully capturing color bars, tone, your own special slate, and countdown and black is just easier for you to edit onto the head of your sequence. If the levels are not representative of the video that follows, this method is not any more accurate than the other methods.

You may find that over time it is very useful to have a premade project that has all of the reference material you use on a regular basis. Store all the company logos, color bars, and other reusable media in one place and lock the media so it is difficult (but not impossible!) to delete. This will make it easier to delete all the media for the current finished project in the Media Tool when you sift by project name and not accidentally delete this reference media.

THE REAL FINAL CUT

When it is time to actually perform the Digital Cut and everything else has been prepared, there is one more setting to check. When you call up the Digital Cut window, note which tracks are highlighted. You can patch the audio or video layers from the Avid to the tracks on the tape, making it easier to customize the master tape without making multiple sequences. Maybe you are going to add titles in a last digital pass in a linear suite because you have a specific look that can be created only by a Chyron character generator. You want to lay off a copy of the show without the video track that contains the titles. If an assistant will perform this function, he or she must be quite clear on the implications of this setting.

What if you are viewing the Digital Cut one last time as it goes to tape and discover a minor mistake? You can perform an insert edit in the middle of the sequence without laying down the entire sequence again, if you are using a professional VTR with serial control. If you are using firewire deck control with Xpress DV you will need to lay down the whole sequence again. Make sure you have the type of edit set to insert in the Digital Cut window first. Then mark inpoints and outpoints in the sequence at existing cut points before and after the change. Then uncheck the Entire Sequence checkbox in the Digital Cut window and highlight only the tracks that need to be changed.

If the change lengthens or shortens the sequence, be aware that there will be some more work to do at the end. If the sequence is now shorter after the change than it was the first time you recorded it to this tape, then you must insert black and silence onto the tape after the sequence is over. There is almost nothing more embarrassing than fading to black and then coming back for a little reprise of the last few seconds! There is a checkbox called Add black at tail. You can determine the amount of black with a pulldown menu. This will not change the length of your sequence but will keep the deck recording for a few extra seconds after the sequence has stopped playing. Make sure this option is turned off if you are just replacing a shot or two and are not rerecording the master all the way to the end.

On older versions, you cannot just add filler to the end of a sequence in the Avid. There are a few workarounds to choose from if this is desired, but I prefer the simplest method. Use the edit controls on the front of the master deck and do a short Insert Edit while playing back filler from the Avid. This will not work if you don't have direct access to the deck or if your deck does not have front panel edit controls, so you may be forced to use one of the following methods.

You can import a black graphic, or capture black and edit it onto the end of your sequence for the length needed. You can also splice any piece of audio on the end and then turn down the volume for that clip. A method invented by Jeff Greenberg is to put in an add edit in the filler at the end of the sequence on an empty track and trim the black on the incoming side to be as long as you like.

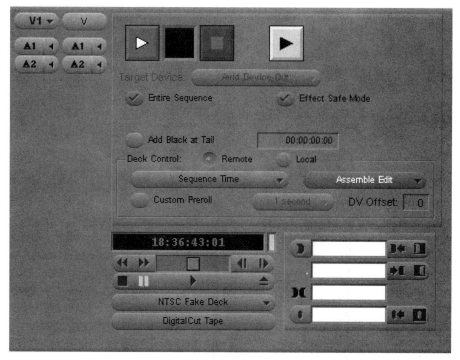

Figure 9.9   Digital Cut Tool Set to Assemble Edit

Other methods to make the system think there should be filler at the end can backfire when you make later changes and the sequence drops the fake filler. This happens if you splice a piece of video on the end and then lift it out. This is fast, but only a temporary fix. Any black or sound added to the end of a sequence ends up in the EDL if you don't remove it, which is why I prefer the method that does not affect the actual sequence.

If the sequence is longer than the first pass after the changes, then make sure the tape has been blacked long enough to accommodate the extra time. If not, change from insert mode to assemble mode in the Digital Cut window, and on the record deck, put the timecode Preset/Regen switch in the Regen position. Make sure you have several seconds of black (as much as 30) after the last frame of the sequence. You never know what used to be on that master tape!

CONCLUSION

This is an overview, although detailed, of all the various considerations to prepare before beginning to finish a project on the Avid. Once these issues are dealt with in a consistent and thoughtful way, many of the annoying quality issues will be absorbed into standard procedures. The next two chapters will go into specific strategies and features designed to speed workflow and to help you cope with flexible and fast-changing situations. Once the basics have been prepared, much of this process can be streamlined to focus on the creative rather than the mechanical.

10

Offline to Online: Finishing Strategies

There are many ways to approach finishing a project on an Avid system depending on length and complexity. Making sure everyone in the process knows what the final result should be means that all carefully laid plans will still be effective in the final crunch time. This chapter covers both simple and complex procedures that allow you to get the maximum flexibility and power from your nonlinear finishing system.

The important factors in finishing are quality control, the speed of graphics replacement, and amount of rendering. The video and audio levels must be monitored at every step to make sure they conform to broadcast specifications. Most people render too much, so Avid fixed this problem with ExpertRender™ and they continue to expand realtime capabilities to someday eliminate rendering altogether. If you are trying to make an air date, then saving even a few minutes becomes crucial, and the knowledge of how to build an effect so that it renders faster or is completely realtime may make all the difference. If you are trying to complete a long and complicated project, then you will especially need the professional controls and level of flexibility that the Avid has evolved over many years of listening to online editors.

USING OFFLINE LEVELS IN ONLINE

It is important that you set the levels for each tape at the offline stage if you are planning to recapture on the same system. You can save these Level Settings attached to the tape name, and they will be recalled automatically at the batch recapturing stage of the online. These settings will be applicable only to the system you are working on now. If you save tape settings and use them on another system, that video board may be calibrated slightly differently. If you are moving to another system for high-resolution capturing or to a Symphony system, you should delete all the Tape Settings from the Project window. If you can get

245

Video Input	001	Project
Video Input	002	Project
Video Input	003	Project
✓ Video Input	Default	Project

Figure 10.1   Tape Settings Can Be Found in the Project Window

the levels correct at the offline stage, you could save huge amounts of time later when the deadline is tight and time is potentially more expensive.

If there are multiple, radically different levels on the same tape, you may end up creating several settings for that tape. Give that tape multiple tape names—a new name for each setting. For instance, you may have tape 005, which contains primarily exterior shots. At the end of the tape there is a change to an interior that requires a different set of video levels to compensate. While logging or capturing for the first time, you might pop the tape out of the deck, reinsert it, and rename the tape 005IN or something similar. Change the Level Setting to compensate for the interior and label the outside of the tape to reflect that the project now thinks there are two tapes and will ask for the second tape during the online session.

After the levels are set correctly for the video, make sure that the audio levels are set correctly and watch them closely. Some facilities mount extra large audio meters above their waveform and vectorscopes so you can see any distortion from across the room. Always keep in mind that you can use this audio as the finished mix. If you are going to end up on digital videotape, you can capture at 48 kHz and go digitally to tape (AES/EBU or optical connections) for the final output. This is generally the best sample rate to use although you will have to perform sample rate conversion on material from audio CDs (44.1 kHz) or DV25 footage that was accidentally recorded at 32 kHz. 48 kHz audio is so common now that it is usually the best choice.

If the audio goes into the system digitally with carefully monitored levels, and you can mix the sound as well as needed in the Avid suite, then you will save a significant amount of money. A MIDI controller to ride audio levels in realtime and the AudioSuite sound plug-ins are excellent tools for the finishing touches or special effects. On the other hand, professional audio designers can be worth their weight in gold on the right job. You may be able to hand those audio media files directly to the audio designers as AIFF, WAV, or SDII files with OMFI or AAF sequence information for ProTools, AudioVision, or other digital audio workstations (DAW). This will save you time and money at the beginning of the mix session.

When graphics and animations are created, you import them and cut them into the show as needed. Import graphics and animations at the highest finishing quality your system supports. Unless your entire show is at 1:1, use 2:1 JFIF on

Meridien so you can mix it with video at other compressed resolutions. High quality graphics don't take up that much more storage space, but they do require faster drives, so make sure your offline system can play back the finished quality graphics. If you don't do this, you will need to reimport them later when you get to the online stage. Reimporting graphics in order to uprez them (replacing the lower quality graphic with a higher quality or final version) is easy with Batch Import, even if the graphics have alpha channels, but this requires a rather tedious workaround on ABVB systems. If you use many graphics when you finish (especially with alpha channels), I would highly recommend upgrading to a system with Batch Import since the time savings on just the first few jobs will make a big difference. If this is not an option, I will outline a method later in this chapter that I devised to batch import graphics and animations without alpha channels.

Edit the sequence together until you get the final authority to sign off. Many people call this *picture lock* and assume that nothing is going to change after everyone sees this final version. (Don't ever believe this, but it is a nice idea.) In the process of getting final sign-off, you need to get this low-resolution version as close to the final high-resolution version as possible. That entails, in many cases, creating all the effects and titles at this offline stage. You may have a specific look that has been mandated by the designers or art directors, and it must be accurate down to the kerning and leading of every letter of every title. This, too, must be created in the offline so that there are no chances for them to say, "I didn't know it was going to look like that!" and start redesigning while the clock ticks. Everyone's approval in the offline stage is important. There may inevitably be temporary effects while something difficult is completed on another system using advanced compositing software, but try to keep these to a minimum if schedule allows. Again, the Avid Conform from offline to uncompressed standard definition or high definition online is one of the main values of using an Avid system like Symphony or DS Nitris. This is a critical competitive advantage for you at this stage.

SIMPLE RECAPTURING

When it is finally time to take the final cut and recapture at the finishing resolution, you have two choices of action. You can choose either to recapture the entire length of the original master clips used in the sequence or to capture just what is necessary to play the sequence. The first choice, capture all of the original master clips, is simple but takes up much more disk space than being a bit more selective. To find the master clips used in the sequence:

- Put the sequence in its own bin.
- Using the fast menu in this new bin or the main bin pulldown menu, choose Set Bin Display.
- Choose Show Reference Clips.

The master clips used in the sequence will appear in the bin. These clips are always connected to the sequence, but they are usually hidden from the user unless you ask to see them.

- Select all of the master clips and choose Batch Capture under the Clip or Bin menu.

Perhaps you need to be more selective about what you actually capture because you need to save time or disk space. Consider one of the many ways Avid allows you to be efficient during the batch capture.

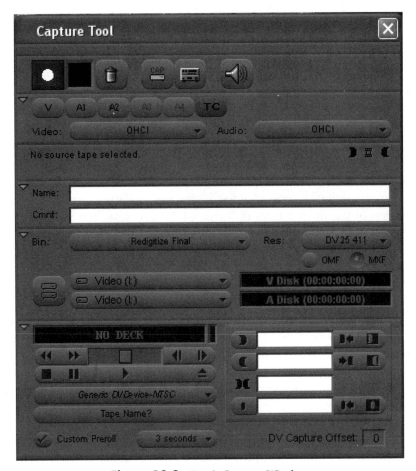

Figure 10.2 Batch Capture Window

Batch Capturing a Sequence

Here is the simplest way to batch capture a sequence:

- Take all of the low-resolution media offline. Do this by either deleting or moving media out of the MediaFiles folder.
- Change the resolution(s) in the Media Creation Tool.
- Highlight the sequence.
- Choose Batch Capture from the Bin menu.

First, the system asks if you want to batch recapture everything or only the "offline media." Because you have taken all the media offline, it is all offline or unavailable. The system can keep track of what has been recaptured, so when you stop and start, it knows what has already been captured (that media will be online). The system skips the available clips, but it ignores video resolution when checking, which is why the lo-res clips must be taken offline.

The system then asks for the required tapes and takes every shot it needs from each tape. The beauty is that the system captures only as much of the shot as it needs to play the sequence with a little extra thrown in as *handles*. The handles give you enough media to make minor changes, like the length of a dissolve, or to allow you to slip the shot a few frames to avoid a dropout. Do not underestimate the importance of handles, but do not forget that you are using up disk space very quickly at the high resolution. At the end of the batch capturing process, you will have a bin full of new master clips that have the original clip name plus the extension .new and a number at the end. This is your new high-resolution media.

Batch capturing the sequence this way is simple, but it doesn't give you very much control. If you have a very complex or long sequence, you are at the mercy of the system in terms of stopping or pausing or even trying to anticipate what the next needed tape will be. Under some circumstances you might want to leave all the low-resolution media online when you start recapturing because you will continue to edit low-resolution versions after making the high-resolution version. You might be finishing a project on one editing station while continuing to edit on another with shared media on Unity MediaNetwork. If you do this, you cannot take advantage of the "capture only media which is unavailable" feature. All shots will be available, and the system does not distinguish between resolutions. If you stop capturing and want to pick it up later, it is not so easy. You have to skip over every shot and every tape already captured, or every tape not present, one shot or one tape at a time. If you have skipped tapes on purpose, maybe because they were not ready, you have to skip over them manually with "skip this tape." You will have to skip shots and tapes each time you stop and start again. With several hundred tapes, this can get old fast.

	Name	Tape	Offline	Duration	Start	End	Audio
	Upper Pool.new.01	001	V1	10;13	01;00;15;20	01;00;26;03	
	Upper Pool.new.02	001	V1	8;19	01;00;33;15	01;00;42;04	
	Upper Pool.new.03	001	V1	8;13	01;00;44;17	01;00;53;00	
	Angel's Landing.new.01	002	V1	13;12	02;15;08;20	02;15;22;02	
	Angel's Landing.new.02	002	V1	12;26	02;15;32;01	02;15;44;27	
	Angel's Landing.new.03	002	A1-2	28;28	02;15;35;17	02;16;04;17	480(
	Angel's Landing.new.04	002	V1	12;11	02;15;50;01	02;16;02;14	
	Angel's Landing.new.05	002	V1	13;14	02;16;05;26	02;16;19;10	
	Angel's Landing.new.06	002	V1	15;04	02;16;25;00	02;16;40;04	
	Interview with Park Ranger.new.01	003	A1-2	57;09	03;14;19;08	03;15;16;19	480(
	Interview with the Park Ranger.new.01	All Interviews Compiled	V1	10;23	00;01;59;28	00;02;10;23	
	Handbook v4 Bin1.02.new.01	Bryce Canyon	A2	15;19	14:43:28:27	14:43:44:16	480(
	Paiute Chant.new.01	Bryce Canyon	A2	11;19	14:43:30:27	14:43:42:16	480(
	Zion Discovered.Copy.01			1;23;18	00;00;30;00	00;01;53;18	

Figure 10.3   New Clips in a Recapture Bin

DECOMPOSE

Decompose is the feature you can use with Media Composer and Symphony to specifically address the need for recapturing sequences. It is designed to give control back to you by breaking the sequence into new master clips before capturing. The basic logic of the system tries to protect you from capturing over material that is already there. This is why, in the first scenario, you should take media offline before you batch capture. When you decompose, the low-resolution media is taken offline automatically as the first step before recapturing.

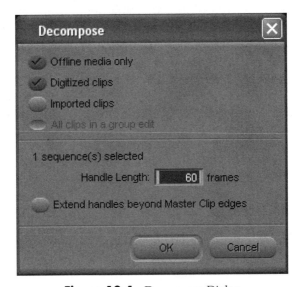

Figure 10.4   Decompose Dialog

Figure 10.5 Decompose Unlinking Warning

Duplicate the sequence and put it into its own bin. If you want to use the offline audio for online, see the later section, "Splitting the Sequence," before continuing. Highlight the sequence and choose Decompose under the Clip menu. When you decompose, you get a warning that your sequence will be relinked to new offline master clips. This message means that, for the selected sequences only, the system will break the links to the media already on the drives. In one step, it creates the new master clip with the extension .new. Since these new master clips have never been captured before, they link to nothing on the media drive, so all media is considered offline. There is no need to take any media offline before batch capturing. Figure out how much space you need on the drives by selecting all the decomposed master clips in the Decompose bin and pressing Ctl/Cmd-I (Get Info). This calls up the Console and gives you the total length of all the clips in the bin.

Now you can arrange the clips in any order you like. If you sift the clips based on tape name you can select all and batch capture one tape at a time with tape order completely under your control. Or you can sort the clips based on tape name and start timecode (Shift-click the multiple headings before sorting), then print out the bin for a guide to see what tape will be needed next. And you can easily skip over tapes that are not ready yet or clips that will be replaced later. When you finish batch capturing, you have all the high-resolution media you need to play the sequence plus handles for minor changes. We will come back to the Decompose option later in this chapter because it is so important as the first stage of some sophisticated variations on this simple recapturing process.

All titles must be recreated, an automatic process that leaves them intact, but changes the resolution. Re-import the graphics at the highest resolution if you need to. All imported graphics with or without alpha channels can be batch imported. Some effects must be rerendered and then everything is digital cut to videotape.

SETTINGS TO INCREASE SPEED

There are several settings that can add up to a significant increase in the speed of batch capturing. They are spread out across the interface, so let's deal with each one separately.

Preroll Method: Best Available

This is the default for this setting under Capture and it should stay that way. This means that the system will always make several attempts to capture over a break in the timecode. This can happen if the original shot was captured on the fly and there is not enough preroll for the recapturing. The system will attempt to shorten the preroll to capture even with the timecode break. If it is unsuccessful then it will use control track instead. It is only when there is also a break in the control track that the system will abort the batch capture. This setting reduces the amount of times a batch capture will abort due to timecode breaks.

Optimize for Disk Space/Speed

This setting is under the Batch tab in the Capture setting. If you optimize for speed the system will look ahead to see if there are two or more clips that are within five seconds of each other on the tape including handles. In that case, the system will continue to roll forward and grab all the shots without stopping, backing up, and prerolling. This saves time, but unfortunately the extra footage in between the shots is captured too and cannot be accessed. If you are really tight for disk space, choose the Optimize for Disk Space. But consider using the faster option since most high resolution footage is deleted or archived shortly after the project is finished. The implications of wasted disk space are not as severe as they would be for a longer term project.

Log Errors to Console and Continue Capturing

This setting, just below the Optimize setting, allows you to walk out of the room with the confidence that the system will not stop capturing until it reaches the end of the tape. You can then call up the console (Ctl/Cmd-6) and see if there were any problems that caused the system to skip a clip. If you have the bin sifted per tape, and you have the Offline heading displayed in the bin, you will be able to see which clips were skipped. You can sort or sift based on offline status, then select the problem clips and try again.

Extend Handles beyond Master Clip Edges

This choice is in the Decompose dialog and is off by default. A problem with creating a preset length handle for every shot in a sequence is that there is no way to know if that extra material actually exists! If you have created a handle that goes too far then there will be a timecode or control track break. Even with the preroll set to Best Available, you really don't want to create master clips that have timecode that doesn't exist. To avoid this possibility you can set the decompose function to look at the length of the original master clip that was created during the offline. If the handle length specified in the decompose step goes beyond the length of this original master clip, the decompose function will automatically shorten the handle for you. In other words, if the offline edit used all of the original master clip then you would end up with 0 length handles. Or if the offline edit used a section of the master clip that started 30 frames after the beginning of the original master clip and the decompose handle was set to 60 frames, you would get only 30 frames—the length of the original master clip is the handle limit. This is really in your best interest in all situations except one: You have been working from a film transfer that has no timecode breaks on the reel. In this case, you can turn on this option and 60 frame handles in the decompose will always mean 60 frame handles on the new decomposed master clips.

MORE COMPLICATED CAPTURING SCENARIOS

Drive Space and Offline Media

Your strategy for recapturing is going to first require a decision about drive space. Do you have enough space on your media drives to keep all of your low-resolution media? This is a critical decision since you may need to recut sections and make multiple versions after the final sequence is finished. Maybe your producer will want to make changes after viewing the final high-resolution master and you will have to go back to the source material to find new shots. Deleting your low-resolution media is a big step toward the point of no (quick) return. This should be done only after serious consideration. It also significantly impacts the way that you manage your media during the recapturing. Let's look at several scenarios depending on whether you keep or discard media while streamlining and minimizing the recapturing process.

Splitting the Sequence

In the first scenario, you have enough space for all the media to be on the drives at the same time. The decision also is to keep all of the audio exactly as it is right now. There has been lots of sound mixing or the mix will be done by someone else. Either way, there is really no reason to recapture audio unless it is distorted or is being replaced with material that is much better quality, say the DAT original instead of a U-Matic dub. If the audio needs to be recaptured, then the mix has to be tweaked because the levels during the recapturing may be just slightly off what they were originally. You can save time and money if you can do anything to keep from recapturing and remixing your sound. Think seriously about changing your offline methods so that the audio does not need to be recaptured during the finishing stage.

This strategy entails stripping the video away from the audio when starting the recapturing process so that you are working only with video clips. There are several easy ways to do this; let's use the easiest first:

- Duplicate the finished sequence and put it into its own Recapture bin.
- Load the sequence into the Record side of the Source/Record monitor.
- Highlight just the audio tracks, and press Delete and OK.

You are left with only the video tracks.

- Highlight the same sequence in the bin.
- Choose Decompose from the Clip menu.
- Uncheck Decompose only those items for which media is unavailable.

This function will create new master clips for everything in the sequence whether it is offline or not. A .new extension is added to the end to make sure there is no confusion with the original (longer) master clip. Batch capture the video only and cut the audio from the low-resolution sequence back into the finished high-resolution sequence.

USING MEDIAMOVER TO TAKE VIDEO OFFLINE

There is another way to take video offline that does not require splitting the original sequence into two separate sequences of audio and video. Chapter 4 discussed a third-party program called MediaMover (www.randomvideo.com). This program can search all your media disks and find all the media for all the projects used in this sequence. You will need the correct version of MediaMover for the type of media you have on disk (and there are several depending on your version of video board and software). MediaMover versions 2.0 and later have the capability to find and then move just the audio or the video out of the MediaFiles folder and into a separate folder with the project name. This effectively makes the media offline or unavailable. If you do this to the video-only clips, when you use the Decompose function, leave the "decompose only clips for which the media is unavailable" checkbox checked.

Decompose will then make .new master clips for the video only. If you have used material from many different projects, then this method is not very effective since you will have to move the media for all the projects involved. Whether you split the video and audio tracks or move the video out of the MediaFiles folder, you end up creating new, offline, video-only master clips.

Handles and New Master Clips

A checkbox in the Decompose dialog allows you to not decompose graphics and animations (imported clips). If this function is off, you will not get a new master clip for your graphics. By default it is off since Avid assumes you are managing your graphics through batch import. If the function is on, you will get new master clips for all your imported files that do not have alpha channels. Use this as a list to gather what you need and then batch import. The choice is up to you: Batch import directly to the sequence or batch import to the bin based on decomposed master clips.

Another reason you may not get a new master clip for every shot is the second part of the Decompose dialog: the handles. The handle length defaults to 60 frames (NTSC) or 50 frames (PAL) and is easy to change. If you shorten it and you have hundreds of cuts in your sequence, you save yourself considerable disk space. One hundred clips multiplied by 120 frames per clip for handles adds up to approximately an extra 6.5 minutes of footage in NTSC. If your one-hour show has 600 to 800 edits, this is an amount of extra storage space at high resolution that must be accounted for. However, occasionally two shots in the sequence that were right next to each other on the source tape may actually overlap handles. The decompose function determines if there is an overlap and combines the two (or more) shots into one master clip and so reduces duplicated frames.

Combining multiple shots intelligently into a single master clip reduces the number of master clips. It speeds up the recapturing process by eliminating the amount of time needed to stop, search, and preroll for each shot individually. Extrapolated over hundreds of shots, this can really add up to a time savings. Depending on your project, occasionally lengthening the handles can add up to faster capturing and less wasted disk space.

Modify Video-Only Masterclips to Add Audio Back

To summarize this video-only technique: When you decompose, the sequence breaks its links with the low-resolution material and links to the new master clips that have not been captured yet. The .new master clips are offline and they are video-only. You could sort and sift, stop and restart the batch capture to a higher resolution and have complete control over the process. However, there is one drawback: What if you are asked to continue to edit another sequence with these new uncompressed video-only clips? Suppose you are asked to make multiple new versions of the same sequence? Where are you going to get the sync sound? From the original audio? And then sync every shot by matching timecode? This seems like a huge amount of work. Working with video-only can backfire and force you to lose the savings created by leaving the audio alone unless this is really the last step in the project. If you are totally sure that you will not be asked to continue editing at all with this material, go right ahead and start batch capturing. But if you know your client too well, there may be an extra step.

You can modify the video-only master clips, add audio tracks to them, and continue to edit using sync sound. This seems totally redundant after you just stripped the audio out, but this new audio will be used for a different step in the process. You will not use this audio for the final mix! The levels do not have to be recreated perfectly; the new audio is for reference only and you will not accidentally cause the sequence to link to this new audio. It is used only in case you are asked to go back to the high-resolution master clip and play the whole thing through. If you match frame on the video track in the final sequence, the system calls up a shot with sync audio. If you match frame on an audio track in the sequence, you call up the original, low resolution master clip with the original audio. You really have the best of both worlds except that recapturing the audio takes up more drive space. But audio really takes up very little space compared to video, and if your sequence is really not that long, the difference in storage is minimal.

Controlling Levels during Batch Capture

The Tape Settings created during the offline are recalled automatically during this batch recapturing process. But you may not want the Tape Levels Setting to be recalled! Perhaps someone set up the tape incorrectly or you are using the original digital betacam reel instead of the S-VHS dub and the two tapes are quite different looking because of a bad dub. You may be batch capturing on a Meridien system, although the offline was done on an ABVB system. You have to create a new setting with the same tape name to replace the existing one or go to the Project window and delete the Tape Setting completely.

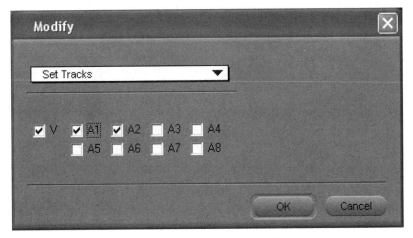

Figure 10.6 Modifying a Video-Only Master Clip to Add Audio Tracks

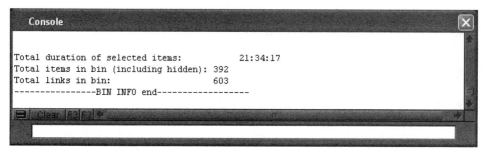

Figure 10.7   Get Bin Info in the Console before Batch Capturing

Some people make the deck preroll time a little longer during this stage so they can quickly correct any little level problem they see before the shot starts to capture. This is why you take the deck adjustment out of preset. Many times the shot requires a tiny raising or lowering of the video gain to even things out.

The tapes are requested by the system one by one and each tape is used only once. The system starts by capturing the earliest shot on the tape. Then the tape moves forward to the end of the reel, capturing all the shots that are needed. You can speed up this process by either rewinding all the tapes beforehand or actually cueing up to the first shot that is needed on that tape. This is easy to do if you have an extra deck or two and a printout of the decomposed bin.

Putting the Sequence Back Together

Once the high-resolution video and reference audio are captured, use the sequence as if it were a source clip. Load the new sequence into the source monitor and overwrite the high-resolution, video-only sequence as the new video track. The new video should match perfectly and have the same starting timecodes. If you press the Home button and mark in on both sequences, the new high-resolution sequence will be in perfect sync with the original sequence. Even the sync breaks work because they are looking for tape name and timecode. So if you continue to tweak the sequence and you knock yourself out of sync, you are still given the white numbers in the timeline. This may be the last step for a simple sequence.

The Low Disk Space Variation

When you have very limited disk space, then all low-resolution media must be deleted and the audio must be reduced to the bare minimum to free up space for the high-resolution material.

Here, you can use the Media Tool to delete large amounts of low-resolution video all at once. Choose to show all the media files and precomputes for this project. Select all the media in the Media Tool and press Delete. A dialog box appears and asks you if you want to delete audio or video. Uncheck the audio choices and delete all the video-only. Scan through the sequence to make sure all the video is gone. If there are any shots left, then they were part of another project. You must decide whether they can be deleted or not. To be completely sure this method will work, you should either delete these leftover shots or figure out how to take them offline. You can do this by:

- Dismounting the media drive that has the other project's media files.
- Unlinking the clips (Shift-Ctl-Relink).
- Moving the media from those projects out of the MediaFiles folder using MediaMover.
- Locating the media file using Reveal File and moving it out of the OMFI MediaFiles folder.

Click on each file in the Media Tool and choose Reveal File under the File menu. The Media File folder will open and show you that specific media file.

This is a variation of the first method and is similar except that you have deleted the project's video media instead of just moving it out of the way. If you are still tight for space, you can consolidate audio used in the sequence. This takes the audio media and copies only the parts you need. To do this:

- Copy the sequence into a new bin.
- Select the sequence and choose Consolidate from the Clip menu.
- Uncheck Skip if media is already on the drive because you are not just moving the media, you are shortening the original master clips to only the parts you need.
- Check Delete after consolidating if you have set the Consolidate drive list to include enough drive space to finish the entire sequence. Be extremely careful with this choice since you may be sharing media across multiple sequences. It is best not to delete media until all sequences are safely consolidated.

Now you have a sequence that is linked to the consolidated audio and you can decompose the video-only. Because you deleted the video as the first step, it is already offline. Decompose Only those items for which media is unavailable and get video-only master clips to recapture.

Mixing Low Resolution and High Resolution

This scenario involves mixing any low resolution offline media with high resolution online media. As the project comes to a close, delete the offline quality media and batch capture only that. Basil Pappas solved the problem with the best method I have seen. I keep this method for ABVB only, although it will work just as well with later releases. The Batch Import feature has made this method less valuable, but still handy as an advanced technique.

If you don't need the low resolution material, you can sift for it using the Video heading in the Media Tool and delete. But if you do need to keep the low resolution stuff around then you need another way to separate the low resolution from the high resolution.

- Copy the sequence and put it into a new sequence bin. Then load the sequence into the timeline.
- Strip the audio tracks from the duplicate sequence and make it a video-only sequence as in the previous technique.
- Open the Media Tool and show master clips from all projects.
- Make sure nothing is highlighted or selected in the Media Tool.
- Highlight the new video-only sequence in the new sequence bin and use the Bin menu choice Select Media Relatives. This highlights all the clips in the Media Tool that are used in the sequence.
- Drag all the highlighted clips from the Media Tool into the new bin.
- Sift the clips in the new bin using the Video heading so that only the low resolution clips are showing (there may be several different low resolutions so sift on all of them).
- Select them all (Ctl/C-A).

Now unlink these clips from their low resolution media and purposely make them go offline. Don't worry, we can relink them later.

- Hold down the Control and Shift keys at the same time, and under the Clip menu (Media Composer and Symphony) or Bin menu (Xpress) choose Unlink.
- All the low resolution material unlinks and becomes unavailable.
- Move the video-only sequence into yet another new bin.
- Select the sequence and decompose. Check Decompose only the clips for which media is unavailable.

The sequence will decompose only the video clips that were linked to the low resolution media. Because these clips have been unlinked, they are offline. Decompose creates new master clips for all the original unlinked clips and leaves everything else in the sequence alone.

- Go back to the bin with all the unlinked low resolution clips.
- Select all, and relink them.
- Check Relink master clips in the Relink dialog. All the low resolution clips should go back to normal.
- Capture the decomposed, video-only master clips that you just created.

If you like, you can also take the extra step and modify the video-only master clips and add audio tracks.

Decompose without Timecode

Many people capture nontimecoded material at the finishing resolution during the offline stage. They know they can't recapture it later in the traditional offline to online process, so they want to capture it only once. This method works well if you have only a few shots without timecode and don't want to use the previous method to isolate them. What happens to the shots with no timecode when you decompose? If they were decomposed along with the rest of the sequence, now they are offline in the sequence. How do you get them back?

When the decompose and batch capturing process is finished, there will be gaps in the sequence. This is where the nontimecoded material should be. You can check to see if you have any gaps in your finished sequence by putting a copy of the recaptured sequence into another bin and decomposing the sequence again. This time leave the box checked for Offline media only. All the media should be online now except those shots not recaptured. If you missed a shot for some reason, it becomes a new decomposed master clip. This is used purely as a simple way to pinpoint what is missing; you don't really want to recapture these clips. You can also use Show Offline Media in the timeline. This will give you a fast visual reference since all the offline shots in the sequence will be highlighted in bright red, even if they are nested several layers deep. When nontimecoded material originally was captured, it was given a time-of-day timecode that cannot be recreated. This timecode is made up by the system and assigned during the capturing and is not really on the tapes. This fake timecode is enough so that the system thinks this material is different from a graphic, video mixdown, or another kind of imported media that has no timecode at all. You have a choice of whether to decompose imported media, depending on the options checked in the Decompose dialog, but nontimecoded material is decomposed just like all other kinds of video from tape. Consider this: The sequence can always be relinked to the original media still on the drives, even after the decomposing. If the recapturing of timecoded sources is complete and you relink the sequence, the sequence stays linked to the new, high-resolution, just recaptured media. The sequence looks to the media drives to try to fill the empty gaps left over from the missing nontimecoded shots. When the sequence tries to link, it is looking for timecode and tape name. It finds the original time-of-day clips and relinks them with the finished high-resolution sequence.

The goal of all these methods is to reduce the amount of recapturing to a minimum. These strategies require selectively removing only what needs to be replaced. Many people are content to spend the extra money on drives and always work at high resolution. If you can afford it, you can skip these techniques, but there will still come a time when you have too much media and not enough storage.

Recreating Titles

If the titles created from the Title Tool or Marquee used in the final sequence were first created at the low resolution then they must be recreated at the finishing stage. This is an automatic process that takes little time and replaces the old titles with new, high-resolution versions. There are two parts to every title: the type information about drop shadows, borders, and alpha channels, and the title media created at a specific resolution when you leave the Title Tool. The type information must be used again when you re-create the title media, and new media is created based on what is set in the Media Creation Tool.

- Set the Media Creation Tool to the highest resolution compatible with the rest of the sequence.
- Mark inpoints and outpoints at the beginning and end of the sequence.
- Highlight all the video tracks.
- Choose Recreate Title Media from the Clip menu.

The Avid looks through the sequences, finds all of the titles, generates new media, and links it to the sequence.

In the process of recreating titles you will remove any nested effects. This refers only to nested effects within the titles. Nesting video in a title is a very cool and easy-to-use effect. You fill a title with moving video by stepping into the title and editing new video over the unlocked video track (V2), and maybe you add some other effects inside. The problem when you are recreating the titles is that the Avid is going back to the original title information. The original type information knows nothing about what you did to the title after it was created. In Meridien versions there is dedicated hardware called a downstream keyer (DSK). Media in the DSK is by its nature uncompressed. Editing video into a title will make the title lose its uncompressed status in the DSK. If your title was at low resolution when created, you will see a big difference in quality of the title's edges. You can preserve the work you did inside the title by copying the nested effects, dragging them to a bin, and then placing them back in the title after it has been recreated. The only way to do this, in later versions, is to use the Clipboard button:

- Mark-clip the entire material inside the nest.
- Hold down the Option or Alt key while clicking on the clipboard icon. This puts the material automatically into the Source monitor. Since the only time I use Clipboard is as Option/Alt-Clipboard,

> I have mapped the Option/Alt button—a little dot in the Command palette with the letters OP or AL—onto my Clipboard button.
> • Use standard subclip techniques to drag the nested material to a bin.

On Meridien systems you have the option to use uncompressed titles and title rolls. You can't promote these uncompressed titles to 3D, but you can play them in realtime while there is another video track below with unrendered real-time video effects. Rolls are as easy as pressing the R button in the Title Tool and typing away. Crawls are one of the few things the ABVB does in realtime that the Meridien does not. In the ABVB system, you will have a C button in the Title Tool for creating crawls. Using the Marquee Title Tool is almost as easy since rolls don't need to be rendered and crawls are a fast render (not an animation).

REPLACING GRAPHICS

The final step for "rezzing up" is to bring all the graphics up to the finished quality resolution.

BATCH IMPORT

After you finish capturing the timecoded material, you must reimport all the nontimecoded graphics, animation, and music (especially if it is from a CD). Of course, the music does not have to be recaptured or re-imported unless it has been deleted, since AIFF files will travel easily between Mac and Windows at full quality. You can consolidate the music to a drive for the online or bring the CDs with you to the online session.

You can bring all of your original material over to the finishing station through any of the methods we will discuss in the next chapter and then import them again using the Batch Import function. This reimport is frame accurate even without timecode. To use Batch Import:

- Highlight the sequence in the bin.
- Choose Batch Import from the Clip pulldown menu.
- If all your graphics are offline, choose to replace all of them.
- Navigate the menus to the folder with all the graphics. They are imported and connected to the sequence frame accurately.
- If your graphics are scattered across different folders, repeat the previous two steps.

There are actually two methods to Batch Import, depending on your desired result. The first method is to highlight a sequence and batch import all the files that are connected to that sequence. The system looks for the names of all the offline imported media and grabs those files from the folder you have pointed it to. Those files are linked only to that sequence, because you may have multiple sequences that are still at low resolution that you want left alone.

The second method also keeps in mind that you might not want to affect all the sequences on your system with the new high-resolution graphics. This method is best if you want to relink multiple, slightly different sequences all at

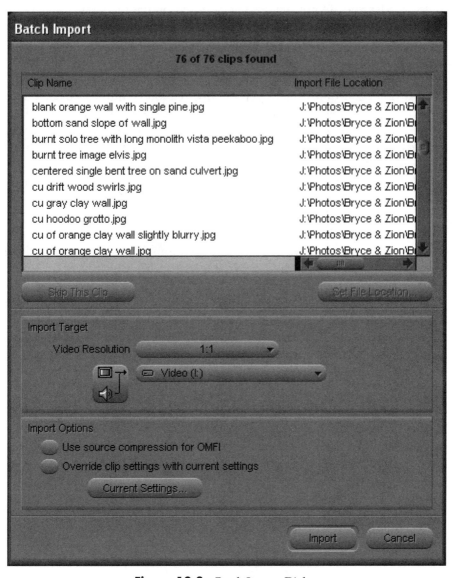

Figure 10.8 Batch Import Dialog

once and they share many of the same graphics. Go to the original bin that contained all the clips representing the offline imported media, select them all, and batch import. You will now have a bin full of online high-resolution media, but this media is not really linked to any sequences yet. Go to the sequence and drag it into the same bin as the new batch imported clips. Select all and choose Relink all non-master clips to selected online items. The specific sequence is linked to the high resolution graphics. Any sequence not relinked in this manner will still retain all the links to the old graphics.

The hidden benefit to the Batch Import function is that you can use it for versioning. Identification of the graphic is based on the filename. You can point the Batch Import file hierarchy to any file and it will import and link the new file to the sequence. If you have an updated graphic, just delete the media from the old one, navigate the Batch Import dialog to the new graphic, and batch import it. The system will link the new graphic frame accurately to the sequence in place of the old graphic. Perhaps you have a new version of the music used and all you need to do is import it and link it to the sequence where the old music was. The user can then change the name of the clip to reflect the version number. If you have been doing many nested layers or editing the music and cutting to the very specific frames on the beat, then this will save you hours.

The following method for fast replacement of material is useful on older systems, and edits directly from the bin to the timeline.

- Use the Mark Clip button to select the low-resolution graphic in the timeline. A quick way to do this is to highlight the graphic with the red segment arrow and then press the Mark Clip button. You bypass having to turn tracks on and off.
- Choose the red segment mode arrow and drag the high resolution graphic or the updated graphic from the bin to the timeline.
- Media Composer and Symphony users will need to hold down the Ctl/Cmd key for the graphic to snap to the inpoints and outpoints in the timeline. Xpress users don't need the modifier key.

If you created keyframed moves for these graphics (including fading up and down), you want to preserve that work for the new high-resolution replacements. To save the keyframes for a regular graphic:

- Go to Effect mode and highlight the graphic with the keyframes in the timeline.
- Drag the effect icon in the upper left-hand corner of the Effect mode palette to the bin. This saves the keyframes without the source material when there is no alpha channel.
- To save the keyframes from an imported graphic with an alpha channel,
- Option-drag the effect icon from the Effect mode palette to a bin.

Batch Importing Animations

In the past, the main strategy for rendering animations in third-party programs for use on both the offline and finishing systems would have required rendering the animation twice using the Avid Codec. You would render it once as the offline resolution and once as uncompressed. Since you can now mix resolutions in your timeline there is no reason not to have the highest quality graphics at all times. Do some experiments to make sure your drive configuration can support uncompressed graphics. If not, consider striping your drives or moving up to SCSI or fiber channel. Even if you want to use a lower quality version of your graphic at the offline stage you still need to render only one high quality version. This is because you can change the resolution of the animation during the import even if it has used the Avid Codec. This method will take longer to import because when you change the resolution during the import, the system must open the file and recompress every frame. The Avid Codec was designed to make this unnecessary, but you may find that the extra time to import a high resolution animation at a compressed resolution is trivial. Creating only one current version of an animation simplifies your media management.

It is not that much extra time to make a high resolution and a low resolution version of an animation if you are rendering overnight or you are continuing to render multiple draft versions at lower resolution until you get it just right. After the animation has been approved on the offline system, you can render the second version for finishing at the higher resolution. This will give you the faster import on the offline system, and when it comes time to import the animation at the finishing stage, you can use the Batch Import feature. Just point the Batch Import dialog to the new file and it will replace it as if it were the original file.

RENDER SETTINGS FOR MOTION EFFECTS

When you go from an offline version of a sequence to the finished online version you will have to take into account the changing quality of motion effects. There are up to seven kinds of motion effects available, depending on your system: duplicated, both, interpolated, VTR-Style, blended interpolated, blended VTR, and FluidMotion™ (see Chapter 7 for more detail). The duplicated fields option is used primarily with single-field resolutions and is faster than the other two, so it is preferred by most offline editors. It is also excellent for previewing motion effects on two-field material because it renders so fast. The problem for the online editor is how to change all of the offline or scratch quality motion effects from duplicated, which look just fine at 20:1, to something that looks good at 2:1 or 1:1. This means changing all the motion effects in a sequence to the highest quality that looks good for that particular type of motion.

To render an entire sequence as a new kind of motion effect, go to the Render Settings User setting. There are two sets of choices: Motion Effects Render and Timewarp Effects Render. Since there may be old-style motion effects coming from an older system, you may choose to use the Motion Effects Render setting. If you have created the motion effects on a newer system using the record side style timewarps, you can choose the Timewarp Render setting and have a few of the newer choices. The one thing you can't do at this time is take a sequence full of old-style motion effects and convert them all to timewarps at once. You can step through the motion effects and promote them one at a time. This could be tedious, so you want to recommend that the offline editor use the Timewarp effects if possible or convert only the old style motion effects to timewarps that really need some help to look better. The render choice defaults to the setting that maintains any motion effect type that is already there: Original Preference. The offline editor may have been a master of the art of motion effects. They may have been working in a two-field resolution and picked many different kinds of motion effects depending on the type of motion in the image. If this is the case, you don't want to wipe out all their hard work by changing all the different motion effects in the sequence to one type. Step through a few of the motion effects in the sequence you have been given to finish and if there are many different types then use Original Preference before you render. Freeze frames currently do not have a render setting, so be careful to make them at the highest quality (Both or Interpolated

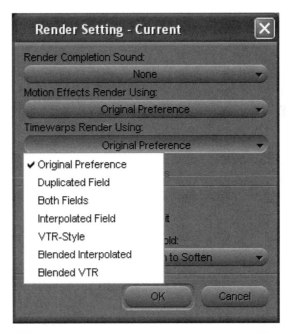

Figure10.9 Render Settings to Change the Motion Effects Type

under the Two Field choice) so that they will look best when recreated with two-field media.

The best way to make any motion effect look better is to pick a speed that divides more easily into the 100 percent frame rate. Although this is definitely a last resort, it should be considered if you are close to that speed anyway and the change would not be noticeable in terms of the content of the shot.

On a Meridien system, while you are in the Render Settings, turn on the Show Intermediate Results checkbox and watch the client monitor the next time you render. It doesn't slow the Meridien system down at all, and it is amusing to the clients. Show Intermediate Results is useful if you see a part of an effect that needs to be tweaked. You can stop and tweak before you waste time rendering the entire effect. In some cases (where you haven't used acceleration or spline on the effect), you may be able to add an Add Edit before the area to be tweaked and then just render from there. This would take advantage of the Partial Render feature that keeps everything rendered until the user makes a change that invalidates the effect parameters. To best use Partial Render, make sure the Render Range choice in the timeline is set to partial so you can see how much of each effect has already been rendered. The Show Intermediate Result choice is not yet available for Adrenaline- or Mojo-based systems.

After all the graphics and animation have been re-imported and linked to the sequence, it is finally time to render. With Adrenaline-, Mojo-, and Nitris-based systems you have a huge leap in realtime capabilities, but you will still need to have a rendering strategy for efficiency and guaranteed playback. There is an in-depth explanation of the uses and benefits of ExpertRender in Chapter 7.

SCRATCH REMOVAL

One of the most important stages to finishing any project that has originated on film is scratch removal or, as it is sometimes called, "schmutz busting." Perhaps this term comes from the ubiquitous Dust Buster and the fact that so much stuff that ends up on the final print is of an unidentifiable origin. Anyway, this process consists of many sharp-eyed people evaluating the final video image to catch any and every defect before it is signed off. This feature is on Media Composer and Symphony only.

As the dirt, or "schmutz," passes by, it is flagged and fixed. With the scratch removal tool this is a one-step process. By stopping on the problem frame, the user can click on the Scratch Removal button. This button must be mapped from the Command Palette and looks somewhat like an eraser. The Scratch Removal button will automatically put the user in Effect mode, put Add Edits one frame back and at the end of the problem frame, and put the Scratch Removal effect on this new segment. In one step, the problem area has been prepared for fixing.

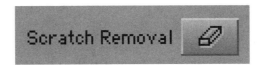

Figure 10.10 Scratch Removal Button

Where should you get the pixels to cover the scratch? You could grab them from somewhere else in the frame if there are many similar-looking surfaces. Or you could use the Paint Tools to clone a small section of the image and move it over the scratch. But most likely, you will want to take the good pixels from the last good frame. That is why the first Add Edit (those little white marks that indicate an artificial edit point) is placed one frame before the problem frame. Now you have the ability to steal the pixels from the previous frame or the previous field. The default for this temporal cloning is automatically set to two fields, or one frame.

Generally, you will get a better looking fix if you take the pixels from a field one and apply it to another field one. This becomes less of an issue in the 24 fps progressive projects because there are no longer any fields, but it is still important in all 2-field projects.

Sometimes the problem extends beyond a single video frame, as when the dirt is on a B or D frame in a film-based project that is being edited in 30 fps NTSC. Then you can trim the Add Edit to make the Scratch Removal effect as long as you like. The default setup is to take the good pixels From Start so the first good frame or field is used automatically. You can also set it to From End or Relative. Set to Relative, it will always change the field to maintain a relative offset. The temporal offset—the number of fields in either direction that can be used—is limited to 10 fields.

However, you cannot use fields that extend beyond the limits of the Add Edits. This limits most fixes to only a few fields. If you need more than that, bypass the automatic features of the Scratch Removal button and apply the Add Edits yourself. If you know that you will be adding a Scratch Removal effect to a long segment, put your mark in and mark out points around the bad area. With your blue position bar set in between the marks, click on the Scratch Removal button. The system will put Add Edits at the marked points, plus one extra frame at the head of the effect for the reference frame. This last feature would be good for fixing the kind of thin vertical scratch that goes from the top of the frame to the bottom and lasts over a long period.

AUDIOSUITE PLUG-INS

The audio special effects capabilities can be amusing for all the times you need robot voices, but the real power of these plug-ins is in fixing everyday problems. They can be used either to lengthen or shorten an audio sound bite

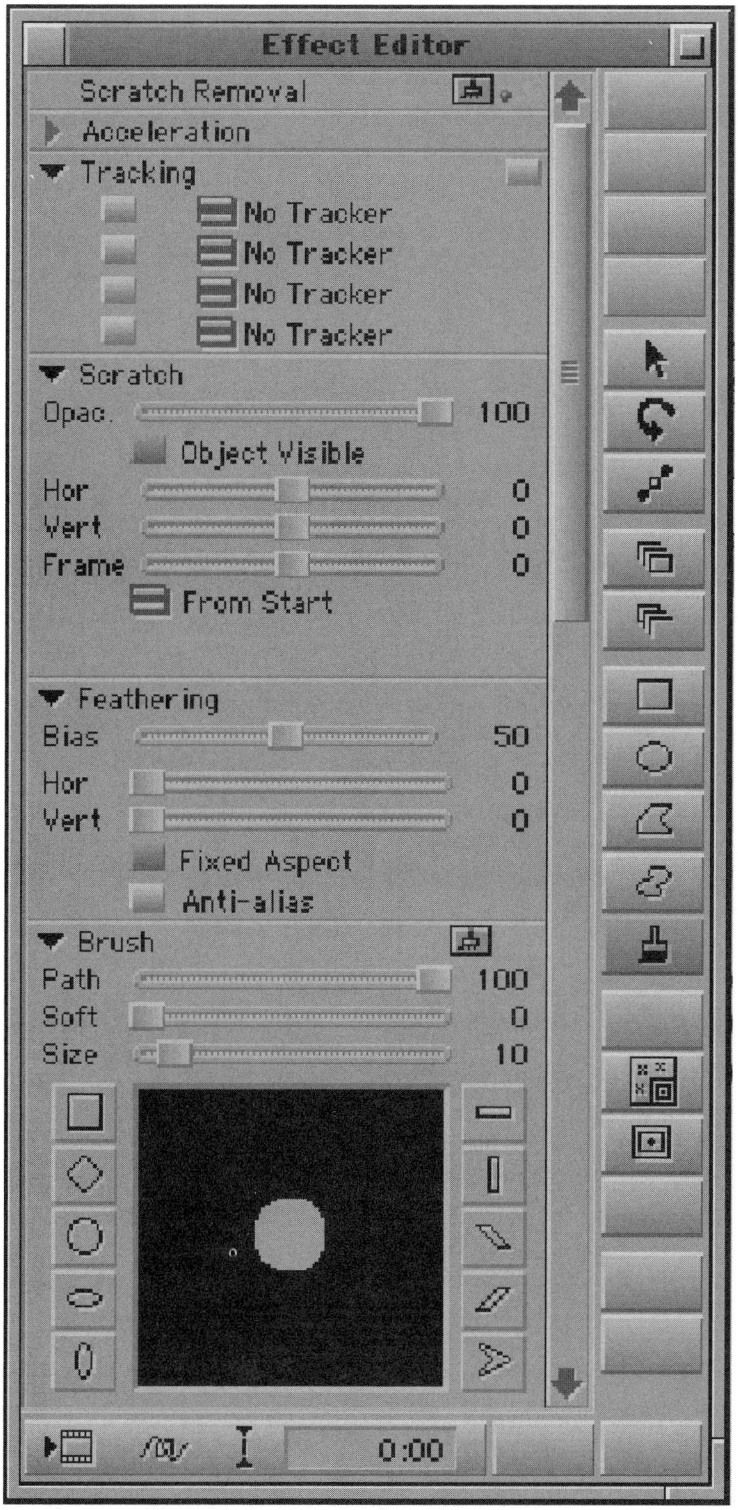

Figure 10.11 Scratch Removal Effect Interface

and keep the pitch the same, or to change the pitch and leave the length the same.

A hidden capability of the AudioSuite Plug-ins is the ability to apply the effect to a master clip and create a new master clip. Thus, if you have a voiceover that is 31 seconds long, you can make a new master clip to be used in all the sequences that follow without having to reapply the effect.

- Open the AudioSuite Tool.
- Take the audio master clip from the bin and drag and drop it onto the AudioSuite interface. A new section of the interface will open up and show you the controls that apply to the master clip.

After you make the basic effect decisions, you create a new master clip and can now use this as if it were a captured source. From now on, every time you use the AudioSuite master clip, you can apply it into the sequence with the correction already done.

There are multiple benefits to applying an AudioSuite to a master clip. The first is that it allows you to apply an effect to multiple audio channels at the same time, or to a stereo pair. It also allows you to create a new master clip that is longer or shorter than the original clip. Instead of applying the effect in the timeline and then trimming to accommodate the new length, you can cut this new clip in without any further adjusting. If you want to apply the effect to stereo pairs, be sure to check Parallel Processing mode. This will ensure that the effect is applied to both stereo channels simultaneously rather than one after the other (Serial Processing mode).

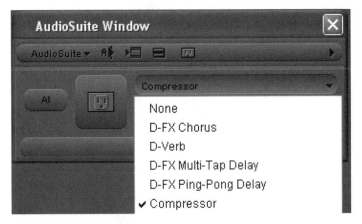

Figure 10.12 AudioSuite Effect Plug-ins

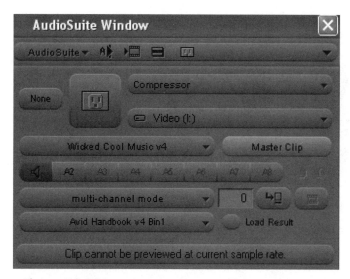

Figure 10.13   AudioSuite Interface in Master Clip Mode

CONNECTING A MIDI CONTROL SURFACE FOR MIXING

An excellent choice for mixing sound in the Avid suite is to connect an external Midi control surface to adjust the audio rubberbanding or Audio Gain Automation on the fly. This is a relatively easy connection to make, although it does take up one of the ports on the back of the computer. There are switches for multiple devices to be connected to these ports if you really need two decks connected and the Midi control. Purchase a serial port switch to minimize yanking cables with a flashlight in your mouth.

Follow the instructions in the manual for the connections for a JL Cooper Fadermaster Professional™. You are ready to start recording audio rubberbanding with all eight channels. When you are done recording, you can go back and filter out the unneeded audio keyframes to make it easier to adjust later with the mouse. To use the pulldown menu in the interface to filter the automation gain, you must first select the track(s) in the Audio Gain Automation window by clicking on the button that shows the track name. If you want to go back and adjust the levels after a first pass, you need to manually position all of the faders on the JL Cooper Fadermaster Pro. Both of the little lights next to the Track Name Selection button should be blue on the Audio Gain Automation window. Now you know that you are starting with the faders in the proper place to begin recording, and pick up smoothly where you left off.

Avid officially supports the JL Cooper MCS 3000X and the Yamaha 01V, which both have flying faders. The flying faders are important because they will automatically reflect the rubberbanding level of any sequence as you begin to

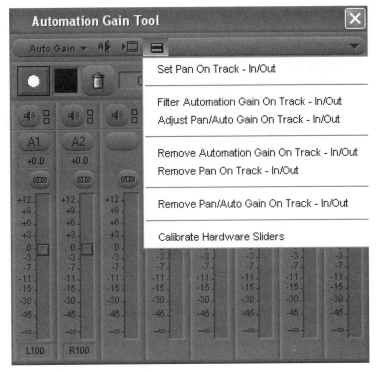

Figure 10.14   Audio Gain Automation Interface

make changes. Start the audio gain automation recording and grab the faders as they are moving. The JL Cooper is touch sensitive, but the Yamaha is not. The Yamaha, however, can double as an audio mixer. You can use it to control all the ins and outs for audio sources and then switch over to become a Midi control surface on demand.

IMPORTING AN EDL

After reading Chapter 8, you should be aware that an EDL is a terrible way to archive a sequence; however, there are times when you take a sequence to an online linear tape suite and assemble a project only to have it return to the Avid suite for more changes. Now you need to get the EDL into your Avid and capture it. This is taking a noncompatible format and translating it to something the Avid can understand.

Using EDL Manager you can open an EDL and either save it as an OMFI file or import it directly to the Avid editing application if they are both running at the same time. Use the arrow icon on EDL Manager that is going back toward the editor icon, and the list should import into an open bin of your choice. This brings in a sequence and a series of master clips that are almost 24 hours long. The master clips are there for reference and are not really to be used.

These imported master clips from the EDL do not have a project affiliation. This is a basic limitation of the import process since EDLs do not contain project information and it is not assigned during the import. Project name is critical to managing media on your drives so you should assign a project manually. Select all the new masterclips and select Modify from the Clip menu. Modify the project name by changing the tape name. Change the tape name from the current name to an identical name. You have just given the clip a tape name that belongs to the current project. If the tape name already exists in the list of tape names in this project then use the existing, identical tape name. In other words, you have changed the tape name for the masterclip from Tape 001 from no project to Tape 001 from the current project.

Do not use these 24 hour long master clips created in the EDL import process for anything else. You will be getting new clips from decomposing the sequence.

- Load the imported sequence into the Record monitor.
- Use Remove Match Frame Edits from the Clip pulldown menu.
- Decompose the sequence and recapture all the new clips.

If your sequence does not import correctly, you must strip some of the complex information out of it.

- Open the EDL in a simple word processor.
- Make sure that everything is in a monospaced font like Courier.
- Look for a setting in the word processor that will show all invisible characters, including spaces.
- Using cut and paste, make all unusual effects into dissolves, strip out comments and GPIs, and basically make the list as simple as possible while maintaining all the original character placement.

Although EDL Manager is forgiving about missing spaces in the EDL, it is still best to leave everything in the original format. There are too many variables to guarantee that importing your list always comes in perfectly to the Avid, but you should get the majority of it without a problem if you simplify as much as possible.

RELINKING MEDIA TO AN EDL

If the captured media from the original edit is still on the media drives, it is frustrating not to be able to link the imported EDL to the footage. The reason that this does not happen automatically is, as you will remember from the previous chapter, the importance of project name to relinking. When you import an EDL and turn it into an OMFI composition using EDL Manager, it then imports as a sequence into the editing program; however, there is no project associated with the source tapes in this new sequence. Without a project name connected to the tape names you will need to uncheck the relink option for Relink only to media from the current project.

The most important aspect of the new relink choices is the ability to link media already on your drives to an imported EDL. In the past, this took many steps and was a rather unpredictable process. It was only made possible at all by following a workaround I published many years ago. It has now made it into a full-fledged feature.

To relink existing media to an imported EDL:

- Open the EDL in EDL Manager.
- Save as an OMFI file or, if the editing software is running, click on the arrow icon in EDL Manager that points toward the Avid editing software icon.
- EDL Manager will ask what bin to put the sequence in if there are multiple bins open; choose one.
- Select the sequence in the bin and choose Relink.
- In Relink, choose Relink by Source Timecode and Tape Name and Offline non-master clips to any online items. We are relinking the sequence, not the master clips.
- Make sure Relink only to media from the current project and Match case when comparing tape names is unchecked.

If there are multiple projects with the same tape name or there are multiple resolutions of the media, you may have to take another step. If you want more control over exactly what media is connected to this sequence:

- Open the Media Tool and sift on the appropriate criteria for linking. This could be a combination of project name and resolution or multiple project names.
- Select all these sifted master clips and drag them into a bin with the imported sequence.
- Select everything in the bin and Relink all non-master clips to selected online items.

There are still some obvious issues with relinking based on the text of tape names in an EDL. What if the tape names in the EDL do not match the names of the tapes assigned to the captured media? There will be no relinking. This could happen because when the original EDL was made in EDL Manager, the type of EDL chosen could not support the original tape name. If you have a tape name that is longer than an EDL format can handle, the tape name will be truncated. In other words, "Exterior Tape from Yesterday" becomes EXTERI, which will never relink. Since we are working from a simple text match, if the text is different, there is no automatic way to tell what it used to be. If this is the case, then you need to go through another step.

To force the link when the tape names have been changed:

- Check to see if any media has not relinked to the sequence through the preceding process.
- Decompose the sequence with the checkbox set for Decompose only those clips for which media is unavailable. This will create new master clips for all the areas of the sequence where the media is offline.
- Select the master clips and choose Modify from the Clip menu.
- Modify the tape name to be the same as the original captured media that is already on the drives.
- If you have many clips from the same tape, you can modify them all at once. Otherwise, you need to do this to each master clip separately.
- Relink the sequence again.

This points out the importance of naming tapes so that they will fit into the restrictions of the most common EDL formats, even though you may have no intention of making EDLs at the beginning of the project.

Some people will use this feature to load sequences that have been created from a tape database. By using the database to create an EDL, they can quickly assemble sequences from stock footage, import the EDL to the Avid system, and relink.

FTFT

Another important method of relinking, FTFT (Film Tape Film Tape) relies on keycode. This method is extremely important when working with multiple film transfers of the same material. Assume that most film-originated productions transfer all of their dailies using a one-light or inexpensive film transfer designed to get as much footage transferred as quickly as possible. Then this one light is captured at an offline resolution to save disk space and to put as much footage on the system as possible at the creative stage of shot picking and offline editing.

At the end of this process, the quality of the image must somehow magically be improved and then output to the master tape. Just recapturing the one light transfer is usually not good enough. The magical bit happens by keeping track of the film's keycode, numbers that are printed onto all raw film stock. First, in the telecine, where the telecine operator transfers film to tape, they record the keycode information as part of the telecine session log file. This log file will eventually be transformed into an Avid bin using Avid Log Exchange (free Avid software) as an intermediate, translation step. This keycode is entered into a heading in the bin for each master clip and this follows the footage throughout the project.

The editor or assistant, after the offline has been completed, creates a *pull list* using FilmScribe, which is a film cutlist generating tool available under the Output menu on systems with this option. This creates a list of entire scenes to be retransferred from the original film with final color correction. The pull list is generated based on the keycode and the telecine operator/colorist does the final transfer on only the selected scenes. This is all basic stuff so far, but we are just about to hit a snag.

The original one-light transfer was captured into the offline system and used to edit based on the timecode of the original videotape. But the latest, greatest film transfer is on a new tape with different timecode. How do you reconcile the two different timecode numbers that apply to the same shots? Video decks don't search and cue up based on keycode! Relinking is a powerful stealth tool, so Avid has created a new capability: relinking by keycode. Go ahead and recapture the new material at 1:1 quality based on the new files from the second color correction. Pay no attention to the timecode on the new tape. Both color correction sessions have used the same keycode, and they will dutifully be listed in the new bin if you capture from the telecine session log converted to an Avid bin using Avid Log Exchange (ALE).

Take the original offline sequence (or a duplicate just to be safe) and put it in the same bin as the new, recaptured 1:1 material.

- Select all the items in the bin (sequence and master clips).
- Choose relink from the Clip menu.
- Choose Relink by Keycode from the top pulldown menu in the Relink dialog (the one that defaults to Relink by Tape name and Timecode).
- Choose the second radio button, Relink all non-master clips to selected online items.

For some reason, the majority of these second transfer bins are opened and used in a new project, different from the original offline. This may be because the recapture session happened on another system different from the offline system and the assistant did not have access to the original project. It doesn't matter, except that if you are recapturing in a new project you must uncheck the setting for Relink only to media from the current project.

The original offline sequence will link to the new, recaptured, retransferred material, and in an instant you have avoided the typical eye matching that must occur with other systems. Especially on longer, more complicated, or short-form multilayered projects, this one procedure will save you hours of tedious work.

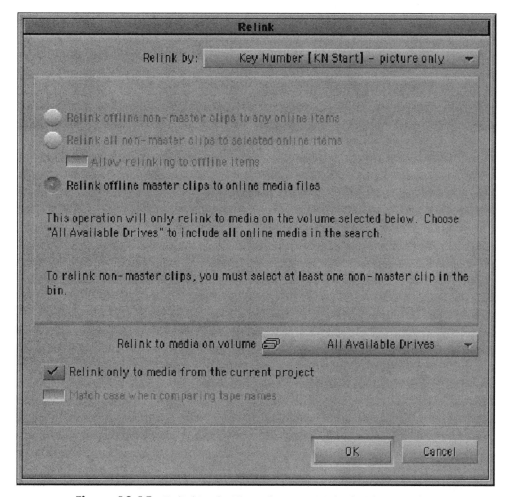

Figure 10.15 Relinking by Keycode across Multiple Film Transfers

VTR EMULATION

There are times when the Avid system does not control a deck, but is itself controlled by an edit controller. This is useful if you want to tie the Avid into an edit suite and use it like a disk recorder or another deck. It could hold bits and pieces that you will be using repeatedly for a linear online assembly. You can also use VTR Emulation to output the sequence time to tape and slave other devices to the Avid. This becomes especially useful when tied to an external, hardware-based MPEG compression system that can control a tape deck. You can compress MPEG directly from the Avid output when you connect it through VTR Emulation.

VTR Emulation requires a special cable that must be ordered from Avid, and then you can set it to behave like any type of VTR that works best with

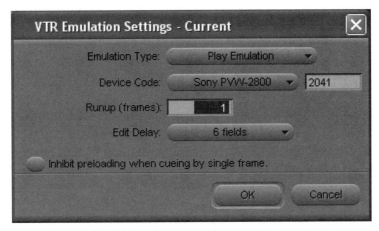

Figure 10.16 VTR Emulation Setting

your system. VTR Emulation defaults to a Sony PVW-2800 but can be set to a wide range of decks if necessary. The H-Phase and the SC Phase also need to be adjusted in the Output Tool just like any deck, so that they match the timing of the switcher. If you adjust the H-phase and the SC phase correctly, you don't get a giant horizontal shift or color change when you attempt to dissolve it with another source through a video switcher.

CONCLUSION

The methods in this chapter are relatively advanced and complicated, but they are created in response to the unexpected needs of the real world. Even if you can't remember all the steps involved in these different techniques, at least you know that these procedures are possible. There is a lot of flexibility when it comes to going from offline to online. This extra power means that you must spend a bit more time and go through more steps when you are working on a complicated sequence. If you find yourself always able to work at the highest resolution, you can save yourself much of this effort, but that may not be realistic. When the time comes that drives are so inexpensive that most people can work at high resolutions all the time, then all of these techniques will be as obsolete as knowing how to thread a quad VTR.

11

Macintosh and Windows: Working Together

THE STATE OF WINDOWS AND AVID

There has been much gnashing of teeth, pulling of hair, and wasting of Internet bandwidth over which operating system is better: Macintosh or Windows. Although the Internet may have plenty of bandwidth to spare, we humans do not, so I would rather spend time discussing specific issues that apply to editors. Avid chose both operating systems and will continue to develop for both. If you look at the decision rationally, the choice comes down to what fits best into your existing infrastructure or what your own personal preference is for ease of use. OS X added new permission functions. We live in a hybrid world so no matter what your feelings are about an OS, you will need to figure out how to make the Macintosh and the Windows systems live happily together. This chapter deals with moving media between the two systems. Traditionally, the workflow has been offline Mac and online Windows, but with the addition of Mac Symphony and many inexpensive Xpress DV systems running on Windows laptops, you will need to swing both ways. This is hardly the bleeding edge, but it is a few extra steps. Once inside the editing applications, you will quickly forget the differences.

EVERYDAY EDITOR ISSUES WITH WINDOWS

Administrator Privileges

Windows is based around the security and hierarchy of privileges. This may be one of the reasons artists and nonconformist types have been attracted to the openness of Apple (OS 9.x and earlier). If you never want to bother with permissions or user IDs or different passwords, then you can make everybody an Administrator. Use a blank space instead of a password. This is quick to get up and running in a facility that never dealt with security before, but you may find that this will lead to more downtime in the long run. If everyone can install anything onto your editing workstation or muck about with important system files, you will eventually find something that conflicts with the Avid software and specialized hardware. You must weigh this possibility against the potential frustrations of a user who needs to do something important and time critical when there is no Administrator around.

In general, one person should be responsible for the upkeep of the system software, and everyone else should get his or her own password and user ID. Although this may take some getting used to, overall, the limited access will help keep billable systems in a billable state. The Administrator can call up records of who used the system and when. This can help for tracking usage and may also allow better troubleshooting since the last person to use a system can answer many of the basic Support-type questions. All user actions are recorded automatically in the Event Viewer (under Administrative Tools). Currently you must be at least a Power User to run Avid editing software on a Windows system (DS users need to be Administrators). You may not like the fact that someone somewhere is always monitoring your actions, but when push comes to shove, it is better that the records are accurate. Integrity is more important than security, but somehow security always wins out.

Fonts between Windows and Macintosh

Be sure that you have the exact same fonts on both Windows and Apple systems. If you have an older offline Mac, make sure it is at least Avid versions 2.1/7.1. This is because Windows and Mac fonts do not play well together. You may find that you actually need to take all the fonts you have in one format and convert them to another format (Mac to Windows and vice versa). There are some shareware programs like CrossFont from Acute Systems (www.asy.com) and, of course, the professional shrink-wrapped application Fontographer by Macromedia, if you need to convert lots of fonts. The utility Transtype is excellent for this since it runs on Mac OS 9 and OS X. If you have Adobe Type Manager® on your Mac system, you will not see the same level of font manipulation (smoothing, kerning, etc.) unless you have it on the Windows system, too. If you find that the correct font is not on your Windows system, then you will be prompted during the Re-create Title Media function to pick another font. Be aware that this font, or even the same named font on Windows, may have different kerning, leading, and point size. You should probably check all of your titles after recreating a project on a Windows system if you began on a Mac. Once replaced, the new font will be used on every title that originally used the missing font. The font replacement information can be copied and moved to another system or deleted since it is only a Site Setting. The file is called AvidFontSub.avt and usually is located on the C drive in the Settings folder. It is created the first time you substitute a font.

If you have used any Avid version earlier than 2.1/7.1, then you will need to open the project in the latest version of the software. Once opened in the later version, you will need to open every title and save it again. Clearly, this could be tedious, so make a special effort to offline your project with the correct, supported Symphony-compatible version 2.1/7.1. If you do not have suitable fonts that match across platforms, as a last resort you can always export a title as a PICT file with alpha and import it to the finishing system as a downstream key.

Using QuickTime

There is a QuickTime Codec for both Macintosh and Windows that will make cross-platform movies that can be imported to either system (Avid Codec version 9.0 and later). Before making any QuickTime file, it is best to know where it is going to be used. If you don't know, then you *must* make it a cross-platform movie. Otherwise, you will need to open the Macintosh-only QuickTime movie in a third-party program and resave the movie as cross-platform.

The Avid Codecs and editing systems will recognize an alpha channel embedded into a QuickTime movie, a 32-bit file. Most paint and animation programs will let you save with an alpha channel so include it if you can (always a straight alpha, not premultiplied if you are given a choice). If the older system you are using cannot recognize a 32-bit QuickTime file made with the Avid codec (version 7.2 until 2.5/8.0 on Mac, and 3.0/9.0/2.0 on Windows), then any movie with an alpha channel made with the Avid Codec will be a very slow import. This is because the system has to discard the alpha channel during the import. Slow import also means that you are changing the resolution of the file during the import. This has implications if your alpha channel is a different resolution than your foreground.

If you know you are going to an older system for import, you are better off using the earlier methods for working with alpha channels. Create two separate files; one for the foreground element and one for the alpha channel (see Chapter 6 for more details). Consider if you are going to use the same QuickTime movie on multiple systems. If some of them *can* take advantage of the new 32-bit Codec, then you may want to put up with the slow import on the older systems just to enjoy th simplicity of moving around a single file.

Avid has made a concerted effort to place all the most current codecs onto their customer Web site (www.avid.com) under the Support section. You m y find it easier to point your colleagues to the Web site regularly to download the latest and greatest versions to stay current with what you are using. The only exception to this is Avid's DV Codec. Because of licensing issues, Avid is distributing the DV Codec only with the installation disks. You can still put the Codec on multiple machines; you just can't download it from the Avid Web site.

In summary, if you are using the latest versions of editing software on either platform, you can import and export QuickTime movies using the Avid Codec and embed an alpha channel. You should make sure all vendors and all workstations used for graphics and animations have the latest QuickTime Codecs for either Mac or Windows and check the Avid Web site periodically for any improvements. Finally, if you are working on an up-to-date system and must move material backward to an older system, you may want to make your QuickTime movies without an embedded alpha channel when using the Avid Codec. Include the alpha channel as a separate movie and combine it with the foreground as a Matte key while in the editing system.

MOVING FROM OFFLINE MAC TO ONLINE WINDOWS

Moving from offline to online on a different system with different video hardware limits what you actually want to bring across. You should be interested mainly in three things: the project (including bins, sequences, and original graphics), User Settings, and the audio.

Use the setting in the General Settings of Avid versions 2.1/7 and later that keeps you from naming a project, bin, or any clip or sequence with characters that are incompatible with Windows.

SENDING METADATA

Technically speaking, bins, projects, and settings are data about data: metadata. They are the most important part of any project that needs to be recreated. Although, at the very least, you need only the sequence bin, having all the bins just in case is an excellent idea. The only question now is: "What is the best way to send that data to the Symphony or DS Nitris system for finishing?" For the DS Nitris system you can only convert your bins and projects to AAF and AFE. Current Avid systems can export as either format (for more details see Chapter 12). There are two major choices for bringing over metadata: removable media or a network. Which one is best for you depends on ease of use and the size of the files.

First, you must decide what else you need to send besides the project and bins. In all projects, after 2.0/7.0, there is a file called Statistics. This is a folder

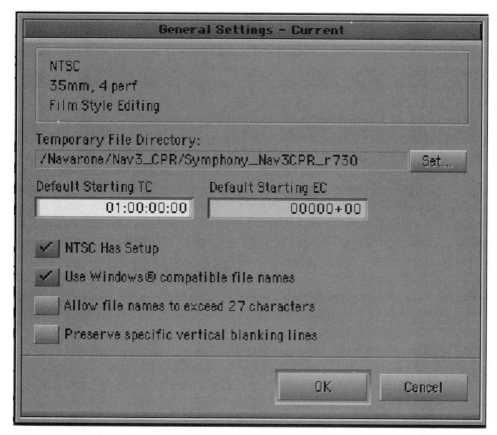

Figure 11.1 Windows-Compatible Setting in the Macintosh General Setting

full of usage records as text files. Consider that if you bring the Statistics file over to Symphony, it will continue to increment the usage amounts correctly. Decide if you need to take up the extra space and time it requires copying them over.

The Attic now saves a separate set of files for each individual project, so you should think about bringing that over as well. Have you been good about making multiple copies of every sequence at crucial points in the edit? If not, then you may have to fall back on the Attic as a last resort when the producer claims, "It was better on Thursday. Can we see that version?" And then hope that you still have the Attic that far back. If you want to copy the Attic for your project, make sure that you put the folder inside the new Attic folder on the Symphony system.

There will come a time when you will need to compress the metadata to make it fit on a floppy or send a large file over the network. The most common way to compress data on the Mac for the Wintel platform at this time is Mac WinZip. *Do not use this program!* It will corrupt your bins and you will have little success opening them. Instead, use Dropstuff shareware from Aladdin Systems on the Mac. Stuffit Expander for Windows is also shareware from Aladdin and can be used to uncompress when the data arrives on the Windows platform (www.aladdinsys.com). If you don't have these installed already, go to the Web and get them now so you won't be flustered at crunch time.

If you are using Unity then you will probably want to use Avid's Media-Manager. This browser-based asset management program is based on the library principle of checking material into the database to make it available to everyone and checking it out to your project when you need it You can drag and drop material to the Media-Manager for check-in (like from a graphics program or directly from the bin when you mount drives) or have it do so automatically in the background as you are capturing. You can then search, sift, and play media directly from the MediaManager interface. Anyone with network access and a fast enough connection can play media up to 3:1 compression (this will obviously get better over time). When you find what you need, either a sequence or a bunch of master clips, just drag and drop the files to your open Avid bin and the metadata is copied into your project. The media is either copied or linked to this metadata depending on where it actually exists. If the media is on an archive or a "Nearchive" (a large group of slower drives with instant access) you might have to wait a few moments for the media to be available.

NETWORK

Let's take a look at setting up a network between Windows and Macintosh. Project and bin information is really not that large. A long format sequence bin with several sequences could be 10 megabytes or more, but compared to the size of the media, this is insignificant. If you are concerned with sending large amounts of graphic and sound files you should consider only 100 Base-T Ethernet or faster.

The best way to set up a multistation network between these two competing operating systems is through a server. A server is just a computer that is used as a central point between multiple systems and is useful for archiving, printing, and searching databases. If you use a Macintosh as the server, you will need to use a third-party application on the Macintosh server, like Dave by Thursby Software (www.thursby.com). If the Windows system becomes your server then you can use PC MacLAN by Miramar Systems (www.miramarsys.com). These are pretty standard applications for connecting Windows and Mac, but technically they are not supported by Avid. This means that the full range of usual testing and documentation has not been completed. If you choose to use them, you must figure out the details yourself using the directions and the manuals provided by the manufacturers.

You can also use a Windows server between the Mac and the Windows system and use the Windows server package for Macintosh. The Windows server operating system is more expensive than the standard Windows work-station; be prepared to learn a lot about domains and administrative privileges if you are a Macintosh user. Again, Avid does not completely support this con-figuration, although it is quite standard for other common connection needs.

If you are on a Unity system then you will have an allocation group (a bunch of striped drives that are seen as one unit by the file system) just for your metadata. This means that anyone with permissions can open bins and play media without copying, importing, or exporting. In fact, many people can be using the same media at the same time. If you find yourself bottlenecked going from offline to online or needing to move projects between systems all the time, consider LANShare (an inexpensive drive cluster that connects with Ethernet) or Unity MediaNetwork (fiber connections to mirrored or intelligently striped drives) as a permanent installation. Your customers may not appreciate how much easier it is for you, but they will notice that things are just simpler when they work at your facility.

The simplest answer is just to connect the Ethernet cable between the two systems and use Transfer Manager (or AvidNet on older systems). Transfer Manager is Avid's Ethernet transfer protocol and network software and allows users just to drop information into the "in box," send it over the Ethernet con-nection, and open it in the "out box" on the other side. If you have no existing network already, this simple Peer-to-Peer connection may be the answer for you. Consider a simple five-port hub between the systems. If you want to expand this network to handle multiple systems, however, you may want to revisit the server solutions. As mentioned in an earlier chapter, the good news is that although Transfer Manager serves as an easy Mac-to-Windows-to-Mac translator, it is also usable no matter how complex your network is. With Mac OS 10.2, it is very easy to link to Windows computers. Just plug in the Ethernet cable and set the drive permissions. You don't even need a crossover Ethernet cable because the Mac OS is self-sensing. You can almost network the systems by accident!

There are certain limitations for Transfer Manager or AvidNet transferring between Mac and Windows, but most of the common needs are handled well. Check the matrix of what is supported in the Release Notes for the details. If you are going between a Mac system with local Mac drives and a Windows sys-tem with local NTFS drives, then you should be just fine. Make sure that the clocks on every system on the network have the same time or you may end up replacing a new sequence version with an old one based on a wrong time stamp. Keep Transfer Manager or AvidNet in mind as the easiest way to get data between your multi-OS Avid systems.

If you are savvy about networks, then you should be able to set up an FTP site and use Fetch free software or a more full-featured shrink-wrapped application. For those who know how, this inexpensive method of putting files on a server for people with permission to download from anywhere is proven and secure. Just be sure you give permissions for the right project to the right customers!

REMOVABLE DRIVES AND SNEAKER NET

Unless you are clever about setting up wide area networks (WAN) over the Internet, you may still need to rely on generic removable formats like floppy drives, Zip and Jaz drives, CD-ROM, DVD, USB memory sticks, and Media-Docks. Floppies are becoming significantly less important in the exchange of data; however, they are especially important getting logged information into the system via Avid Log Exchange (ALE), FLExfiles, or making EDLs. A single floppy can hold hundreds of ALE files. USB memory sticks are quickly becoming a popular way to exchange files up to a few hundred megabytes, but you still can't use them on a CMX linear editor made in 1991!

Unfortunately, when you launch Windows right out of the box, you will find that it is clueless about reading drives formatted on its operating system rival. There are several shareware and shrink-wrapped choices to easily rectify this apparent oversight by Microsoft. The first is MacDrive by MediaFour Corporation (www.mediafour.com). After installing MacDrive, the Windows operating system will recognize Mac-formatted drives and floppies and attempt to recognize Avid bins and projects. An Avid bin should have an .avb extension, and an Avid project should have an .avp extension at the end of the filename. All Avid bins and projects coming from the Mac versions 2.1/7.1 and later will be handled correctly. A Windows system can recognize what kind of a file it is dealing with only by the extension at the end of the filename. If you ever have an Avid bin that has come across to your Windows system as a generic file icon, you can add the .avb extension to the end of the filename and the Windows system will recognize it and treat it correctly as an Avid bin. You can always open a bin from inside the application, so even if the extension is missing, you can still use it.

There is one area where even MacDrive may not help you when going from Mac to Windows, and that is when you use the dreaded illegal characters. Never start or use these characters anywhere in a bin name: _, /, \, :, *, ?, ", <, >, or |. You will be given a truncated, modified filename. For obvious reasons this is as undesirable as it is easy to do. If you find that you have done this by accident or on an old project, there are shareware renaming applications like SigSoftware Name Cleaner (www.sigsoftware.com). The last ABVB versions and everything later have a new checkbox in the General Settings (defaulted to off) for maintaining Windows legal names. It takes only a few attempts to open a bin called Scenes 22/23/24 on Windows before you realize the value of this precaution.

The other choice is to format all removable media on the Macintosh with a DOS format. You can do this easily using PC Exchange 2.0.2, a Macintosh extension that ships with Macs 7.5 or later. You need the version that ships with

Mac OS 8.1 or later to be fully compatible with Windows filenames. PC Exchange 2.2 can read names longer than the 8.3 format of Windows, up to 31 characters. If you do not have the option to format a floppy with a DOS format or you get a message that your DOS disk is unrecognizable, you probably do not have PC Exchange enabled. Move PC Exchange to the correct Extensions folder in your System folder and restart the Mac (OS 9 and earlier only). If it has been thrown away in an earlier attempt to clean out your system to the bare minimum, then it must be reloaded from the Mac OS installation disk or borrowed from another Mac. This issue is no longer important with OS X and later.

MOVING MEDIA ACROSS PLATFORMS

If you are moving from offline to online, why would you want to move media? What needs to come over from an offline project that will not change in quality? Digital audio. Since audio is not captured at an offline quality, copy your audio media over to the Symphony system. You do not want to play from a Mac-formatted media drive on a Windows system, even with MacDrive, so the first step after connecting the drive is to copy audio to the Windows system audio drive. You can copy the media to any large, inexpensive drive and just use it for transport or actually copy it to a fast enough drive to do the job and use it directly in the online as your audio media source.

To move audio easily to a Windows system, you must first capture it on the Mac as an AIFF or WAV file, not an SDII file. This setting can be found in the Audio Project settings in current versions, and in the General setting in some earlier versions. This Audio setting must be correct before you begin to capture the offline project. If you plan to bring a project from an offline system directly to a Windows Symphony, then you should always capture your audio as AIFF or WAV. If you are going first to a ProTools system or another digital audio workstation (DAW) that uses SDII files natively, then you should stick with SDII. The newest versions of ProTools do handle a wide range of files, so ask the audio post facility what they want. The audio professionals will have to send you back the finished version as OMFI or on digital audiotape. If you make a mistake and capture as SDII when you really want to go to Windows Symphony, then you can export your audio as OMFI with media. By exporting as OMFI, you are secretly converting the SDII audio files to AIFF.

There are some reasons to move video from ABVB to another system, but not many. The Meridien board does just about everything better with signal quality than the ABVB and later versions also have high quality signal integrity when ingesting. The only real reason to bring video from an older offline system would be that you have lost the original tape or you have spent a long time color correcting something at AVR 77. Opening AVR 77 as 2:1 doesn't make it look better, but it shouldn't make it worse. If you want to move video, you can export it as an OMFI file and import it to the Symphony system. Or you can just copy the files and import as OMFI (the media files are already OMFI native). You must change the Import Setting on the Symphony (or any Meridien or later system) to change the field ordering to odd. This is important because the field ordering of the video frames is different between the earlier systems and the current ones. Meridien and Media Composer Adrenaline systems start displaying the odd field (field 1) first in NTSC, which follows SMPTE standards. PAL field ordering stays the same between ABVB and later systems.

Since the compression schemes with the older systems are completely different, copying media and importing it will be a slow transfer because all the video will need to be uncompressed and then recompressed on the Meridien systems. The Adrenaline systems can play ABVB media without any conversion. You may find that it is faster to recapture in many cases if you have the source tapes. If not, beware of going past the two-gigabyte file size limit for any file on the Macintosh when you export as OMFI composition with media. You can perhaps export your sequence in smaller chunks. Remember also that an OMFI export with media takes the entire original file. If you want to take only the media needed for the sequence, then you should consolidate the sequence first to create smaller media files. You may even want to hide the original media files so you don't create a huge OMFI file with original embedded media by accident. Some audio designers do not want the huge media file created using this method, so ask ahead of time.

So to move video from the Macintosh offline system to the Windows Symphony, consolidate the sequence and export as an OMFI file with media. This may be your only choice when you are working with nontimecoded video sources, material restored from a DLT, or material for which you have lost access to the original sources. Of course, you can always just capture the digital cut from the offline session for scratch video.

IMPORTING CD AUDIO

If you have used the CD audio directly from the CD during the offline on the Mac and you are not bringing all your mixed audio files over for the finishing stage, then you can use the Batch Import process to reimport the audio directly from the CD just like with graphics.

ROUNDTRIPPING

Roundtripping is the term Avid uses for bringing a project back from an online system to an offline system. There are some basic problems with this process that have to do with different versions of the software and different ways that media is handled. Generally if you are going from an older system to a newer system, you will have good success with metadata. If you are going from a newer system back to an older system, you will find that many things will be stripped out. However, if the core of the sequence can be roundtripped, the media on the older system should just link up. You can also take an EDL from an online system (especially if you are onlining in a tape suite) and the media on the offline system should relink to the EDL (see the section on Relinking in Chapter 10 for more detail). Bringing video media from a Media Composer Adrenaline or a DS Nitris system backward to Meridien or ABVB, however, is not supported today. The media from the older systems (v7.0 and later) for the most part should be compatible on the newer systems.

CONCLUSION

There are lots of industry standard answers for getting material between Macs and Windows systems. Networks are clean and simple once set up, but unless you are on Unity, they may be too slow for moving video media. With the procedures and network connections in place, the main workflow should be designed to reduce recreation of creative work, recapturing and allowing flexibility if the online project must go back to offline for more versions or tweaks.

12

Finishing on Avid Systems

Avid Symphony and Avid DS Nitris are Avid's top of the line finishing systems. As such, they receive Avid's biggest, coolest innovations for the high-end editorial space. They are designed specifically for the kinds of projects that demand the best quality and the most streamlined workflow. However, software-based Avid models like Media Composer Adrenaline quickly are becoming almost as capable of finishing these same projects. When you finish on any of these models, you are getting the best conform from an Avid offline with features designed specifically for the fast-paced world of the professional front-room edit suite.

Even if you are working primarily on DV25 material, you will still benefit from using these systems for your final online. Although you may be using the DV25 material throughout the process, you still should consider the final pass to be all about quality. Tweaking color, sound, graphics, and keys will pay back in a big way if you take the time to focus on the final pass using some of these more advanced tools.

This chapter will go into detail about the benefits of Avid's Symphony and DS Nitris as it relates to the finishing of offline projects begun on other Avid systems at low resolution. Since the features of Avid's Media Composer Adrenaline are fast approaching the hardware-based features of Symphony, much of this chapter will apply to these new systems, too.

DESIGNED FOR FINISHING

At the finishing stage, all the pieces of a project are reassembled from the offline process. Audio is laid back and graphics are inserted at their highest quality and final version. Color correction is applied shot-to-shot, and blemishes and mistakes are covered up or corrected with scratch removal. Of course, you can still edit during the finishing stage — one of the real hidden benefits to finishing on a nonlinear system is the ability to put in last minute changes. Here's an example of what this flexibility means in the real world. A major PBS series received a large donation from a new sponsor at the last minute. Using Symphony, the editors were able to shorten the program to accommodate the funding credits and show the producers multiple choices as to how the programs would be shortened without affecting content. On each of the four one-hour programs, they could see where those cuts would be and what they looked like even after the online was completed.

Another movement to get to faster-better-cheaper in the online process is toward streamlining the batch capturing process. Avid has added the timeline choice to show offline media and the clips that are missing from the sequence highlighted in bright red. As the capturing process proceeds, each of the offline elements indicated by the red highlights is eliminated, eventually leaving the timeline color looking normal. This is a great way to keep an eye on your progress or to pinpoint a problem area if you think you have finished, but the display shows you otherwise. Showing offline media is a good feature to use just before an output or Digital Cut since it will show even if nested media is offline. You can step into the red elements in the timeline down through the nested layers to the individual element that is offline.

AVID CONFORM

Symphony and Media Composer Adrenaline take all the bins and projects from the Avid offline—which could be a Windows *or* Macintosh system—and reproduce them exactly. The DS Nitris will take an AAF version of the sequence or an AFE version of the bins and convert a large amount of effects, AVX plug-ins (currently v1.0 AVX effects only, but soon AVX 2.0 effects), and the parameters of those effects.

If you have been using EDLs to conform your program to take advantage of the uncompressed quality of a D1 linear tape suite or another NLE, then you have been working too hard. All the effects that could not be represented by an EDL need to be recreated by eye. Every layer needs to be meticulously tweaked with effects, compositing, and drop shadows all started from scratch. Audio levels, pan and scan, motion effects, DVE and AVX plug-ins are all automatically just there if you conform on an Avid finishing system.

If you have an older offline system you may not have the option to export as an AAF or AFE directly from the system. You will need to send the bin to a new version of MediaLog. This will allow you to open the bin and save it as an AFE. Everything in the bin will be able to be opened in the DS Nitris system. Just drag the sequence to the DS Nitris Timeline and you're away. If you have a recent version of MediaLog (it came with your editing software) then you should be able to convert an OMF sequence and bins to an AAF or AFE file. But you can count on the DS Nitris operator having a copy and being able to convert your bin for you. They can download a copy of MediaLog from the Avid

DS Support Web site if they don't have one handy. You may want to decompose the sequence and include all the new master clips with the sequence in the bin you hand to the DS Nitris operator. It could make the recapturing smoother. Of course, your system may be able to export the sequence as an AAF if it is up-to-date. Choose this as the method to import directly to the DS Nitris system to relink or recapture the media for the finishing or effects stage.

With a little planning you can create templates of commonly used segments, like an opening animation, if you are assembling a program the same way every week. Just drop the new video inside the existing effect or composite. Now you have more time for creative tweaking to get the program just right. Remember, every speed efficiency increase doesn't always add up to getting the program out the door faster—it can give you more time to make the project better!

Nonlinear systems make better use of expensive video tape decks; especially with the price of HD deck rentals, using them only for the final day or two of batch capturing is cost effective. Using Unity MediaNetwork you may even find it more *time* effective to capture with multiple decks at the same time for the same sequence! A linear tape suite needs at least three decks most of the time, but with a majority of effects-intensive online compositing being spent without any decks rolling at all, this seems like quite a waste. Hang on, though. Don't try to take any of the decks away! Who knows when you might need them again? A nonlinear system generally uses one deck for a specific period of time and then doesn't need it again until it is time for the Digital Cut. You can take the deck to other Avid suites for dubbing, viewing, and logging other projects.

ADVANCED COLOR CORRECTION

The Avid products slowly have assimilated the color correction capabilities of Symphony—a breakthrough in nonlinear color correction. The new Advanced Color Correction mode sits alongside the other, more established modes of Source/Record, Trim, and Effects. The advanced color correction engine has a new interface and many new capabilities to address two types of users. The editor now has easy-to-use, powerful controls, and the more experienced colorists have enough of the established types of controls to feel comfortable. If you have done any color correction on another system for film to tape, desktop publishing, or on a high-end compositing workstation, you should see something that seems intuitive to use. Because of this wide, existing user base appeal, there will be very few people who use all the controls all the time, and many of the controls are meant to be redundant. If you find the one or two windows that perform the job you need, then look no further until you have a very difficult problem.

The Advanced Color Corrector mode handles two streams of uncompressed video in real time and still has enough bandwidth for another realtime effect and a DSK (either a title or a graphic with a static alpha channel). This mode is a primary color corrector and is designed to be highly efficient when matching shots scene to scene or making badly lit images look good. For more effects-style correction, you may want to use the Spot Color Effect that combines the vector-based drawing capabilities of the Paint Effect with the Color Effects controls. There are also eyedropper functions with NaturalMatch™ capabilities in this interface. Anything that requires field-accurate shape animation (keyframed moving objects) must be rendered, however, so even with the new color correction hardware, the Spot Color Effect is a software or blue dot effect.

If you are an offline editor, you may not be immediately interested in Advanced Color Correction, but perhaps you should give it another look. Many times there will be a discussion about whether a particular shot can be used at all because of on-location problems. In the past, you may have discarded a marginal shot or used it and crossed your fingers that someone else could make it work. With many of the easier-to-use capabilities extending all the way to Xpress DV, you may want to crack the manual. Give some of the easy-to-use graphical tools, like the eyedroppers, another try.

Technical Overview

The capabilities of Advanced Color Correction mode are a direct result of a new 10-bit RGB color correction chip that has been added to the Meridien board with Symphony version 2. They were extended to work in software on Adrenaline and use the Nitris DNA hardware to work at 12 bits on 10-bit source material. The higher bit depths allow a wide range of adjustment with higher accuracy (10 bits is 1024 units vs. 8 bits at 256 units) and a very high-quality color space conversion from native YCrCb to RGB and back again. With many such color space conversions, there is a chance that the color in the new type of color space (YCrCb vs. RGB) will look different. The Avid system purposely minimizes this difference. There is no rounding of the color values, and there is, at most, a difference of one 8-bit unit (1 level out of 256) between the color in YCrCb and the same color in RGB when working with 8-bit sources. To further reduce the impact of this conversion, the system will pass the YCrCb signal all the way through the hardware without change if no adjustments require the extra power of RGB color space.

There has been some debate about the benefit of higher bit depth in the processing than in the source material. In other words, what is the good of 10-bit processing of 8-bit material? Think of any effect as processing the image (running various mathematical calculations) and imagine the computer as a giant calculator (which, after all, it really is!). Now imagine that your calculator only has the ability to compute a number with four decimal places. The calculation you want to do (say, making the image brighter) goes to five decimal places to get the look you want. You would have to truncate or round the final computation to something your calculator could handle. This adds a basic inaccuracy to the calculation. Now imagine that you need to do 30 calculations to the same image and every one of them had to be rounded separately. You would get quite a cumulative error! Now if you were using 10 bits for the calculation, you could do all 30 calculations at once and then round down to eight bits once for the final result. The rounding would be done only once and the difference would be very small between what you needed to do to the image and the final result. Do the calculations in 10 bit and round down to eight bits or do the calculation in 12 bits and round down to 10; you will always get a better result.

NONLINEAR COLOR CONTROL

What is it about nonlinear color correction that could possibly make it more efficient than a stand-alone color correction system? It's certainly not in the manipulation of controls. Working with the $250,000 dedicated keyboard and control surface is more productive than using a mouse and a generic keyboard for color correction, but consider flexibility and efficiency of overall process when comparing the two approaches. First, when a dedicated color correction system is used in the film-to-tape telecine process, all the original material is corrected and transferred. This may be hundreds of hours on long projects and requires the specialized skills of a full-time colorist and the most efficient interface money can buy. But don't try to do anything else in that suite! On the Avid finishing systems, there is a different approach that views color correction as the last step in the process of a final master sequence. The editor is generally not responsible for correcting dozens of hours of footage and may even be doing a second pass on footage a colorist has already handled once. The pressure to crank through huge amounts of footage at an expensive hourly rate is not appropriate at this stage. Certainly, the Avid system is not controlling a telecine device, so that initial stage will not be eliminated until the telecine step is completely eliminated from production.

The nonlinear color correction stage more likely will be used to replace tape-to-tape color correction. It will also be used to add that little extra value to projects that have originated on multiple sources like VHS, hi-8, stock footage, and different types of film. The overall evening or grading of the images from scene to scene subtly helps to reinforce the message of the program by reducing distracting influences. Conversely, color correction can help to accentuate sharp transitions to different looking original sources if that is part of the storytelling. The color correction on Avid systems can also be used to improve an inexpensive or quick "one-light" film transfer by adding contrast and saturation with a set of premade templates.

Using different types of source corrections, the Avid colorists can adjust large amounts of the program material at once. By choosing to adjust source color based on master clip or source tape, they quickly can affect every time that master clip or source tape is used. Although there may be some more fine adjustments required, this one step will help make up for some of the limitations of using a more generalized editing system to do color correction. At this time, only Symphony and Media Composer Adrenaline have the separate source and program or the live linking features that will be discussed later. There is plenty of opportunity for skilled, professional colorists to participate and add value throughout the finishing process, even when color correction is handled nonlinearly. The colorist's importance comes not from their ability to run a dedicated piece of equipment, but from their ability to instantly analyze, problem solve, and add input to the creative process. All of these areas of expertise can be translated to a nonlinear color corrector.

COLOR CORRECTION STRATEGIES

You could spend a lifetime mastering the subtle art of color grading and correction, but there are some basic strategies that will point you in the right direction of improving most of your material. There are three areas to concentrate on: contrast, hue, and correcting technical faults.

Contrast

Unless your image has been color graded by a professional it probably has some lack of contrast. This means that the full range of possible whites and blacks is not being fully utilized. Chapter 7 went into detail about how to use some of the automatic functions to adjust contrast. Not all images will benefit from increased contrast, but most do and you will hear it referred to as "making the image pop."

Use the Levels tab for best contrast control. Look at the input side, the left histogram, on this interface. You will see how the image is spread between the extremes of white and black. If you see the energy of the histogram bunched up in a particular section, you will benefit from evenly distributing it. Move the black point and white point sliders so that they bracket the energy of the image. This tells the system to use the blackest blacks available for the blackest part of the image and make the whitest whites for the brightest part of the image. You will quickly see the output side, the right histogram, show the even distribution of energy as the image spreads out.

The next critical point, which will take some skill, is to move the center slider, the gamma triangle, to the appropriate midpoint in the image. You can do this by watching the image on the video monitor to make sure you are positioning the gamma based on what the image really looks like. Do a quick split-screen comparison before and after the Levels adjustment to make sure you are

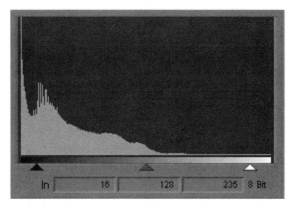

Figure 12.1 Levels Sliders before Correction

not losing some critical detail. You should be able to see a sharper and more defined image that is much less washed out.

Hue

There are many reasons why the hue of the image is not accurate or pleasing. Surrounding objects can reflect light onto skin tones to create a color cast. Shooting in the shade many times will create a blue cast. Shooting under different natural and environmental lighting conditions will also add undesirable colors. This can happen from street lights, fluorescent lights or a mix of natural and artificial light. If you have a reference of excellent skin tones or whites, grays, or blacks somewhere in the image you can use the color matching and cast removal eyedroppers. This was also detailed in Chapter 7.

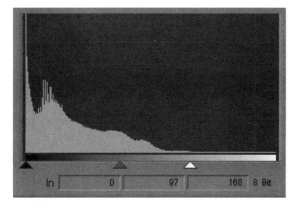

Figure 12.2 Level Sliders after Correction

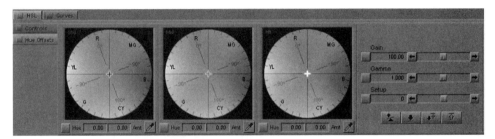

Figure 12.3 Hue Offset Wheels

If you want to take a more creative approach to adjusting hue then you should look at the Luma Range and Hue Offset controls. Go to the Luma Range window and click the Three Tone button to see how your image is distributed among the three areas of shadows, midtones, and highlights. Is the problem limited to one area like the blue shadows of midday? Then adjust only the shadow areas with the hue offsets. In reality you will probably adjust at least one other area of the image if the problem overlaps across the luminance range.

Pick the Hue Offset range and move the crosshairs away from the problem color to the other side of the color wheel. Holding down the Shift key will give you more control over fine adjustments. If this still appears to affect too much of the image you will want to go to the Curves windows. By clicking in the Red, Green, or Blue curve window you create a very selective control point. Each point maps the input color (the color before adjustment) to the output color (where you want the color to be). The control point has an automatically smoothing spline curve so that you affect a wider portion of the image when you adjust the point. You can create some protection points on either side of the area you want to change so that you limit the effect on the entire image (see Fig 12.4). This will also allow you to get more extreme with the correction if the problem is especially severe. Remember that you can fix an image by removing the offending color or sometimes by adding in more of the other two colors for a better balance.

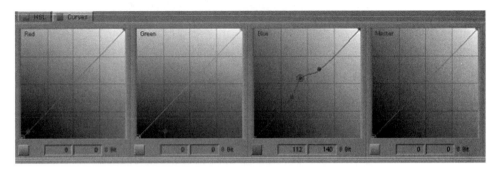

Figure 12.4 RGB Curves with Protection Points in Blue

Technical Faults

There are a wide range of things that can go wrong on a shoot. And you should be glad for it because that gives you an opportunity to show how valuable your skills really are. White balancing is something every camera operator should do when they set up a shot. They hold a white reference sheet in front of the lens and press a button to neutralize any color cast that would be created by a color temperature shift since the last shot. Color temperature can change quickly during sunrise or sunset and certainly when moving between artificial interior light and exterior light. Sometimes white balancing isn't done enough, or is done incorrectly, or not at all. White balancing can be fixed quite easily by removing color casts as documented in Chapter 7.

There are other technical faults that are harder to fix, like a missing blue channel. These can be better fixed using the Channels tab. In the case of a missing channel you can reconstruct that information by substituting the blue channel with the luminance channel. You may even want to blend the red and green channels together until you get something that looks normal. You will still need to tweak the hue to get it right, but you will have saved an unusable image.

There is keyframing of parameters on the DS Nitris system so if you have a problem with oscillating levels you can attempt to match the changes over time and balance them out. You can set the DS keyframe interface to loop a series of keyframes, which will be a perfect answer to a problem that continues for a considerable length. Nail the frequency of the recurring problem and loop the counteractive keyframes until the shot ends. With the Symphony system you can put add edits at the points where the image shifts and put in dissolves to ease into the change.

Every time you are asked to fix a technical problem, be sure you are watching the image through the waveform and vectorscope so that you don't make the problem worse. Also keep any successful settings as a template to be used again or shared with colleagues.

THE INDIVIDUAL CONTROLS

The rest of this section will go into detail about how to use the full range of controls available on Avid color correction interfaces. As you look at the interface of Color Correction mode, you will see three screens across the top for scene-to-scene comparison. The center screen is the active image. With a right-mouse click (or click above the monitor), any center screen image can be turned into a reference image, or on Symphony you can use the reference image as the second half of a split screen. The split screen will default to the usual before/after comparison without the right-mouse click.

Figure 12.5   The Center Screen Interface with the Center Split to Show Reference

Below each of the three screens is an individual timeline for each of the panes, and the controls interface. Notice that there is a green half (the left) and a blue half (the right). The green controls apply to the source correction and the blue to the program, or timeline, correction. Whenever you choose one side or the other, the colored lines in the timeline will reflect this choice once the Correction display is turned on in the timeline. On the source side, you can pull down the menu to decide what source type will be affected. The default is Segment, which restricts any correction to the active clip in the center screen. The other choices are Source tape, Source Clip Name, Master clip, and Subclip. On the program side you can choose Segment, Track, and Track "in to out" (this is a User Setting). I will go into more depth about these

Figure 12.6   Different Kinds of Nonlinear Source Corrections

choices when we discuss the power of the nonlinear aspects of this interface. The main importance of having two separate sets of controls on the timeline is that you can adjust each shot individually, in context of the sequence, as you compare scene to scene. Then you can add an overall correction to the entire sequence. The parameters of the two adjustments are summed, and one set of instructions is sent to the dedicated color correction chips so they do not create some kind of additive distortion of the image. This means that you can first make an adjustment that will make all the shots look even as the sequence plays, and you can then go back and add a creative "look" to a section.

There are lots of adjustments you can make to this interface to customize it. You can activate the controls you use most or optimize it for speed. The four icons in the upper right of the Color Correction Controls window give you direct access to a range of controls. The Correction Settings, the ability to save color correction templates, SafeColor Settings, and the ability to leave comments attached to a segment are all represented by icons. Comments show up in the Timeline and will carry through to an EDL. You can also save four different temporary templates by Alt-clicking the row of empty buttons (C1 through C4) on the right of the interface. Normal clicking on the temporary templates will apply them, and Alt-dragging the Template icon to a bin will save it as a premade correction to be applied later and carried around with the project and bins. This is especially useful if a product shot needs to maintain a consistent look across a series of spots, for example, the red of a Coke™ can.

THE FOUR MAIN WINDOWS

The four windows of the interface are Controls, Channels, Levels, and Curves. The Controls window is what most editors will use immediately since it is close to what most basic color correctors do. Using the Controls window is the next step up in sophistication from the existing Color Effect. Hue Offset is the most powerful control in this window and perhaps the entire interface. Beginners can go straight to the Master Hue Offset, whereas experienced colorists will know exactly what to do with the other three hue controls for Highlights, Midtones, and Shadows. The luminance range will choose the part of the image adjusted with Highlights, Midtones, and Shadows (see Fig. 12.8). This can also be used as a very limited way to do

Figure 12.7   Controls for Advanced User Configurations

Figure 12.8   Luma Ranges

secondary color correction based on luminance values (somewhat like creating a notch filter with the EQ tool). The Controls window will be the most used and easiest to learn part of the interface for common correction.

The Channels window will be used mostly by experts or those most familiar with desktop publishing techniques. Rather than try to use this interface for color mixing or balancing, this mode is more appropriate for stealing detail and cleaning up noise. Desktop graphics gurus know that when a scan does not reproduce well, many times it is because one of the color channels lacks detail. With a little experimentation, you can find a channel with lots of luminance information, and that channel can be blended with a problematic channel. This same process can be done in the Channels window (see Fig. 12.9). First, view the channels to decide where the most detail may exist in an image. Then use that channel to blend with the channel that is slightly underexposed or overexposed. Using percentages of each channel, you can experiment until the overall detail of the image is restored.

The Levels window is the most advanced way to adjust luminance levels and the levels of each individual channel for optimum use of contrast (see Fig. 12.10). The histograms allow the user to see the distribution of energy across the range from dark to light. Two graphs represent the range of levels that will be affected: the levels that will be input into the processing, and then the final or output levels after processing. By graphically representing the energy of the image, you can see whether the blacks are all bunched together.

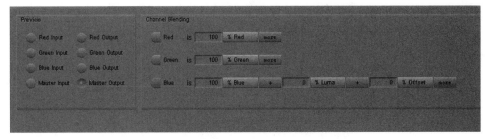

Figure 12.9   Channels Window

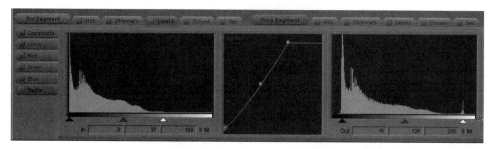

Figure 12.10   Levels Window

They may need to be better distributed across the available range for a more even contrast. You can also see whether an overly bright image can be improved by moving the midrange or gray point.

The Curves window is the first window devoted completely to RGB adjustment. It shows the curves for each channel that can be manipulated with control points. Since this is a primary color corrector, you will eventually add more control points than required to improve the image. Try to keep the adjustments simple and you will have better results. This is also the main window for NaturalMatch, since that function works completely in RGB colorspace.

NATURALMATCH

Eyedroppers are an easy and graphically intuitive method for matching colors from one image to another. The Eyedropper icon is available in many of the different windows in Color Correction mode. Usually, eyedropper controls apply specifically to the settings in that window. In other words, in Levels mode the eyedropper will allow you to pick the black point, gray point, and white point of the bad image and replace it with the corresponding points in the good image. This will extend the contrast range of a dull, washed-out image. The RGB values sampled with the eyedropper can be saved by Alt-dragging them to the bin, where they will be named automatically based on a database of standard color names.

Many times an exact match or absolute match does not return the results you really wanted. This is because you wanted only a part of the color values of the good pixels and didn't want the entire image to change based on those levels. What you really wanted was a relative match, or a match that gives results that are more natural. To address this, Avid invented NaturalMatch. This scheme of matching RGB values takes advantage of complex mathematics, which takes hue and chroma levels from the good image and applies them to the bad image. This makes an excellent starting place for more advanced matching, but many times it may be enough for the critical parts of an image to match from scene to scene.

NaturalMatch is especially useful for quickly changing a video-originated image that was not "white balanced" correctly. Pick a skin tone from a similar person in another shot and apply it to the skin of a non-white-balanced person (easily recognized as having a heavy color cast). The luminance values of the scene are maintained, but correct reference for hue and saturation is applied. The result is the right color in the target lighting condition.

SECONDARY COLOR CORRECTION

Secondary color correction can be used for the outrageous and for the subtle. Any correction that affects the entire image is considered primary so secondary color correction is for adjusting individual colors and leaving the rest of the image alone. You can use this to turn the yellow car purple or everything black and white except the hero. You can also use it more conventionally to make the bricks more red but leave the product the same Pantone™ color. In the Symphony color correction interface it is the last tab in the color correction interface; the other models don't have it (see Fig. 12.11). Basically, you select a color in the image using one of the eyedroppers and this becomes the input color. This is represented in the input vectors on the left of the interface. You then manipulate the output vectors to change the selected color to something else. On the Meridien Symphony systems this process is performed by a different chip than the primary color correction and so you are limited to changing only the hue and saturation of the selected color. You can't change the luminance value just yet. Since you are using a chip elsewhere in the hardware, the secondary is considered a real effect and uses up the one realtime effect you have available at any one time. You cannot have a transition or a DVE and a secondary color correction in realtime at the same time.

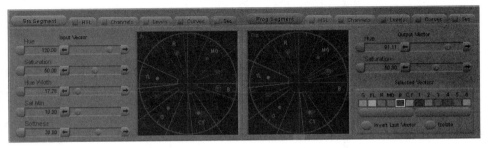

Figure 12.11   Secondary Color Correction

SAFECOLOR

Another benefit of color correction in Symphony, from an offline editor's point of view, is the SafeColor™ feature. SafeColor guarantees that any level that goes through the video board will be corrected and brought within the specification created by the user and saved as a template (see Fig. 12.12). You can create Safe-Color templates for composite, luminance, and RGB levels and easily save them as Site Settings, or you can move them from system to system to ensure a level of consistency between editors and different edit systems, or as targets for different network requirements.

Any video that has been captured or any graphic that has been imported without an alpha channel will be corrected using a clever method of pixel replacement. The standard system in most facilities to ensure broadcast standards is a processing amplifier, or "proc amp." Facilities may install a proc amp at the output of their video switcher to make sure that all levels that go down on tape with illegal levels above and below a preset standard are chopped off. This "chopping" is sometimes referred to as a hard clip. It does not make the image look any better. In fact, a hard clip will cause all parts of the image above the preset level to lose detail. So where you may have seen a bright yellow object with shadow and edges, you will see a slightly darker yellow blob. Many of these proc amps will work to eliminate luminance or chrominance excursions above the broadcast standard. Sometimes the image will become darker, or less saturated. And sometimes you will lose all detail in the problem areas.

There is also a setting on some proc amps referred to as a soft clip, which starts to reduce the illegal levels at a preset point and gradually, over a wider range of signal, begins reducing the problem. This soft clip is less noticeable to the eye, but either begins to affect signals well before they are illegal or does not contain them completely within the legal standards.

The Avid SafeColor system is smarter than this. Instead of compromising the signal, under many conditions, the SafeColor feature will improve the

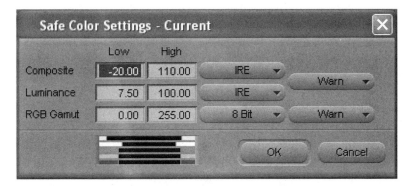

Figure 12.12 SafeColor Template

image quality while making a correction. It does this by identifying all the pixels that are outside the standard and replacing them with their closest neighbors that are within the standard. So bad pixels are replaced with the closest good pixels—many times adding back detail that would have been otherwise lost.

Does this mean that all editors unfamiliar with how to set a proper level on a shot will never have to learn? Of course not! That would be too easy and then anyone could do it! Setting the levels correctly in the first place will minimize the chance of illegal levels and will certainly produce better looking images in the long run. Also, keeping the image within the digital realm of 16 and 235 (for both PAL and NTSC) will guarantee that your image is in the center of the optimum range for broadcast and reproduction. The security of SafeColors is to help those editors who may miss an occasional bright glint off chrome, the bright blue police lights, or other unexpected luminance or chroma levels that are transient. Generally, online editors do not like to adjust the entire image for such tiny but critical-level problems. They rely on other tricks to keep the image within the proper range, but always optimized for the majority of the screen area. With SafeColor turned on, these problems will be fixed automatically.

There are two kinds of images that the Avid editing and finishing systems do not treat as real video, titles and imported graphics with alpha channels, so SafeColor does not work the same way with them. Usually, they are treated as downstream keys (DSK). Both titles and graphics with alpha channels can be treated as a non-DSK by unchecking the DSK checkbox in Effects mode. This turns all DSK graphics into normal video for SafeColor to be effective. Rendering any of these types of graphics will also correct illegal colors since the rendered precompute file generated by rendering is also treated like standard video.

The main benefit of keeping graphics as DSK is that they will key in real-time and remain uncompressed in quality. So if you want to use DSK graphics then you need to be careful when creating them. If the SafeColor limiting is turned on for graphics and titles, then they will be limited when they are imported or created. The Title Tool has a setting for safe levels where it will cut off the composite signal based on the SafeColor template active at the time the titles are created. Imported graphics with alpha channels will also be limited to the levels active at the time they are imported.

If your levels requirements change after the graphics and titles have been created, then you will need to change the SafeColor template and import the graphics into the system again. Fortunately, this is easy. With titles, mark in and out around the entire sequence, turn on all of the video tracks, and use Recreate Title Media (it takes only a few seconds). With imported graphics that contain alpha channels, batch import them again using the original files. Either of these answers could be tedious if you had, say, 20 sequences that needed to be fixed, but otherwise they are fast. If you keep your level conservative to start with, then any change will likely be to make the requirements more liberal and all existing graphics and titles will fall within the new tolerances. After all, it is rare to have your broadcast requirements change during the finishing stage.

USING NONLINEAR TOOLS

When the Symphony operator is going to create a source color correction, there are multiple, powerful ways to correct a large amount of material at once. The default setting for source color correction is Segment, which will restrict all corrections to the segment that is active in the center monitor of the three-window interface. The colorist may know that everything on that particular tape is slightly overexposed, incorrectly white balanced, or part of a "one-light" transfer that could use a premade template (see Fig 12.13). A single correction or application of a template can then be applied to everything on that tape, master clip, or subclip. The same can be done on the Program side to correct an area that affects every clip between a marked inpoint and a marked outpoint. Check the Correction User Settings to make sure Use marks for segment correction is enabled (default is off), since it is considered a bit of a specialty usage with its own set of implications.

From that point on, there is a live link between the image on the screen and its subsequent uses. The power of a single tweak to ripple through a project is an extension of the power of a nonlinear system. If one of those changes sends a small part of one shot out of the defined correct levels, then SafeColor will kick in. It will quietly make the image conform without the user having to check out the entire sequence one more time. Not all users are comfortable with the power of these live links and one little tweak by the wrong person trying to make a shot look "cool" may result in widespread changes that are not desirable. You may want to take steps to ensure that the feature is used correctly. You can flatten the live links throughout the sequence so that all corrections default to being Segment only by loading the sequence and right-clicking in an empty part of the color correction interface. Choose Flatten from the right-click menu and feel confident about handing the sequence off to the next step in the process (see Fig 12.14). You may want to keep a version of the sequence hidden away in a secret place before you flatten it (or check Create New Sequence on the Flatten Choices window).

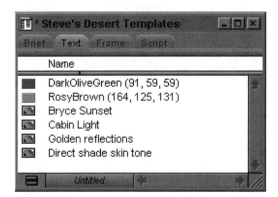

Figure 12.13 Saved Reusable Swatches and Templates

There may come a time when the sequence is kicked back to you, and you will need to continue to make corrections with all of the live links intact.

Another tool in the nonlinear stage of color correction is Update, which takes any new shot added to a sequence and checks to see if it has been used before. If the shot has been used before and you want to apply the color correction added to it using one of the live links, you would choose Update. You can then decide if you want to update everything from the source tape, master clip, or subclip. If you have flattened the sequence, then all of the live links have been eliminated and Update will not make any difference. In the case of a flattened sequence, you will need to create templates and apply them to the new shots in the sequence.

The system cannot always assume that every time you want to add a specific master clip you want it to look the same. You may have added a creative look to the master clip the first time it was used (in a dream sequence or flashback, for instance). The next time the shot is used, it is meant to represent reality and blend in with the surrounding shots. This is why Update is a manual, controllable step taken by the user after a shot has been placed in the sequence.

SHARING MEDIA

With the growing adoption of media sharing systems like Avid Unity Media Network, there can be many people working on the exact same frame at the exact same time. This type of distributed, parallel workflow is revolutionizing postproduction and is a definite step forward for nonlinear color correction. The colorist or the online editor who is skilled in color correction can step into this

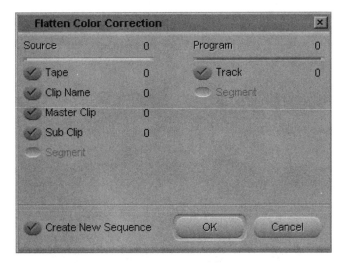

Figure 12.14 Flatten Dialog for Breaking the Live Links

parallel process in many stages, and the Advanced Color Correction makes this as seamless as possible.

There is another control called Merge that, although it seems somewhat similar to Update, is meant to be used in a completely different set of circumstances. Imagine that while you are batch capturing a project onto Symphony at an uncompressed resolution, your colorist is looking at the same images—the exact same media file—on another Symphony. While you are examining the shots, fixing dirt and scratches, your colorist is looking at the sequence for scene-to-scene correction. Perhaps while you are examining the sequence, your producer realizes she can make some changes, or some last-minute replacement footage arrives and you begin to make some new cuts.

What happens when you finish editing and the colorist finishes correcting and you discover the sequences are very different? This is when you Merge the color correction information from one sequence to the other. Take the colorist's version and the editor's version and merge only the color correction while leaving all the edits untouched. If there are conflicts where you have done some colorizing and you don't want it to be wiped out by the incoming changes, you have various choices for such conflict resolution.

Load both sequences, one in the source window and one in the record window. Right-mouse click or Shift-Ctl-click on an empty part of the Color Correction interface. Choose Merge. Make sure you know which sequence is the final color correction version and use that to merge with the final version of the edited sequence. You should probably choose to make a new sequence during this process.

Advanced Color Correction is a first step toward the merging of nonlinear editing, media management, and professional color correction tools. We can look forward to uses that are more advanced and significant improvements in workflow as more editors and assistants become skilled at color correction and take advantage of media sharing. Symphony addresses these situations through the live links, the ability to share correction information across multiple sequences, the security of SafeColor, and the straightforward controls for everyday needs.

HDTV AND 24P

With all the discussion of how best to prepare for the coming of HDTV, many people fail to look at the most important aspect of all: the workflow. If you've read this far in *The Avid Handbook*, then you know that one of the most important design considerations at Avid is, "Does the new method improve workflow?" In a technological world where sales are made by trumpeting the next breakthrough that "changes everything," we often find that this method of advancement is one step forward and two steps back. If you look at the industry as a complex ecosystem, then some of these breakthroughs are like landfill for the wetlands. Sure, you can build the shopping mall, but what is the real cost? 24P is the emerging format for postproduction that combines the frame rate of film with the quality and convenience of progressive frames. There is flexibility to use 24P for a wide range of aspect ratios and frame sizes for television broadcast, and an accurate film cut list for theatrical distribution.

The beauty of 24P is that it is fast becoming the most important aspect of workflow for HD; however, it is not new. 24P is really an extension of the film work that has been going on for over 100 years and has a fully developed support system. Also of benefit to the Avid editor is that Avid was one of the first, the most successful, and the most recognized leaders in nonlinear editing in 24 frames per second. Although there is some new technology to get the progressive frames to work, the 24 fps part is an extension of Avid's existing Oscar™-winning Film Composer system.

Taking into account all the possible different types of HD, you can choose from 36 formats based on the tables published by the ATSC (Advanced Television Systems Committee) including all the frame rates. So which one will people really use? Already many networks have chosen to use a specific format that is somewhere in the middle of this range of choices.

If you had to pick one HD format that would be the easiest to convert into any or all of the others, you must take four important criteria into account: Consider image scan type, frame rate, frame size, and aspect ratio. What one format would convert each of these four items with the best possible quality into all other formats? If even one of these criteria is not adequate, then the final product will be rejected and the conversion will be useless.

Let's take a quick look at the issues surrounding each of these criteria. Remember that we are discussing origination formats, which may be very different from the final distribution format.

Scan Type

The requirement for standard-definition digital signals today is the ITU-R.bt.601 specification, commonly referred to as 601. This standard for digital images specifies an interlaced signal, and interlaced has been the standard for analog images as well. As we discussed in Chapter 6, the interlacing is a type of bandwidth and flicker reducer that was extremely important in the early days of television broadcast. Today, it causes many types of image degradation with computer systems that work in a progressive format. The interlaced image projects only a single field out of two at any one time. This is *half* the resolution of the image. Because of the persistence of vision and the latency of the RGB phosphors of the television screen, the two fields appear as a single frame.

Sony's HDCAM is capable of recording a form of 24P called PsF for Progressive segmented frame. This records two half-frames that are not interlaced.

It is not quite completely progressive or completely interlaced. Although both halves of the image should be identical, they will not play at the same time. This way you get the benefit of no intraframe motion but not the full quality of having all the frame information displayed at once.

There have always been problems with representing the film frame with interlaced images, especially with NTSC, so that you can have video frames with half of two different film frames. But much of our technology for editing, effects, and transmission assume interlaced frames, so any switch to progressive frames is a major undertaking. Having both fields onscreen at the same time with no temporal change between them will increase image resolution, which is what progressive scan promises. Beware of the claims of nonprofessional systems who are trying to break into the film and HD market by claiming they are progressive. If you look carefully, you will see that they are still throwing away one field and interpolating or duplicating the remaining field. If you want to convert from one video format to another, you will get a higher quality image when you go from progressive to interlaced than if you go the other way around. When you take progressive HD frame and convert it to SD, you are taking a high-resolution progressive frame and splitting it into a lower resolution, interlaced field. Because of the higher resolution, ease of frame-based effects (no jagged lines with motion), and interoperability with computer systems, if you are given a choice of scan type to be converted to all other formats, progressive is the better choice. This is the "P" part of 24P.

Frame Rate

You would think that the higher the frame rate, the higher the quality of the image would be, but this is only partially true. If you consider that the highest quality image most of us ever see is projected at 24 fps in the movie theater, then it is clear that there is more to quality than frame rate. There is some advantage to having a faster frame rate for sports coverage or other types of programming where motion effects are critical. Some broadcasters have decided to go with 60 progressive fps for very specialized needs. At this point, we are considering what would be the best universal choice for conversion during the transition to HD while still satisfying PAL and NTSC 601 broadcasts.

The main issues involved with choosing a frame rate have to do with downconverting from HD or cross converting between PAL and NTSC. Also important in the frame rate consideration is how accurate a cut list would be to cut film negative that has been edited on video and will eventually be projected in a theater. If you are making programs that have a wide distribution, the conversion from 29.97 fps (59.94 fields) to 25 fps (50 fields) has been expensive, time consuming, and never quite as good as starting the whole project over again in PAL. But most people live with the compromises as long as the resolution of the images remains in the standard-definition state.

If you want the best possible cross conversion, however, you need to take into account that the original PAL frame is larger than the NTSC counterpart (576 vs 486 vertical scan lines). Converting from a larger frame to a smaller frame will look better than the other way around when working with pixel-based media. This means that the absolute best starting point for multiple-format output would be to shoot or transfer film to PAL. However, the 25 to 24 frame rate conversion is not as clean as the ability to extract 24 real film frames from a 30 fps film to tape transfer. Even though the frame size is bigger and the frame rate is closer to film, it is still not as good as going back to film. The conversion of 24 fps to 25 fps is done by speeding up the film 4.1% and pitch correcting the audio. The continuing trend will be to transfer film to 24P 1920 × 1080 (HD) and downconvert to either PAL or NTSC.

As illustrated by the NTSC-to-PAL conversion issues, going from a high frame rate to a low one means blending or dropping extra frames. The best answer would be a lower frame rate that can be stretched with redundant fields from one format to another. Since the conversion from 24 fps to 30 fps or 25 fps is already commonly accepted, why not choose 24 as the universal origination standard? This also has the important side effect of being perfect for feature films that need theatrical release as one of the deliverable requirements. Starting at 24 fps means the possibility of a perfect film cut list using an Avid finishing system or Avid FilmScribe™. Finally, because the conversion from 24 fps to everything else is done to the final sequence, you can be guaranteed that the 2:3 pulldown in NTSC is consistent from start to finish. This becomes extremely important for high-quality, problem-free MPEG encoding. Since most network and cable broadcasts go through some stage of compression during transmission, this is an important time and money saver for the rest of the production process. Now we have the "24" of 24P.

Aspect Ratio

If you could just shoot everything on film or 24 fps HD, then all your frame rate problems would be gone and distribution would be easy, right? Don't you wish it were that simple? Not so fast! There are still two more important criteria to consider. The next is aspect ratio.

Standard definition television aspect ratio is 4:3, sometimes referred to as 1.33:1. The ATSC decided on the most common alternative format of 16:9 for HD. This format was already being used in Europe and Asia for standard definition digital betacam and is currently broadcast in an anamorphic format today.

When you shoot standard definition television (SD) with 16:9, you are still restricted to the standard 4:3 frame size. The camera squeezes the image as it is being shot or transferred to tape from film (see Fig 12.15). This is called 16:9 anamorphic (squished, skinny, and funny looking) to distinguish it from 16:9 letterbox (black bars on the top and bottom, smaller actual image) or true 16:9 for

Figure 12.15   16:9 Anamorphic Image Squeezed to 4:3

HD. The anamorphic image is then unsqueezed during editing and playback as a choice for monitoring (see Fig. 12.16). Those with the dual-aspect ratio monitors need to switch over the 16:9 mode to see the image unsqueezed. As you go out to tape for the final master, you output it in the original squeezed format for copying and broadcasting as 16:9 anamorphic.

If a program is shot in 16:9, many times the center action is "protected" for 4:3 or (just to confuse things more) 14:9. Protecting for these different formats means that the director of photography must make sure that all of the important action happens in the center of the frame. If the 16:9 master is broadcast as 4:3, then you don't want to piece together the story from sound effects as all the major action happens just outside the frame of your television set. Nor do you want to watch a conversation between two noses. 14:9 is a compromise used to minimize letterboxing when an image is shrunk (but not squeezed) to 4:3. On the other hand, the director of photography, having lots of horizontal area for compositions, needs to fill it with something!

Television programming is still primarily 4:3 and will become 16:9 as more people buy home television sets that can switch between the two formats. A 16:9 television playing a 4:3 broadcast will have lots of empty black space on both sides (see Fig. 12.17). This is being referred to as "pillarbox." It may be set incorrectly to play 4:3 as 16:9 and stretch the image horizontally. This is very unflattering and generally considered a mistake except by the salespeople at the electronics store. Some DVD movies offer a 16:9 version for those who have the

Figure 12.16 16:9 Anamorphic Image Displayed Correctly

right monitors, but in North America today there is precious little 16:9 programming unless it is high definition.

The Avid editing systems can switch over to 16:9 by checking that mode in the Composer Setting or by right-clicking or Shift-Ctrl-clicking on the Source/Record monitors. All titles, effects, and imported graphics need to be in the 16:9 ratio as well. The Avid system will create the proper-sized titles when you create them in the Title Tool while in the 16:9 mode. If you forget to do this or change your mind later, you must recreate the title media while in the proper mode. The Universal output features in Symphony have a slightly different approach and allow you to create both formats at the same time.

Effects are already created perfectly in 16:9 when you are in that mode with Avid editing systems. This becomes critical when you have circle or diagonal wipes. Keyframing of DVE becomes slightly different as well because you are composing the graphic positioning based on a different use of horizontal space. Be sure any compositing program you use for animations and advanced special effects can handle the correct 16:9 aspect ratio. You don't want to stretch or crop a 4:3 composition to fit!

All graphics must be created in the right aspect ratio when working in Photoshop or other graphics creation programs. Refer back to Chapter 6 for the proper methods for creating 16:9 graphics and importing them.

So if 16:9 is the new emerging standard and it is already possible to edit and finish 16:9 programs on the Avid editing systems, then how do you output 4:3? Again, consider that two of the most common answers, center crop and letterbox, are quickly becoming unacceptable to most viewers. The center crop would be to unsqueeze the 16:9 image and chop off either side in order to

Figure 12.17   4:3 Area in 16:9

output as a standard 4:3. This is clearly not optimal for taking advantage of the original composition. This is why in Europe they protect for 14:9 because they figure they can actually broadcast to a public of mostly 4:3 television sets with a little bit of black letterboxing. They can also broadcast 16:9 without making any changes. This is a workable compromise if (and this is a big if) all of your material was protected for 14:9 during the shooting. This would not apply to theatrical feature releases that are broadcast. It is still not great for graphics and titles. Letterboxing may be required as a deliverable, but it is still not an optimal use of a 4:3 aspect ratio. If the original composition is that critical, then 16:9 will not be good enough either. Most people still don't buy that a smaller image with big black bars is progress. But widescreen films of 2.35 aspect ratio will have less letterboxing when viewed on a 16:9 monitor, thus using more pixels for better picture resolution than viewing 2.35 letterboxed within a 4:3 frame.

The best answer is a process called pan and scan. Up until recently, pan and scan was a very expensive session where the director sat in a telecine facility and chose the framing. Shots were moved and repositioned on a scene-by-scene basis. It could also be done in a tape-based online suite using a DVE, but this would still be very costly to keep the quality acceptable. No one likes to make creative framing decisions when the clock is running *that* fast. The expense and time involved meant that this technique was out of the question for many productions.

With the release of the Universal output options on Avid systems, pan and scan became a very simple, dedicated process using a high-quality, realtime resizer chip. By using the Universal Offline option on Media Composer, the director can sit in an offline Avid suite and choose all the framing and reposition moves at an offline rate (see Fig. 12.18). Once all those decisions are made, they will translate

Figure 12.18   Pan and Scan for 4 × 3 Delivery of a 16 × 9 Master

easily and completely to a Symphony system or DS Nitris for uncompressed finishing. The pan and scan can still be adjusted at the finishing stage if time is short and people will not make important decisions until the very last minute.

With the Avid implementation of pan and scan, the proper-sized grid is displayed over the larger original frame and the editor can keyframe repositioning scene by scene. The editors can also create their own pans to include vital parts of the scene when the point of interest shifts with the actor's blocking.

As broadcast moves toward a 16:9 aspect ratio, you will need to intercut legacy material that exists only as 4:3 with new material that is 16:9. Now you can use the pan-and-scan tool to correct just the 4:3 material so that it may intercut with the 16:9. In general people try to preserve a "common top," where the top of the 16:9 image is the same as the 4:3 image. However, in order to frame an image properly, occasionally you may have to create vertical reposition or tilt effect. This process has been quite common in Europe for several years.

Frame Size

Frame size is another problem in dealing with HD. The problem is really rather simple: bandwidth. HD takes almost six times the bandwidth of SD. This means all existing systems designed for SD are inadequate and need to be significantly upgraded to handle all the items we have come to expect from an editing system. We are discussing only a single stream of HD and no audio! Imagine realtime effects and eight channels of digital audio with rubberbanded levels, DSK titles, and color correction! This dilemma is what spurred the design of the Nitris DNA hardware to replace Meridien.

Let's take a step back and look at the big picture. You will need to deliver HD at some point in the near future. For a chosen few, it is required today. For most, it will be in two or three years and, for others, perhaps even farther into the future. You can plan, archive, and shoot for it, but you are probably actually required to deliver an SD master first and certainly many variations including 16:9, 4:3, DVD or streaming video.

The DS Nitris system was designed specifically to finish Avid projects with lots of layers and effects at HD resolutions. It was also designed to take in a wide range of frame sizes and frame rates. There is a full-time, high-quality resizer on the output of the Nitris DNA so that you can output every HD production as either letterbox, anamorphic, or center crop. The system is also capable of realtime pan and scan. The Nitris DNA is Avid's answer to the issues surrounding the four issues of 24P.

CONCLUSION

The marketplace is changing very quickly in film and television. Formats are proliferating and standards are evolving. For those who need to stay at the fore-front of the professional nonlinear editing market, it is important to understand what these changes mean to you and for companies like Avid to continue to innovate. Keeping an eye to the future, you can take advantage of Avid's leadership role and maintain the competitive edge in your market.

13

Troubleshooting

Troubleshooting is a detective story, an Easter egg hunt, and a test of method, memory, and patience. It is rarely an excuse for panic, since it is only by calmly tracing problems to their root that you will ever achieve the solution. The end result is to get back up and running to finish the job and complete the vision. You cannot separate the medium from the tools, even if you prefer to think of the Avid as a cloud of magic that exists at the far end of the cables.

BASIC TROUBLESHOOTING PHILOSOPHY

You will never know how a system works until you know how it breaks. This means you need to understand signal flow and basic connections. When a part of your toolset is missing you need to know where it comes from and be familiar with the points of failure. For instance, if you are trying to capture video, but receive nothing but black you need to work your way backward to each stage that might be wrong. Is the capture tool set right? Is the video source type correct? Is the firewire cable to the computer connected? Is the Avid hardware turned on? Is the correct video cable connected from the deck to the Avid hardware? Is it connected to the outputs of the deck? Is the cable bad or not fully connected? What is the signal supposed to be on the tape? You get the idea: Trace backward to isolate the single point of failure in the signal flow.

As you form the hypothesis of the area that should be investigated (i.e., video flow from the deck), be sure to check only one area of suspicion at a time. This means you have two approaches depending on the nature of the problem. In the case of a loss of video, you could use a subtractive procedure of checking one area before moving on to the next, replacing a problematic cable or trying another setting (like trying different video monitor inputs). Eliminate a potential failure point and move on to the next.

Or you may want a more additive approach where you disconnect everything attached to the computer and add them back one at a time. This additive method would apply to problems with drives, networks, and other peripherals.

It is also useful if you can't boot the system at all. Attach a peripheral and reboot. If successful then add another peripheral and reboot again. If changing something doesn't fix the problem, then go back to the original state before you move on—there is no need to complicate the search by eating the breadcrumbs back to where you started.

Keep track of every piece of software that is added to the system and when it was added. If someone has added software that loaded a driver (or another innocent piece of secret software) and now the Avid software won't launch, you should know what to uninstall. Advanced troubleshooters should know how to get into the Windows registry to delete recalcitrant .dll's, but for now just be knowledgeable about uninstalling software and launching to the Last Known Good Configuration.

There are actually some very easy ways to get back to work without much effort. First, if you encounter an error message, but you're allowed to continue, then quickly save and keep going. If there is odd behavior after the error, you may be better off either quitting and relaunching the application or rebooting the machine. If you are on Windows then check the event viewer. Right-click on My Computer, choose Manage, and go the Event Viewer. You will see if there have been numerous errors in a particular area that may focus your troubleshooting.

RTFM

There are many everyday situations where just a little knowledge of troubleshooting can keep you going forward, give you a bit more confidence, and, maybe, help keep your job. The first thing that can help you instantly is to read the Release Notes. This falls under the category of RTFM (Read The Manual, Please), but some people skip it because they just want to be up and running with the newest version no matter what. Stop, smell the roses, and read the known bugs. Avid is pretty good about listing what they consider to be the bugs you need to know. Your definition of a bug and their definition may differ, but you will definitely benefit by seeing that, for instance, one small part of something you need to do all the time doesn't work under certain conditions. It also helps to know if that procedure has been replaced by something faster, better, and simpler. Even if you haven't memorized the Release Notes, they can generate a little thought bubble over your head if something seems familiar. So don't toss the Release Notes; keep them handy and even scan them quickly before you call Avid Customer Support. Over time the base of Avid users has migrated to a larger percentage using Windows-based systems and now the balance is about 50/50 with the Apple systems. Current Avid systems are shipping on OS X and Windows XP, but there are plenty of users still on Windows NT, Windows 2000, and Macintosh OS 9.x and earlier. For all of these versions there are still some universal basics for troubleshooting.

Customer support has improved significantly over Avid's early days of explosive growth into a new industry and Avid takes support very seriously, but for you, that's not really the point. There is time involved in figuring out that you have a problem, realizing you don't know how to fix it, telling the other people in the room that maybe they had better get some coffee, and then dialing for help. Better to say something like, "Hmmm, did you know you are missing the active SCSI terminator?" and subtly imply, "Aren't you glad you hired me?" However friendly, competent, knowledgeable, and good-looking Avid support may be, you want to avoid talking to them until you have a serious problem.

TECHNIQUES FOR ISOLATING HARDWARE FROM SOFTWARE

Part of the trick of troubleshooting is to determine, before you go very far, whether the problem originates with the hardware or the software. If you have just installed new hardware (like RAM, or replaced an old board with a new board) suspect that first. Image distortion, with Meridien or ABVB systems, was typically hardware failing instead of software (though of course there's always the exception). No hardware can function unless it has the correct software to interface with the operating system. So it might not be the hardware itself that has a problem, but the driver for the hardware. This can be a device driver, and Avid uses .sys files for device drivers in Windows. Good examples of Avid .sys files are the APMPHXNTDRIVER.sys (Meridien Digital Media Board) and Pgenie.sys (3D DVE), which are in Windows winnt/drivers/system32/drivers. For OS X, Avid uses kernel extensions (even though they are not technically drivers, for the most part they are the same as a driver). The phoenix.kext (Meridien Digital Media Board) and zanskar.kext (3D DVE) are for OS X and can be found in System\Library\Extensions. These drivers should be installed in standard updater software as soon as the hardware is in place, but if you are replacing a board you may have to load a new driver. Hopefully, this little bit of software came with the new board and will install itself with a simple double-click and a restart. Don't pull out a "defective" new board until you have installed any new software updates that may have been included.

HARDWARE PROBLEMS

There are four major areas of hardware problems with Avid systems: PCI boards, drives, monitors, and keyboards/mice. All these areas include good cables and connectors that go bad, and they can all go bad if yanked hard enough. You can isolate drive problems by shutting down or disconnecting the drives. You can isolate monitor problems by switching the cables to another monitor. Unless you have spilled sugary brown liquids on your keyboard or mouse, then cables and drivers are guilty until proven innocent.

PCI board problems can be diagnosed by running the Avid System Test or Avid System Test Pro. This is a utility that ships with Meridien systems to isolate a problematic PCI board. Most boards need to be installed in very specific slots, and that map of the system should be inside the case to the computer. Follow this map or get lost at your own peril. The System Test may indicate that a bad board may need a driver loaded for the correct board version or that the board needs to be reseated. But it also could mean that the board is bad and needs to be replaced.

If you are very careful and not violating any warranties, your problem might be solved by shutting down the system and reseating the board (perhaps more than once). Put on a static discharge wrist strap, open the computer case, unscrew or unsnap the holding device for PCI boards, and gently pull the board out of its PCI slot. Then gently push it back in and power up the system to run the Avid System Test again. This may be enough to force the board to spring back to life, but should be attempted only as a last resort before replacing the board. Be aware that if you have an extra board lying about, it must be the same version number of the original board. You may make the problems worse by installing mismatched versions of hardware!

If your image has gone green or you are otherwise having problems with 3D effects on a Meridien system you can disable the 3D effects PCI board to see if that solves the problem. Power up the system while holding down the F and X keys. Answer OK to the question about disabling the 3D effects. The board will be reenabled the next time you reboot. If you find this solves the problem you can more permanently disable the effects until another board can replace it. Go into the Console (Ctl/Cmd-6) and type disable 3d. The system will stay disabled until you type enable 3d or load a new version of the Avid software. Let someone else know you have done this.

On the Mojo and Adrenaline systems there are no PCI boards associated with video I/O or playback. There are only SCSI, Fiber, and Firewire boards that connect the media drives. There is an OGL card in the monitor display slot (AGP) and it is responsible for monitors and DVE. This should simplify your troubleshooting adventures wonderfully.

SOFTWARE PROBLEMS

There are two major areas where you could have problems in the Avid application, the project files, bins, and User Settings or the media itself. Try to separate problems with the media playback from problems with the media itself. Media can always be recaptured, but I'll bet you don't have a few extra video cards hanging around. The best way to tell if a particular problem rests with the media is to take a close look at it. Step through the problem areas frame-by-frame. That will show if the problem is there when the drives are not working so hard. Any corrupt images or crazy colors that are visible when you are looking at a still frame can be solved by recapturing that shot. If you can't see the problem, you must go one step deeper.

Is the problem playback related? See if the sequence plays back without any of the media online. Go to the Macintosh Finder level and dismount your drives. On the Mac you can do this by dragging them to the trash (the drive icons, not the folders inside). On all systems you can change the name of the folder on the media drives from 5.x, 6.x, or OMFI MediaFiles to anything else. I just add an x to the end of the name so I don't get into trouble trying to spell the correct name of the folder again when I am in a hurry. As discussed previously, any change to the name of the OMFI MediaFiles folder causes all the media inside to go offline. Now does the sequence play? If there are no errors, you know that the problem is related to the media, the drive, or the SCSI/Fiber board. If you continue to get obscure errors (my favorite is BadMagic), then you need to look at the media databases, sequence, or the media itself.

Media databases are small files in every OMFI MediaFiles folder. As mentioned in Chapter 4, currently they are called msmFMID and msmMMOB on Windows and msmOMFI.mdb and msmMac.pmr on Mac. Sometimes these files become corrupted and do not update correctly. This may result in files not appearing in the Media Tool or media appearing offline when you know for sure the media is there. You must force the Media Databases to be updated and you can do this by forcing the update with the pulldown function Refresh Media Directories. If this doesn't work you may be forced to delete or move the media database files to another folder. This will force a recreation of the file from scratch. Moving or deleting the media database file will force a rescan of all of your media drives, which may take a few minutes, depending on how much media you have. After the rescan and recreating of the media database files, you may have successfully herded all of your lost media back online.

If the sequence or the media is the problem then you must practice Divide and Conquer to find the offending media file or element in the sequence. If you suspect that it is bad media, not just the drive, then bring all the media back online before using this technique. Then split the sequence in half and try to play the first half. If it plays then try to play the second half. If you have problems with the second half then divide that in two and repeat the procedure. Continue this until you isolate the area of the sequence that is corrupt. It may be a single graphic, master clip, or effect. Delete the effect, replace the graphic, and either extract or overwrite the master clip. Consider deleting related precomputes and rerendering. Recapture the master clip to see if it is just that media. If you do not have access to the original tapes or files, consolidate the media to another drive and see if it plays. If the problem remains, you may have to relog and capture the master clip to eliminate the problem.

There may be something corrupt in your bins, your project, or your User Settings. This is so easy to test that many times it is one of the first things an Avid Support Representative asks you to do. Generally, I suspect this particular problem when something that has been working fine all day stops working or features that should be available are suddenly not there. Create a new bin or project and drag the clips and the sequence you were working on into the new bins.

The next step is to remake your User Settings. This should be a fast check to see if creating a new default User Settings will clear up the problem. If it does, then spend the time to recreate your settings. Better yet, call up an archived version of your User Settings that has been hidden where no one can get to it. You should always have a backup of your User Settings, and as a free-lancer you will want to carry them with you. This User Setting may also be corrupted, but chances are pretty good that it is not. Again, make sure you have new User Settings for every major change in the software. If you call up a really old User Setting, it may be incompatible with new menus and functions in subtle, but important ways.

If the User Setting is causing the problem, delete the old one from the User Settings folder. Do not throw away the AvidDefaultPrefs in the Settings folder since this file holds all the standard bin headings and will not be recreated automatically. Make sure you know what files are in the Site Settings before you delete those; the files may contain vital standards settings that are required for every project. You can cover yourself by moving these files into a temporary folder rather than deleting them. If moving them does not solve the problem, move them back.

AUDIO PROBLEMS

In general, if you have an audio playback problem check the audio meters, check the sample rate, then check the cabling. You may solve the problem by powering down the computer and cycling the power on the Avid audio hardware. Check obvious issues like mixer power or speaker power. Trace the audio flow through the whole system, making sure you have a signal at the beginning, and go step-by-step to the end.

Quickly check audio meters by using the Ctl/Cmd-1 to call up the Audio Tool and playing the media. If there are no levels at all then the audio media may be offline or at a different sample rate than the Audio Project Setting. You may want to click on the PH button in the Audio Tool and choose Play Calibration Tone. If you can see the levels, but can't hear the tone, then you have a cabling problem with output. Turn the volume down before you try this method—it will be greatly appreciated by everyone else in the suite.

A sample rate is the amount of audio samples that are played back per second to reproduce digital sound. There are four common sample rates in video editing today: 32 kHz, 44.1 kHz, 48 kHz, and 96 kHz. There are higher sample rates but they are used mostly for high-end digital audio workstations like Pro-Tools HD. Sample rate mixing in ABVB system was a problem never addressed until Meridien systems. You can't play back multiple sample rates in the same ABVB project and you can't convert from one sample rate to the other within the Avid software. Both of these issues are solved in Meridien-based systems and later. Use SoundAppPC for the Macintosh, a freeware utility that can be used to convert audio sample rates for ABVB systems, or recapture the audio at the correct sample rate.

To track either problem, isolate the sequence into its own bin and through Set Bin Display, choose Show Reference Clips. Choose bin headings Offline and Audio Sample Size, then sort (Ctl/Cmd-E) to find which clips have audio offline and which clips are at the wrong sample rate. With a Meridien system or later you can convert the individual clips to the proper sample rate using Change Sample rate under the Bin menu. Alternately you can go to the Audio Project settings and choose Show Mismatched Sample Rates as Different Colors and look at the sequence. If you have different colors in the timeline then you should consider converting the sample rates or changing the Audio Project setting Convert Sample Rates When Playing to Always.

You may experience clicks when editing some digital audio because the edit cuts in the middle of an audio sample. This might be especially prevalent on imported audio CDs. You can fix this quickly by adding two frame dissolves between the cuts.

An ABVB audio problem can occur when you are working along and suddenly the audio starts to sound distorted. If you have heard this audio before, you know it was not captured at levels that were too high. If you call up the Audio Tool, you can make sure the captured levels do not go into the red area of the audio meter. It helps to isolate this problem if you have captured this material yourself and you know that the sound was also recorded correctly in the field. If you have access to the original tapes with Media Composer or Symphony, use the Find Frame button. After pressing Find Frame, the system asks you to insert the original tape and cues up to the exact same frame you are viewing in the source or record window.

If the audio levels are good on the tape and have been captured correctly, but still sound distorted when you play them back in the software, you may have a corrupt file called the DigiSetup. This little file is created automatically, is always there, and sometimes, when it is being updated, gets corrupted. The best thing to do is just to track it down and delete it. DigiSetup is in the Macintosh System folder. You can find it with Find File or by opening the System folder and pressing the letter D. Later versions of the editing software make one for you automatically when you launch the application the next time. Otherwise, you need to launch a utility called Pro Tools Setup and click OK. Another sign that DigiSetup is corrupted is if you can play audio normally for a few seconds, then it just stops or goes silent, yet if you stop and play that section again, there is audio. If the distorted audio is also slightly slowed down, like listening at 95 percent of speed, trashing DigiSetup is always the answer. If all this does not solve your distortion problems, you may need to look at the audio board itself.

ERROR MESSAGES

If you are not used to working on complex professional software, you may not be used to generating error messages. Write down the ones you see and, if they do not keep you from continuing, call Avid Support after the session and get the official explanation. Saying, "I got an error" isn't enough to let Avid Support help you figure out the problem. It may be something systemic or it may be operator error. But if it is a Fatal Error, you should be on the phone immediately if you don't know what caused it. After working on the system for a while, you will learn what causes the most common errors, and if you keep a log of when they occur and to whom, you can identify the pattern that applies to your own system, facility, and way of working.

The real trick to error messages is deciphering them. They may be colorful but essentially meaningless until you figure out what the function is that has gone wrong or what kind of pattern they follow. A section of the programming code always generates a specific error message when something goes wrong. Essentially, error messages tell you what happened, not why. This is because the same error could have been caused for 10 different reasons. The computer cannot look outside of itself and say, "You have too many unrendered effects at 2:1 and the wrong terminator on Media drive 4A and that last sound effect on audio track 6 just put me over the edge!" It will say "Audio Underrun" because what happened was that it could not continue to play every frame of audio and video through that segment of the sequence. Many times this is the only explanation that the computer can confidently produce for why it cannot play.

THE IMPORTANCE OF CONNECTIONS

Let's look at the basic world of connections. You may have enjoyed playing air guitar to Molly Hatchet, but air SCSI doesn't work as well unless it's connected to something and connected correctly. Connections are one of the first things to look at, especially if you have just moved the system. "I just moved my monitor and now it is broken," will lead most support reps to check your cables to the monitor.

Many of the connections to the computer have electricity running through them, and connecting anything "hot" can cause a component to burn out. A good rule of thumb is not to change connections while the computer is running (not including audio or video to an external source like a deck).

Cables that are screwed in tight don't come loose so quickly. This seems obvious, but many facilities decide it is easier to have them not screwed in so anyone without a Phillips head screwdriver can move things and make changes quickly. Do yourself a favor and buy a screwdriver with one end Phillips head and the other end flat. Tighten everything on the back of the computer and anywhere else you can tighten things down. Of course, if you yank really hard, you still have problems because now you have loosened the computer board it was attached to or even ripped the wiring out of the connector!

Make sure that things are fastened down away from big feet and spilling coffee, but don't pack them away so tightly that you can't get at them to take a look. Get a few bags of little plastic tie wraps from your favorite electronics store and wrap cables together in logical groups. You have to cut the tie wrap to pull them apart at some point, but better to discourage the unauthorized, and you can always wrap them again when you are done. The cost compared to downtime is negligible.

Good engineers leave a little slack in all their carefully tie-wrapped suites so that any piece of equipment can slide forward enough to see around back. Large, easy-to-read labels that are also easy to understand make any phone call to Customer Support less embarrassing. "Is the video input connected?" "Uh, you mean cable VI649?" If you need to call your own video engineers, at least you will be sure that the problem is not software related. Also contemplate moving everything in the Avid suite a few feet away from the wall and mounting a small, clip-on lamp back there. This gives the system some more air if the room tends to get warm and will give you easy access.

MONITOR CONNECTIONS

To the chagrin of some editors of a certain age, most people are using high-resolution monitors these days. Yes, it was easier in the past to see basic icons and functions when they take up most of a 20-inch monitor, but you also spend more time arranging your views if you perform many different tasks quickly. More open bins (if they each contain a small amount of well-organized material) keep you speeding along under certain projects. But more interesting is that when you are capturing, you can have the waveform/vectorscope, the audio meters, several bins, and the console all open at once. On the other monitor of the Media Composer, the Record monitor, you can get a huge amount of real estate for your timeline with high resolution monitors. And there are ways to increase the size of text and color code buttons so they are easier to read (see Chapter 3).

There are two common problems with monitors. The first is getting the RGB and sync cables connected wrong in the back if you use the older Mitsubishi monitors. How can you tell? All your colors will be horribly wrong.

The monitor may be completely black if the sync cable is not connected correctly. Newer monitors have a simpler VGA single connection.

The second problem occurs on monitors that allow you to loop signals through to another device. If set incorrectly, your image will be washed out and way too bright. Many people foolishly live with this because they didn't expect the signal to look great in the first place! They accommodate it by cranking down the brightness controls on the front of the monitor. To see this, look at the white type on the splash screens as you launch the software. If the letters are a little too bright, check the termination. Look in back of the monitor, near the input connections, for the termination switch. It has an icon that says 75 Ω, which stands for 75 ohms. Make sure your monitor is terminated since it is very easy to whack that little 75 Ω switch while moving or unpacking. One quick check may save you hours of work if you are doing much color correction to the project.

ADB AND USB CABLES

One of the silliest problems that seems to happen often is with the keyboard cables. USB is the newest version, but older systems use serial (PC) or ADB (Mac) cables. You don't want to have too many USB devices connected at once and be careful about using USB extenders. This is a good place to look if peripherals are acting strangely. First, let me caution everyone who is thinking of pulling this harmless looking ADB cable and reconnecting it while the system is running. You may have done this a hundred times, but the next time you do it, you can fry your keyboard or even your Mac motherboard. Any cable that has power running through it has the capability to create a power surge or damage sensitive electronic parts by "hot-swapping" or changing the connection while the power is running through it. The ADB connects directly to the motherboard and the connection to the board is not replaceable. If that ADB connection fries, you need a new computer. If you don't know which cables have power and which do not, change all connections that are not standard video and audio only when the system is shut down. I know this is overkill, but simple and safe is a general policy. The ADB cables definitely have power running through them and on PC systems, don't attempt to hot-swap mice or keyboards either; USB cables are designed for hot-swapping so this is less of an issue. Unfortunately, all keyboard and peripheral computer cables are always about an inch or two too short and so are under a certain amount of tension all the time. Many times just a slight pull is all it takes and your system appears to crash. You try all the keyboard reboot commands and nothing works and finally you go to the CPU itself to press the reset switch. When everything comes back online, surprise! You still don't have control because what really happened was the keyboard cable had come just slightly loose.

SCSI CONNECTIONS

ADB or USB connection is basic and straightforward compared to the scariest of all computer connections—SCSI. The term "SCSI Voodoo" may not be completely foreign to you and for a good reason. Even though you may follow all of the complicated rules of SCSI or simplify your system to minimize them, you may still encounter situations that just don't make sense.

Basic SCSI Rules

There are some basic rules to keep in mind no matter what your SCSI configuration is. If you want peak performance and as many streams of realtime as possible, you should consider still using SCSI drives. In many cases fiber drives will be sufficient, but firewire drives will not be much good for the higher resolutions and full amount of advertised realtime streams. Firewire drives get faster and larger all the time, so check the specifications to make sure you can get the throughput you need. If you are cutting long form, single, or dual stream projects like feature films and documentaries, you may find the firewire drives sufficient. For everything else, SCSI and fiber are recommended and for uncompressed high definition, SCSI is still the best. Of course, when we see 10Gbps fibre switches all bets are off!

On older Macintosh systems there was an internal SCSI card with every system. It was really meant for things other than video streams. Connect up graphics, audio and backup drives, removable media, scanners, or the ancient external 3D effects "pizza box" (the Pinnacle Aladdin) to this internal SCSI connector. The real power was in the SCSI accelerator card that Avid included to connect the media drives. Most high-end PCs have an internal SCSI connector that is quite fast and may be enough for your standard definition needs. If you have another card, however, you can add more drives to the system. And if you use a dual port SCSI card you can get even more speed from your drives through striping (which we will cover later). Keep in mind, the more things you have connected at once, the more complicated your SCSI troubleshooting will be.

Keep the length of your SCSI connections as short as possible. The cables that ship with your drives are meant to be that short because anything longer than the maximum length of normal SCSI chain causes serious voodoo behavior. There are many supported types of SCSI drives: Classic SCSI-1, Fast and Wide SCSI-2, Ultra SCSI-3, and Ultra2 LVD (a variation of SCSI-3). Each type uses different cables and different cable lengths. So rather than try to outguess the manufacturer, stick to the length that comes with the system. The length of the cable used for the entire SCSI chain must take into account the length of cable inside each drive. That can add up pretty fast at over a foot per drive. A common mistake is to take one look at the length of the cables that came with the drives and rush out to the electronics store and get the longest cable you can find. Avoid the temptation to use these and keep the cables short.

Make sure all your SCSI cables are of the same brand, type, and style—different cables may have different internal configurations that may cause some devices to just never work right. This is why it is so important when mixing different types of drives to get all the cabling and termination correct, as we will cover later.

SCSI cables are very sensitive to twisting and must be handled more carefully than any other cable on the system. Just bending a SCSI cable back and forth a few times can significantly reduce its functionality. Don't strain, kick, stomp, or bite them.

Finally, turning drives on and off has its own set of rules. Make sure all SCSI devices are turned on and have clearly come up to speed before powering on the CPU. Listen for each drive to make a single "click" sound when it has finished spinning up. Keep all SCSI devices on if they are connected in the chain to ensure consistent behavior. Turn all peripherals off only after the computer has shut down.

There are lots of complicated rules about SCSI, especially when you are combining the newer ultra drives with the older narrow drives. These four terms, *ultra*, *fast*, *wide*, and *narrow*, refer to the speed and capabilities of passing larger amounts of data through the SCSI chain. The drives themselves don't look all that different.

Drive Striping

You can get better performance from any drive by combining it with other drives in a striped set. Drive striping is a way to make many drives act as one large, fast drive. There are five main types of drive striping, but the two to remember for Avid systems are RAID 0 and RAID 1. RAID 0 (Redundant Array of Independent Disks) is what most people mean when they stripe drives together for more realtime effects. You use the drive striping software that comes with the Avid, mount all the drives, and stripe them into one large drive with multiple read/write heads and thus faster seek time to find or write media. This process will erase all the media on the drives so do it only when they are new out of the box or when you have backed up all important material. You can even stripe drives across multiple SCSI connections if you have a dual channel SCSI PCI board. If you have four newer drives and a dual-channel SCSI board, you can create what is called four-way striping with two drives on each SCSI channel. It is highly recommended to do this on the Adrenaline systems for a maximum of realtime effects based on many streams of video. Four-way striping is a good way to continue using older, slower drives too, but there is a drawback. If one drive fails, you will lose all the data on the other three drives, too.

RAID 1 is also called drive mirroring and is used in Avid Unity MediaNet-work. This means that you have a duplicate drive for every drive full of media and the duplicate is created during capturing. You can decide to turn this on or off on Unity depending on your needs and the importance of the project. If you choose to use it then you may not even notice when a drive fails. You will con-tinue to keep editing and can replace the failed drive when you have some time to spare. Avid is working on systems that fail even more elegantly, without duplicating drives. These systems will eventually heal themselves by spreading redundant data across all drives so that when a drive fails it can be recreated on a spare in the background without effect on performance.

SCSI-D

There is yet another kind of SCSI 2 on older Macintosh editing systems: SCSI-D or differential SCSI. Differential SCSI is no longer a shipping configuration, but you may find it on older systems, especially Film Composers with SCSI MediaShare. The differential SCSI is for drive towers and storage expanders that allow you to connect many more drives than either the standard seven or eight and are capable of the much longer cabling lengths of 75 feet (compared to 18 feet for single-ended SCSI). The important thing about a differential connection is that it has a noncompatible connector to keep you from accidentally connect-ing regular drives and burning them out because of the difference in the amount of power flowing through the cables. Of course, you can still flip the connector and jam it on if you really try! You might not actually see a puff of smoke, but the effect will be the same. Don't confuse MediaDocks with drive towers. MediaDocks do not require the differential ATTO card or the differential cables.

Termination

All SCSI chains must be terminated. Termination is the way the computer knows where the end of the chain is and which drive is the last device. It keeps the signal from bouncing back to confuse the computer with false signals. Tech-nically, the chain must be terminated at the beginning and at the end, but on most CPUs (all the newer ones and all the ATTO cards), the termination at the beginning is internal. This means that you need to attach the terminator that comes with the drives to the last drive in the chain. Don't worry about termina-tion at the beginning of the SCSI chain (although occasionally that internal ter-mination can fail, too). Always be sure to use an active terminator, the purple terminator that came with the narrow drives or the blue one that came with the termination kit for the wide drives. With the new LVD drives the terminator is beige and has an LED that indicates whether it is in the LVD mode (green) or single-ended (amber). We'll discuss the difference when we deal with mixing drive types. Don't use the generic gray terminators from the electronics store—they are generally not active.

SCSI ID

The most basic SCSI rule is that the SCSI ID, the number associated with this drive in the chain, must be unique to the chain. This usually becomes a problem when you add a new drive to the SCSI chain or bring a drive over from another system. Always check to see that the new drive does not have the same SCSI ID number as another drive already on the system. The SCSI ID is set on the back of standard drives with a pen or other pointed object; on MediaDocks it is set on the front panel. The fact that this fundamentally important piece of information is a small number on the back of the drive again points out the importance of having access to the equipment after it is installed. How do you know which numbers are being used unless you can stick your head back there with a flashlight and read them upside down and backwards? The best way may actually be another piece of software—the Avid Drive Utility (ADU) on the Mac or the Computer Manager on Windows XP. These utilities show you the correct SCSI ID even if the ID number on the back of the drive is broken and displaying a wrong number! ADU has a facility to flash the lights on the front of the drives so if you physically label the drives, this can help when troubleshooting. You can also cause the drive lights to flash by copying a small file to that drive or striped set. Of course, you need to be able to boot your system with the drives attached in order to use this software. The new LVD drives finally have the SCSI ID on the front of the drive.

Getting the wrong SCSI ID may cause the system to not boot correctly or even to damage data. On a Macintosh system, the desktop may also indicate that you have many more drives connected than you really do. When there is a problem with the SCSI chain on startup with the older Macs, you may get the flashing question mark icon. This is because the system cannot find the startup drive with the System folder on the SCSI chain. You have somehow confused the SCSI chain that has the internal Mac startup drive.

Take into consideration that some devices come already terminated internally and must always go at the end of the chain, like the "pizza box" external 3D DVE (the Aladdin was discontinued many years ago, but still pops up occasionally as a SCSI problem). If you have two such devices that terminate internally, they are going to fight with each other until you eliminate one or figure out how to unterminate it. Always shut down the system first if you are having a problem with SCSI IDs. Do not try to change the SCSI ID while the system is running!

With narrow drives, you can use only the numbers 0 through 6 for a SCSI ID. A wide drive can use 0 through 6 and 8 through 15. Never, ever use SCSI ID 7 on any SCSI device since this is the number used by the SCSI card itself (which is also technically a SCSI device) or by the host computer. On the older, internal Macintosh SCSI chain, avoid SCSI ID 0 because that is the ID used by some internal system drives. Internal CD-ROMs generally are set at the factory to use ID 3, so when attaching devices to the Macintosh SCSI chain you should avoid SCSI ID 3 as well. Scanners and Iomega Zip drives may use SCSI ID 5 or 6. When you are adding and subtracting drives from any system, the ID is the most important factor in making sure the drives work happily together.

Most new computers use the IDE internal drives and IDE connectors for CD-ROM, DVD-RW, and other internal devices so you don't need to be concerned about an internal SCSI chain. However, you will still need to take into consideration that a SCSI board (either one that ships with the basic computer or an Avid installed accelerator) will take up a SCSI ID, which usually will be 7. If you have a choice about adding peripherals other than drives like scanners, CD-R burners, or anything else you want to connect to your Avid system choose USB or firewire over SCSI to help eliminate conflicts.

Connecting Wide and Narrow Drives

We have already discussed the difference between narrow and wide drives and, when you upgrade to wide or ultra drives, you need to figure out how to connect them to the same SCSI chain with the narrow drives. The slower (narrow or wide) drives must come last in the SCSI chain.

When connecting wide and narrow drives on the same chain, you need a special termination kit from Avid. Wide drives use a 68-pin connector and narrow drives use a 50-pin connector. The difference is more than cosmetic, however, since there are more active pins inside the 68-pin connector, and these extra pins must be terminated before connecting to the narrow drive. You get a new wide cable to go between the last wide drive in the chain and the first narrow drive, a small blue terminator/adapter that allows this cable to connect to the narrow drive, and a blue terminator to go onto the last drive in the chain.

Use the standard wide cable (68 pin to 68 pin) to go from the computer to the first wide drive. Then use the termination kit cable (also 68 pin to 68 pin) with a special blue terminator adapter (68 pin to 50 pin) added to it from the kit when you connect the wide and narrow drives together. Finally, use the special blue terminator on the last drive in the chain.

Cables to connect the wide SCSI card to the narrow drives (68 pin to 50 pin) are very different from cables that appear similar for connecting wide drives to the narrow drives (68 pin to 68 pin with a 50-pin adapter/terminator). Even though this narrow cable may connect, you will have nothing but problems connecting wide drives to narrow drives. It is a good idea to label these cables "for narrow use" exclusively. Better to lock up the 68-to-50-pin narrow cables somewhere after upgrading to wide drives.

Connecting LVD Drives

Ultra LVD (Low Voltage Differential) are the fastest SCSI drives offered by Avid. They can easily play back 1:1 resolution with a two-way stripe across two LVD disk controllers. LVD drives require the same 68-pin cables as the Fast and Wide, but if you connect an LVD drive to a SCSI chain with a slower drive, they drop their speed to match. In other words, an LVD drive that is connected with an iS Pro drive will perform only as fast as the iS Pro. With the cost of drives plummeting these days, it is a good idea to use only your fastest drives for media playback on the SCSI chain. Take the older drives and connect them to a server for backup over a network or for graphics and compression stations.

There are still many complications to the SCSI chain, but these are the basics. Even if you follow all these rules, you may find that a particular drive works on ID 4 but not ID 2. You may never find a good reason for this (although there is a reason). You may also find that the order of the drives in the chain makes a difference even though all of the termination is correct at the end of the chain. And above all, you may have to juggle extra devices on the internal SCSI to find the best order. Slowly rebuilding the SCSI chain one device at a time and rebooting is often the only way to isolate where the problem is occurring and identify the problem device, cable, or terminator.

MOVING STRIPED DRIVES ON WINDOWS NT

For all of Windows NT's high-speed, memory-efficient, stable architecture, it still lacks an easy way to move striped drive sets from system to system. This has been improved in Windows XP so that it is no longer an issue. However, some unfortunate folks may be stuck on an older operating system for reasons beyond their control. You must use Disk Mounter before moving drives from an NT system. If you are moving drives from NT to Windows 2000 or Windows XP you must save the drive configuration. This can be done by going to the NT Disk Administrator; under the Partition menu choose Configuration and Save.

Here is the recipe that you will need to move striped drive sets between systems. Stripe and format the drives following the User Manual using the NT Disk Administrator. (You will need to have Administrative privileges to do this.) You cannot have any extended partitions to stripe drives, so you may have to delete those partitions before you proceed.

- Run the Avid-supplied software Disk Mounter.
- Click the Register button.

- Click Do It to confirm the procedure.
- Exit Disk Mounter.

Information about the striped set is stored on the drives. This information will be important to the registry of the new system when they are mounted. Make sure that the drive letters or the SCSI ID of the striped drives will not conflict with the drive letters on the second system. If you don't know what drive letters are being used already, then change the drive letter of your striped set to something high in the alphabet.

Check to see if the original striped set used two SCSI channels. If they use two channels, then they must be matched up to the same channel on the new system. If the first drive was connected to the A channel of the SCSI controller, then it should also be connected to the A channel on the new SCSI controller. You may want to label the outside of the drives to make sure they are matched correctly by the person making the final connections.

After the striped drives are connected to the second system, then you should run Disk Administrator to make sure the drives are recognized. Then run Disk Mounter and use the Mount button. Restart the computer after mounting the new striped set.

Occasionally, you may be trying to mount striped drives on a brand-new system that has never been used to stripe drives. You must activate a Devices Control Panel to recognize the striped file system for the first time.

- Go to Devices Control Panel under the Start menu and Settings.
- Find the device driver Ftdisk in the list of devices in this window. If this is your problem, then the Ftdisk is probably disabled.
- Highlight Ftdisk and click the Startup button.
- On the Startup Type window, choose Boot and OK.
- Back at the Devices window, click Start.

You may have to reboot the computer for the changes to take effect. If not, then you should now have access to the striped drives.

Another problem with drive registry data is that it may become corrupted on the second system and not allow you to see the new drives. You will have to delete the corrupted registry data and create new data. First run Disk Mounter and register the data for all the drives before you start this procedure.

- Go to the Start menu and choose Run.
- Type regedit32 and OK to get the Registry Editor.
- Find HKEY_LOCAL_MACHINE and open this window.
- Open the System folder.
- Highlight the DISK folder and use the Delete key to delete it.

- Say yes to the Warning and close the Registry Editor.
- Open Disk Administrator again and allow it to write a signature to each disk.
- Use Disk Mounter again to mount the drives and say OK to each striped set.

Never try to hot-swap drives on NT whether they are striped or not. NT will not see new drives until the system has been restarted.

AUDIO CONNECTIONS

Audio connections have multiplied in type and format over the years. There are now digital stereo pairs (AES/EBU XLR connections), S/DPIF, Optical, and analog (multipin octopus cables on the Adrenaline and Nitris and RCA connectors on Mojo). On Nitris there are two microphone inputs for live audio punch-in for voiceovers and multiple language overdubbing. On Nitris and Adrenaline there are also quarter-inch phono jacks for monitoring so that you no longer need any splitter cables for audio outputs.

With this cornucopia of choices, you may want to simplify everything with a small, inexpensive audio mixer. This will allow you to permanently connect many audio sources like tape decks and DVD players. Also make sure that all of your digital devices are calibrated to the same reference level. Many older systems will be calibrated to –14dB and this caused a problem with inputting audio from Digital betacam decks. See Chapter 9 for a full discussion of audio calibration.

If you have an older system, check to see if your audio card is an Audio Media II or an Audio Media III card. As I pointed out in an earlier chapter, be sure to use the Avid-supplied audio cables going from any source, like a mixer, directly into these specific audio cards. A built-in attenuator in the cables makes sure that the level going in matches the level going out. You cannot compensate for the lack of these attenuating cables unless you have a mixer that outputs at –10 dB.

If you have self-powered speakers, connect the speakers to a power strip with everything else and turn them on through that strip. This is a good idea for most of the equipment since it lessens the possibility that a piece of equipment will be left off accidentally. If the speakers are on and you still have no sound, check the Audio Tool (Ctl/Cmd-1) to make sure you actually have audio playing back. Change the timeline view to show Media Offline and see if the audio segment in the sequence timeline lights up bright red. If you have an older system, you can change the Text in your timeline to show the media filename and see if it says Offline. If it gives you media file numbers in the timeline audio tracks and you still can't hear any sound, then, on the older systems, check the sample rate switch on the ProTools external hardware. If the sound was originally captured at 44.1 kHz and the switch is set to 48 kHz, you will hear nothing. If you have an older system, you cannot mix sample rates in the same sequence and so you must recapture the mismatched material. Also make sure you are set to .99 or 1.00 to match the original capture setting. This external switch on the ProTools hardware can be hit by accident or when switching back and forth from a film project that needs the .99 pulldown rate. If you need only one type of sampling rate all the time, I suggest using a little piece of black gaffer's tape to permanently prevent these switches from being accidentally changed.

More recent systems have this change of sample rate and pulldown as internal software switches in the Audio Project setting and the Capture Tool. Using the Audio Project settings, make sure that the sample rate is set to the same rate the audio was captured at or set so that it will convert between the two sample rates when playing. The sample rate of the audio can be confirmed with a heading in the bin or by looking in the Console, which clearly will tell you that you have a sampling mismatch. This means you will not have to recapture just to hear the audio play but you will still need a roll of black gaffer's tape for something, so keep it handy!

THE BLACKBURST GENERATOR

A blackburst generator (BB Gen) is like a synchronizing clock for video or audio. It provides a steady source of a perfect video signal: black. If you do not use a blackburst or sync generator, your audio and video sync may drift apart over time. This is especially obvious when capturing or outputting long takes or sequences with analog sources and decks. I highly recommend getting a blackburst generator if it doesn't come with the system you ordered.

Connect the BB Gen to decks, monitors, your video card, and the Pro Tools video slave driver to synchronize your audio. Looping it through several devices is OK, but don't overdo it. Eventually you will attenuate the signal too much through multiple devices and it will fall below the minimum signal level to be effective. Digital decks like Sony's Digital Betacam will supply sync to the Avid during capture. When you have set your audio project settings to accept a digital signal then the audio sync is provided by the deck. If you shut off the digital deck you will get an error message complaining that you have lost sync. You should switch the Audio Project setting back to analog sync.

Many decks require a stable signal to their composite video input when they are playing back, especially the Sony UVW-1800. If you are using the component inputs and outputs, consider connecting blackburst to the composite video input permanently. It makes it easier to black tapes when you are not using the deck for anything else. Just set the deck input to composite, set the correct timecode, and start to black and encode a tape with no other input. Just remember to switch it back to component input before you leave the room!

You should definitely attach blackburst to your Mojo and Adrenaline boxes since Avid has invented a unique way of sending blackburst sync through the firewire cable that connects these boxes to the computer. The Nitris system can even handle two blackburst signals: one standard and the other tri-level sync. Tri-level sync is necessary when you are editing at the following HD frame rates:

```
1080i        30.00, 29.97, 25.00 fps
1080Psf      30.00, 29.97, 25.00, 24.00, 23.976 fps
720p         60.00, 59.94 fps
```

(Psf is the progressive segmented frame used by Sony's HDCAM. This is considered progressive by most people, but there is a slight technical difference.)

It is called tri-level sync because there are three sync pulses instead of the usual one: at 0 volts, −0.3, and +0.3.

Connect signal generators like blackburst, color bars, and tri-level sync as permanently as you can and design a system that requires as little connecting and disconnecting as possible. Consider patch bays, mixers, MediaDocks, networks, and removable media like Zip or Jaz drives. The less wear and tear you put on cables, the more reliable and long lasting they will be. A little more initial investment during the planning stages can positively reduce troubleshooting downtime in the future.

STANDARD COMPUTER WOES

Even if you successfully eliminate the potential for problems with connections, there is still the potential for standard computer-type problems. Unfortunately, these kinds of problems are not easily solvable by the typical IT department, even if you are lucky enough to have one. This is one of the reasons it is a good idea to be able to take care of your own computer problems. Standard computer support personnel are going to be at a loss with the Avid problems unless they have been through some training for the specific requirements of high-resolution video.

Extension Conflicts

An extension is a small piece of software, required by Apple operating systems 9.x and earlier, that launches when you start the computer and runs in the background. They are necessary for the operation of certain software applications. The Avid systems have a handful of extensions that are installed automatically in the Extensions folder in the System folder when you use the installer disk. You can run Adobe Photoshop, After Effects, and some other third-party programs that require their own extensions on your edit system computer without problems and, although many times this is efficient and practical, it is sometimes asking for trouble.

If you use programs that add extensions when you use their installer, eventually you will have an extension conflict. Sometimes the conflict is obvious and you cannot boot the system and run the Avid software after you have installed something. That problem is pretty easy to fix. In fact, there is a program for the Mac OS 9.x called Conflict Catcher by Casady and Greene, Inc. that systematically cleans out your extensions folder and puts extensions back in a way that allows you to quickly zero in on the culprit. You can boot the OS 9 Macintosh with the Shift key held down to bypass all the extensions and get the system past the problematic extension. You can also hold down the space bar and be launched immediately into the Extensions Manager. You may find your life simplified by making several sets of extensions and labeling them for the particular functions they are used with. Have a set of extensions for graphics work, one for cruising the Internet, and one that is stripped down for editing. The best set of extensions to create is one that allows you to work with the three or four most useful programs that you run all at once. Test to make sure there are no conflicts, strip the extension folder so that only those applications are supported, and then save it as an extension set. (Back it up in case you are forced to reinstall your operating system.) Your system will run faster on less RAM and give you many, many fewer mystery crashes.

The most difficult extension problems to track are intermittent or related specifically to particular functions. They may occur only when you go into Capture mode or only when you are making a Digital Cut. These may occur because some third-party application has changed some important function in your system. Another video application may reset the frame size of the video board or change the media drive firmware. There are many small but important settings that may need to be reset after running another program on the Macintosh. You may be able to find these programs only after they cause serious problems, but if those problems are intermittent, you may never connect the two as cause and effect. Simple is better, and if you can afford another Macintosh to run third-party applications, you can reduce the amount of crashes when you are running the Avid software.

There is a set of general rules that you should follow when contemplating adding any software to your editing system that is not directly related to editing on the Avid. Most of these rules revolve around the requirements that the Avid system has to be uninterrupted during critical operations like capturing, playing, or rendering. Try to close the Explorer window on Windows since it will attempt to update continuously in the background. In general, don't leave other applications open that might run background processes, like Microsoft Office.

The Find Fast background process is another common application that will try to update in the background. Some e-mail applications will periodically check the network for new mail and may interrupt some other more critical function. Calendar programs are just as guilty of trying to schedule and update at preset times. Screen savers can start to play at inopportune times and may cause the system to hang. Keep your system fonts small and do not use large graphics for the desktop wallpaper display. Keep your editing system primarily for editing and, unfortunately, many of the things a user can do to customize the computer, the fun stuff, may also pull valuable resources away from functions that are already pushing the limits of the system.

Working from a Floppy

Do not ever open a project, a bin, or a User Setting directly from a floppy. The moment you remove the floppy, but continue to work with the bin, you will see very strange errors. The Avid system periodically needs to save back to the original bin file on the floppy. Even worse, you will not be able to close the project or save the bin you are working on if it grows too big to store back on the floppy. All bins automatically save themselves when they close or when the project is closed. You must allow for the extra space if the bin has grown larger. You should always copy bins and projects from removable media to the main storage systems before continuing work.

Access to Original Software

One of the last solutions to difficult, intermittent, or unusual problems is to reload the software. Avid Support may ask you to reload the Avid software, the Macintosh or Windows operating system, or both. All users of the system should have access to these disks. A big mistake, commonly made, is to lock up these disks safely away from anyone who might need them, probably at 3 A.M. Make sure the disks are the most recent and correct versions of both the operating system and the Avid software. You may cause even more problems by loading an unapproved version of the operating system!

Access to the Hardware

None of these techniques does you any good unless you actually have access to the computer itself. Some installations have cleverly hidden the system away in another room or rack-mounted it in a machine room. If you are completely forbidden to touch the hardware because of facility or union rules, then forget about it. You can just hand the phone to the appropriate authority. Otherwise you must, phone in hand, be able to look around back and see that all the cables are tightly connected, that the power to everything is on, and read the disks' SCSI IDs.

This means that you should look into getting a good engineer to set up your suite. Good engineers are worth much more than their salaries when they save you the embarrassment of your first several jobs going out the door with bad levels because you were monitoring the audio in the wrong place! Make sure, before the engineer leaves, that you have a thorough understanding of the cabling and get a wiring diagram you can refer to.

Never Enough RAM

With the cost of RAM these days plummeting as fast as hard drive space, how much you install becomes a trivial decision. Buy a gigabyte and never worry about it again. However, if you have an older system and do not want to purchase more RAM, or are having RAM-based issues with no way to run out and buy more, you will need to look at this next section. More RAM is always better, but there is one thing you must do after buying and installing (or getting a certified technician to install) your RAM. On the Macintosh version 9.x and earlier, you must change the RAM allocation of the application. On Windows and Macintosh OS X and later this is handled automatically. On the older Macs, find the icon of the original Media Composer software and click it once. Then choose Get Info (Command-I) from the File pulldown menu at the Macintosh Finder level. Change the minimum and preferred amount of RAM to use all the new RAM you installed except for about seven to ten megabytes for small programs and enough for the Macintosh operating system to expand when it needs to. To be really sure you are not shortchanging the operating system, go to About this Macintosh under the Apple menu in the upper left-hand corner of your screen while you are at the Finder level. See how much RAM the system is using now and add a few megabytes to it. Subtract that amount from the entire amount of RAM available and give it to the Avid application. If you have a large amount of RAM, well over 1 gigabyte, and you don't have a huge amount of media online, over 200,000 objects, then you might want to leave a larger chunk available for a graphics program or EDL Manager. With versions 7.x/2.x and later, you also have the DAE (Digidesign Audio Extension), which will launch the first time you use an Audio Suite plug-in. This extension may take up to 20 megabytes of RAM. EDL Manager for very large sequences may need 20 megabytes as well, but most of the time it should work fine with around 12 megabytes.

STARTING UP FROM ANOTHER DRIVE

If you have problems when you boot your system you should try to start up using another drive. It is easy enough to bypass all your Apple extensions if you think that is causing the problem by holding down the Shift key during the initial moments of the startup. But if the problem is truly with the internal Macintosh drive, you can force the Macintosh to look for another drive with a System folder on it and see if you can boot at all. Once your system is up and running, you can try to track down the extension conflict or run a disk repair utility to try to fix the problem.

The Macintosh looks to see if there is a floppy disk with a Startup folder already in the floppy drive before it looks to its own internal drive. This is where Norton's Emergency Disk comes in handy. It contains a stripped-down System folder and Norton's Disk Doctor, which you can then launch from the floppy and use to repair the internal drive. If you don't have an Emergency Disk, you can force the Macintosh to look for another drive before the internal drive, like the Disk Tools floppy that came with your system software (if it came on floppies). If you have a floppy in your Windows machine, the system will automatically try to boot from it. If there is nothing to launch from, then remove the floppy! You can boot from a CD-ROM that holds the installation of the Macintosh operating system, too. If you simply hold down C while booting, you will default to starting up from the CD-ROM. If you hold down Command-Option-Shift-Delete, you can boot from a CD-ROM, a Zip, or a media drive if they have a viable system folder. You may want to carry a bootable disk with you to insure you can solve a problem at a customer site if you are a freelancer.

Again, having access to a clean Macintosh operating system and the CD-ROM of the original Avid application may be an easier and faster answer than really trying to pinpoint a problem. You can always go back and try to recreate the problem when you have more time and no one is threatening you with deadlines. Some facilities that regularly work under tight air dates will keep a mirror image of their internal drive on another external drive attached to the Macintosh SCSI chain. If there are any internal drive or system-related problems, they can force the system to boot from the external drive and be on their way. Another quick way to deal with internal system problems is to do a clean reinstallation of the Macintosh operating system. This is as easy as loading the system software installation disk and holding down the keys Command-Shift-K (for klean, I guess) while rebooting. This automatically puts the old operating system in a folder called Previous System Folder and installs the system software again. With OS version 7.6 and later, there is a checkbox for the clean install since Apple finally realized how important this operation is in a pinch.

You can force a Windows system to boot from another drive by changing the BIOS. This is the software setting used by the system during boot. There are different ways to get into the BIOS depending on your PC manufacturer. During boot you will hold down F1 for IBMs, F10 for Compaqs, and F12 for Hewlett-Packards. Then you can choose the order of drives the system will look for the next time it boots. Change the boot order to another drive that you know has a good copy of the operating system, like an emergency boot drive just kept around to get back up and running. If that drive is disconnected later, the system will skip over it and go to the next drive in the boot order list. You will notice that on Windows systems, the PC will always look at a floppy drive first. This is why if you have left a floppy disk in the floppy drive without the bootable files, your system will stop the boot process until it is removed. Leaving a floppy in the floppy drive fools even experienced editors so look for that first if you are having a boot problem.

ROUTINE MAINTENANCE

By far the most important day-to-day steps you can take to keep your system running and happy are just routine maintenance. These techniques are so fast and easy that as a freelancer, I used to come in a little early on the first day of a project to run them on any new machine before I started.

Rebuilding the Macintosh Desktop

The first procedure is called rebuilding the desktop. Do not underestimate the power of this procedure. You are telling the Macintosh to look at every file on all the drives and build a new database of exactly where everything is. There are so many important pieces of software that the Mac must find instantly to run smoothly that having everything accounted for before you begin can prevent all kinds of nasty little surprises. When you start up in the morning, or if you are restarting because you have had a problem, hold down the Option and Command keys during the startup process while the last extensions are loading. It takes longer to start up as the machine looks at everything on all the drives, but it is worth the wait in the long run. You don't need to do this every day, but if you have been moving, deleting, or adding large numbers of files, then it's like chicken soup: It can't hurt. You can rebuild the desktop using Norton's Disk Doctor, which we will discuss shortly. When in doubt, ask Avid Support about versions of any third-party programs that you use on your system for maintenance.

Disk Diagnostic Programs

The next important step of routine maintenance, a disk diagnostic program, takes a little longer, but if run once or twice a week, maybe more often in a facility that never sleeps, it allows you to catch serious problems before they become fatal. Disk diagnostic and repair programs have been around for a long time and people may have had a bad experience with them and media files, but they are a lifesaver to run on internal drives.

Norton Disk Doctor is the most popular and ships with the Avid Macintosh systems. Disk Warrior 3 is also approved for basic disk maintenance. In the past you may have received the entire shrink-wrapped Norton Disk Utilities package. If you did a full install of all the options, you installed a few extensions that can interfere with the operation of the Avid system. All this correct installation information, of course, was in the Release Notes, but . . . just install the Disk Doctor and leave the rest for a Macintosh that doesn't need to play back gigantic video files with realtime effects!

There have also been the rumors that you should never run Norton Disk Doctor on media drives. Part of the problem was that in trying to repair a gigantic media file, old versions of Disk Doctor occasionally left the file corrupted. More recently, the controversy was over the fact that Disk Doctor couldn't deal with striped drives. With Disk Doctor version 3.2 and later, that is no longer true. I run Disk Doctor on the media drives now, but with one exception: I never repair a media file. If Disk Doctor wants to repair header information and other minor directory problems, it can go right ahead, but I don't want it to mess with my 1 gigabyte media files.

Speed Disk is a utility that Norton ships with Norton Disk Utilities that defragments, or reorganizes, a drive so that there is no wasted space. It optimizes disk space by moving files around and, in a very orderly way, filling in gaps on the drive. This is great if you are concerned about space, but not so great if you are concerned about performance of gigantic audio and video files that must play back simultaneously. Avid has determined how to record files to the drive in such a way that video and audio stay together to maximize efficiency when the drive head is reading the file. That way, the drive doesn't have to work so hard at the high speeds that are necessary to keep it from underrunning.

Try to capture audio and video to separate drives when you can. Note that if you put audio and video that are meant to run simultaneously on different partitions of the same drive, you are forcing the drive to work even harder. With the popularity of striped drives (and the repeated assurance that you will not lose half your drive space), capturing audio and video to different drives is less of an issue. Are you really pushing the system, trying to get many realtime effects at high resolution, for instance? If you are pretty organized, then you may benefit from splitting your audio and video to separate drives or even separate SCSI cards.

NT systems come with a utility called CHKDSK, which will check a drive for problems. You can get to CHKDSK in the Disk Properties window. Right-click on the drive icon to get the drive properties, then choose Tools. Have CHKDSK look for and fix file system errors. With Windows XP right-click on the drive icon, choose Properties and then Tools to get to the operating systems' defragmentation and disk checking software.

A new tool to help determine if the drive is really dying is DiskEx or StorEx, or Avid Storage Manager. This utility is available from Avid Support and ships with new systems. Avid Storage Manager exercises your disk by making it run through a series of stress tests to see if it will fail completely. Do not run any of the destructive tests unless you don't care about any of the material on the drive. A destructive test will destroy all the material on the drive, even if you run it only for a few seconds. Never run a destructive test on your boot drive (usually the C drive on Windows). If you run a test with one of these utilities and your disk does not fail, then you can confidently search elsewhere for the problem instead of waiting overnight for a new drive that will not solve your problem.

FONTS

Even though Apple has tried to make dealing with fonts as simple as possible, somehow it is still complicated. When you try to use the wrong kind of font in the Title Tool, it may look absolutely terrible although all the other fonts look fine. The main complication has to do with what kind of font you have and where it is located. Standard TrueType or Adobe PostScript fonts installed correctly should be available to use in the Title Tool. You should have a copy of Adobe Type Manager (ATM) loaded to deal with the Adobe PostScript fonts.

The confusion comes when people download a font from the Internet and install it or take a single-sized font and try to resize it. These are two different problems. The first situation, a font of unknown origin, generally means that it is a bitmapped font. It is meant to be only one size, and if you try to change it in the Title Tool, it quickly looks chunky with ragged edges. You should really discard this font; it is of no use to you with the Avid.

The second situation is similar, but results from a different cause. When you use Adobe PostScript fonts on a Macintosh, there are two parts to every font—a screen font and a printer font. You may find that one of these, usually the screen font, has not been installed correctly on the system. This means that you no longer have a font that rescales for screen display. Time to go back to the original installation disk and reinstall the font. Sometimes good fonts just go bad. They become corrupt and they must also be reinstalled again. You should always have access to the original software, especially fonts, in case you have to reinstall your Macintosh system software from the CD-ROM.

Incorrect installation may account for fonts not showing up at all. All fonts must go into the Fonts folder in the System folder in Mac OS 9. You can simplify this by dropping them onto the closed folder icon for the System folder; the Macintosh asks if you want them to go into the Fonts folder. When you say yes it does the rest after you restart. The problems occur when you just drag them into the open System folder and drop them inside that open window. Then they are not put away correctly and are loose in the System folder. They must go into the Fonts folder. In OS X there are several locations where fonts are stored including System:Library:Fonts (where System is the name of your boot drive) and System:Users:Current User:Library:Fonts (where Current User is the name of the logged on user.

Another problem with fonts is that too many of them take up unnecessary amounts of RAM. If you want to slim down your required RAM for the operating system so that you can give the RAM to another application, removing fonts helps as soon as you restart. Another way to simplify font selection, but complicate your life, is to use font suitcases. Many programs allow you to keep a particular set of fonts together and load them all at once so you have just the ones you need.

MEDIA MANAGEMENT

Another day-to-day concern mentioned before is the management of media on the drives. You may think that you are efficiently deleting all media as you finish each project, but you may be surprised to find all sorts of odd bits of precomputes and imported graphics and so forth floating around on your media drives. The best way to keep a handle on this is to delete media from the Media Tool and not through individual bins. That way you see the project from a big picture point-of-view and can evaluate on a project-by-project basis what must go and what must stay. It also keeps you from accidentally deleting media from another project just because you dragged a duplicate of someone else's master clip into the bin you are now deleting.

The real problem, not only with having many unnecessary small objects on your drives slowing down performance, is that the drives may accidentally become too full. You should always keep a minimum of 10 to 15 percent on any partition or any drive completely free. There are all sorts of files, like the Media file Database, which occasionally must get larger to accommodate the changing nature of the media on the drives. Don't worry about defragmenting media drives, but overfilled drives corrupt media and may crash and take everything with them.

A more difficult problem to diagnose is when you actually have too many files in a single folder. If you are working with nine-gigabyte partitions and capturing offline resolutions, you may find that on long or complicated projects you are exceeding the limit on the number of files that can be in one folder on a Macintosh. This occurs because Avid must restrict the disk cache to the minimum figure in order to reduce digitizing underruns. This can be 96 k on older systems and 128 k on recent systems. This limits the practical number of files in the OMFI MediaFiles folder to around 1200. Above 1200 (or approaching 1200 depending on the size of the files) you may have difficulty booting the system or opening the OMFI MediaFiles folder. If your system is exhibiting these symptoms, you may want to raise the amount of the disk cache (in the Memory Control Panel), move items to another folder, and change the disk cache back again later. If you are always working with offline resolutions, you may want to limit the drive partition size to a maximum of four gigabytes.

SERIAL PORTS

Printers and modems can be connected to your system, but they take up valuable serial port connections. This may be the best reason to use Ethernet for printers since it frees up a serial port so that you can use it for deck control. The modem port and the printer port are identical and interchangeable. It is just that certain software expects to find the modem connected to the modem port and the printer to the printer port (on Windows these are just numbered but are still interchangeable). If the software can't find the right device, tell it to look at another port either in the Macintosh Chooser (OS 9) or in the software itself.

A common serial port issue on the Macintosh occurs when serial port 1 returns an error that it is already in use by another program. This is usually because the Keyspan software used to control video decks has a setting wrong. Go to the Keyspan Control Panel and turn off Emulate Printer Port.

Sometimes the serial ports get stuck in an open or closed setting and appear not to recognize that there is a device connected to them. This might happen if you had a printer or a phone connection problem and the last thing you did with that device did not shut it down or turn it off properly. You may notice this in the Avid when you have lost deck control, although everything else is set up correctly. You have the Serial Tool under Tools and you can change which port you are using for deck control or change the port to None and change it back to the one you want.

If the Serial Tool doesn't clear the serial port, you must either use a utility called CommCloser or zap the PRAM (pronounced pee-RAM). The PRAM is parameter RAM (OS 9 only) and is a part of the computer memory that holds important information that keeps things like the clock going when the system is shut down. PRAM remembers important settings so that when you boot up, everything is the way you left it. Sometimes you must purge this memory to clear the stuck serial port. TechTool, a piece of freeware from Micromat Computer Systems, is the best utility for this because it zaps the PRAM quickly. It is also good for rebuilding the desktop. Otherwise, you must restart and hold down the P-R-Command-Option keys and wait for the restart chime to sound eight times. This can be pretty tedious if you have a lot of RAM because the restart takes so much longer. You then just restart and reset all the control panels that may have changed back to their default state.

On older systems, the most important settings after zapping the PRAM to put back are Memory Settings. Modern Memory Manager must be on. With a Quadra, you must turn the 32-bit addressing back on in the Memory Panel. On newer systems (OS 9) you should check the Memory Control Panel and turn virtual memory off. If you have a Quadra with a Daystar Power PC upgrade card, you must turn it back on through the Control Panel and shut down completely (not just restart). It can take some time to both go through the procedure and to reset all the control panels, so zapping the PRAM is generally considered a last resort.

OTHER DECK CONTROL TIPS

There are several other reasons why you might not have deck control. When you open the Capture Tool, you may see the message No Driver. The first thing to do is to try to reload the deck configuration or force the system to "check decks." Both of these choices are at the bottom of the Capture Tool under the Deck Model pulldown menu. If this doesn't work, then you need to do some digging. Obviously, check the cable connections first. You may have a V-LAN or VLX, which is an external deck control device from Videomedia that gives you a wider range of deck control choices. If so, then make sure you are using the V-LAN or VLX cables and not the Avid-supplied deck control cable.

Check your Mac extensions folder to find Serial (Built in) on a PCI system. On an older system, like an 8100, Avid Media Processor, or Power Quadra, check for the Serial DMA extension.

If none of this works, you may want to check your Release Notes or user manual to make sure you are using a supported deck. Even if the deck is not supported, you should be able to get some limited control using the Generic deck choice in the Deck Template window of the later versions. With Xpress DV, the only way to get your deck recognized may be to do an Autoconfigure. This will usually result in a Generic Device, which can then be switched for a particular template using the Deck Configuration settings.

WORMS AND OTHER INVERTEBRATES

Lately, the spineless, dirt-eating individuals who like to prove how much damage they can do have been rather busy. There have been many auto-replicating worms, Trojan horses, and viruses on the Web. It seems they are just as interested in making sure you do not have a nice day as they are at stealing personal information for use later as fraud. They may also be using your system to bounce spam to the rest of the world. If you have broadband connections to the Internet make sure you have a secure firewall and have used all the security software that is appropriate for encrypting important information. Always use virus protection software; if you find that the constant checking of your disks slows down your realtime performance, you may want to disable background features. But continue to check for new problems on a regular basis. See if your virus protection software can be updated from the manufacturer's Web site to protect you in the future. Check this Web site for updates anyway until the human worms wise up and get a life.

BASIC WINDOWS TROUBLESHOOTING

The main objection I hear many editors use to avoid switching from Mac to Windows (besides the politics of "monopoly" versus "insanely great") is the fear of troubleshooting a more complicated operating system. The ease of access to the Macintosh System folder is a double-edged sword, and moving vital resources in and out on a whim is the cause for many editing problems. It is also something that many editors have spent time to learn and so can do some basic troubleshooting on their own. Switching to Windows means you must relearn the top ten things to do when a problem arises.

A serious problem with Avid Windows editing systems is when somebody changes something from the original shipping configuration and the system will not boot. This is handled easily by the Windows OS during the boot process when it gives you the opportunity to invoke the Last Known Good configuration. Press the space bar and then type L when prompted. This will bypass any changes the last user may have tried and will load a configuration that worked the last time the system booted successfully. You may then try again to get the configuration to be the way you want; however, if the present configuration did indeed boot correctly, then it will be the Last Known Good configuration and you will need to look elsewhere to find the problem.

Your Meridien Windows system probably came with three hardware profiles already loaded onto the system. The Avid Configuration (Network Enabled) is the default profile and has all the network information already configured and ready to go. The Avid Configuration (Network Disabled) profile can be chosen during the boot process if you feel that the network is causing your problems (or there is no network connected). It also simplifies the system to make troubleshooting that much easier. And finally, Original Configuration will omit any special Avid hardware drivers except for the screen display driver for the Meridien Display Controller. Again, this would be used for troubleshooting to simplify the system to its basic Windows components. This profile must also be chosen during the boot process if you suspect that it is the Avid hardware or hardware drivers that are causing the problem. If your system does not have these three profiles already loaded you should spend the time to create them yourself.

Another mode that can be engaged during the boot process is the VGA mode. Choose it if you have problems with the display card or suspect that the Avid display drivers were causing conflicts with something new loaded onto the system. This mode is also extremely useful if you have been messing around with the configurations and suddenly your display becomes unusable. By rebooting and choosing the VGA mode, at least you will have a usable image on your monitor to continue to troubleshoot, reload the Avid display drivers (EDCInstall), or undo what you just did! Your Avid software, like most software installers, will default to installing on the C drive, which is generally the boot drive. The C drive may actually be a rather small partition (as small as 2 gigabytes) and you should consider forcing the Avid installer to put everything on the D drive. Pay attention during the installation process to where the software will go and change it when prompted.

WINDOWS XP RECOVERY

Windows XP is designed with a feature called System Restore. This keeps track of the system configuration and saves them at crucial times as known reference for when the system was running well. If you are having problems starting your system you will want to take the choice of starting in Safe Mode, which disables part of the operating system that may be causing problems. Then you can use System Restore, uninstall problematic programs, or run diagnostic programs on your disk drives.

If you are extra cautious about restoring your system you may want to look into a third-party program like Norton's Ghost, which has many options for creating mirrored copies of your system drive. You would be able to switch over to the mirrored drive if you were under pressure to continue working without time to troubleshoot.

WINDOWS NT RECOVERY

Every basic NT troubleshooting guide will walk you through two major procedures to make sure you can get your system up and running quickly in the case of problems. These two procedures are creating the NT Setup diskettes and creating an Emergency Repair disk.

Creating Windows NT Setup Diskettes

Your system should have arrived with these diskettes already created. You may want to make another set or you may have safely stored these important diskettes in the locked office of your company's accountant (where no one can get them). These are the disks that you will use to boot the system in case of a crash and you are no longer able to boot from the internal C drive.

Use three high-density diskettes and format them. Then load the Windows NT CD-ROM that was used to install the original operating system. Click the Start button and choose Run to bring up the Run dialog box. Know the drive letter of your CD-ROM drive; it is usually F with a standard Avid configuration. If it is F, then type the following text in the Run dialog box:

```
F:\i386\winnt32 /ox
```

Notice the space after winn32 and be sure to substitute your CD-ROM drive letter instead of the F if it is different. Follow the instructions from that point to load the three floppy disks one at a time and label them correctly. Save these floppies in an accessible but safe place. They should be able to get you to a stage that will allow you either to continue troubleshooting or to reinstall Windows NT in the case of an emergency.

If you boot the system using Setup disk #1, then you can use the Emergency Repair disk to repair a problem that might have been caused by a corrupted operating system, user files being deleted, or someone interfering with the configuration.

Creating an Emergency Repair Disk

There should have been two Emergency Repair disks (ERDs) with your Avid system. The first would be the bare bones Factory Default ERD that will help bring your system back to the configuration before any Avid hardware or drivers were installed. This ERD will also help you to troubleshoot a simplified NT system and is an important tool if you need to completely reinstall the NT OS.

The other ERD is the Avid Operational ERD, which is the disk that should be used most often. Generally, there will be something minor that needs to be repaired and you don't need to take the system back to the factory defaults to fix it. This disk (or another disk just like it) should be updated when new applications are added to the system. To create or update an ERD, you should go back to the Start button and choose Run again. In the Run dialog box type:

```
rdisk /s
```

Notice the space after rdisk and press Return. This will ask you to insert a high-density floppy, which will be reformatted, and then record the registry information required to reproduce your current setup. Remember that in order for it to return the system to the current configuration, you will need to update the ERD every time you make a change.

Windows 2000 Emergency Repair Disk

In Windows 2000 the procedure is very similar but when you create the ERD you need to go to the Start Menu and then to Programs/Accessories/System Tools/Backup. Choose Emergency Repair Disk and back up the registry to the repair directory.

If you need to repair the system because of a booting problem, then boot from the original operating system CD. Choose to Repair, not Install when given the choice and insert the ERD when requested. This will restore the registry of your system to the state when you last made or updated your ERD. This is why you need to update the ERD whenever you change the software or other configuration information on your computer.

Creating a Bootable Disk

If your problems are serious enough, you may just need to boot the system far enough to reinstall the Windows operating system from the distribution CD-ROM. A bootable disk has been included with all of the Avid installation software. You can also boot from the three set-up disks discussed earlier for NT.

Reinstalling Avid Software

If you are missing a file or a file has become corrupted, you may want to reinstall only those problem files. By loading the Avid software installation CD-ROM, you can do custom installations to replace drivers, Codecs, or anything else you may suspect as being a problem.

You may want to reinstall the entire Avid application just to start from a clean slate. You may want to move all your AVX plug-ins and AudioSuite plug-ins to another location so they don't get erased; however, if you suspect that the plug-ins may have something to do with your problem, you may want to reinstall them as well. Be sure you have all the registration information for plug-ins that require being registered the first time you use them.

VERSION NUMBERS

You should always have, either in the back of your head or written down someplace accessible, the version number of almost everything associated with your system. First and foremost is the version number of the current Avid software. You should know it down to the last digit because every small change in the software has a reason. Avid puts out what they call the gold version of the software, the most tested and most stable version they can achieve within the time allowed before it must be released. Version 11 is an example of a gold release. Release dates are based on complicated interrelationships all lining up at the same time. If, after the gold version is shipped, some features didn't make it into the software, even though they were planned, there may be other releases with an extra decimal point. Release 2.1 is an example of that since it included the color correction interface for Symphony. There may be some procedures that were not tested and appear to have problems, so there may be another revision called a patch release. A patch release is meant to fix one or two small problems. This would be version 6.5.1v2. The important thing to remember about all these releases is that you might not need any of them except the gold version. The other versions have not been tested as thoroughly as the gold release because of the importance of getting out a fix in a timely manner. Most of the time this is not a problem, but asking for a patch release if you don't absolutely need it pushes the envelope unnecessarily.

It is always best to have all the systems in your facility running the same version of the editing software. Most versions are forward compatible, but not backward. This means that a bin created in version 5.5 opens fine in version 6.5, but not the other way around. Once a bin has been opened in the higher version, it has been converted forever to that version. By just opening the bin once in version 6.5, you may not be able to open it in version 5.5 again. You have to export the bin full of clips as a shot log and import it into the earlier version. Unfortunately, this does not work well for sequences. A bin converter program goes between versions 6.5 and 5.5.1, but it is best not to depend on such complications.

If you are forced to change systems in the middle of a job, always insist on the same version of the software or later. You should know the version of the operating system you are using. Some versions are not approved for some models, and you should make sure that your computer can run the latest software before you install it. There will be a time when a new Macintosh will not run OS 9.x and you will need to upgrade your computer to run the latest version of Avid software that is optimized for OS X.

Always make sure you know the latest version of Macintosh operating system Avid has tested. If at all humanly possible, try to get all your Macs to run the same Macintosh operating software. Even experienced editors can make their Avid system unstable by loading an unapproved copy of the operating system. Do not do this casually, and make sure someone else doesn't do it for you just to keep you up-to-date on the latest must-have features. Keeping everything interchangeable is a valuable goal and should not be complicated by an IT person or an especially enthusiastic editor who wants to put the newest software on the machine as soon as it is available.

Version numbers also carry over to the hardware. Each of your boards has a revision number, which should be considered during troubleshooting. You may have an old version of a board or a version with a known conflict. You can check the revision of your hardware in several places. If you have Avid software running, look under the Hardware Tool (in the Tools menu) to see configuration and drive use information. A utility called Avid System Test allows you to get more detailed information about each PCI board (if you have any!). This is the better answer if you can't actually launch the Avid editing software. If Avid System Test (called the Avid System Utility on older systems) can't see the PCI board at all, calling it an unknown board, or if the slot is empty, then that tells you the problem is with the board or the way it is mounted in the slot.

Every utility that ships with your Avid system has a specific version number. The ones that ship are the ones that are meant either for that hardware or that software. Grabbing a version from another system just because it is newer may get you into trouble. Certain versions of the Avid Drive Utility, for instance, are designed primarily for four-way drive striping. The rule of thumb is: If the new version has a tangible improvement, completely compatible with your system and supported in the Release Notes, only then should you load it onto your computer.

ELECTROSTATIC DISCHARGE (ESD)

The sneakiest and hardest problem of all to diagnose is one that is very easy to create: damage related to electrostatic discharge (ESD). You may not realize it, but the human body can store and discharge frightening amounts of static electricity. Shuffling across the carpet with a relative humidity of 10 percent generates 35,000 volts! Compare this to the smaller, faster devices that are needed to achieve the kind of performance necessary to keep the system running happily, and they have a range of susceptibility of several hundred volts.

You can zap a component with static electricity by touching the outside of an ungrounded device or, more probably, by opening the device to do some simple troubleshooting. You may be asked to remove or add RAM or to reseat a troublesome board. Any time you are going to open a case, be concerned about voiding a warranty or causing ESD damage to the sensitive components inside.

The key to touching anything inside the system is to be grounded. Being grounded ensures that any buildup of static electricity is channeled off to a ground and dissipated. This is best done by wearing a grounding wrist strap and connecting it to a ground or to a metal component inside the computer. Most important, make sure the computer itself is grounded. The best way to ensure this is to plug the computer into a grounded outlet. An even better solution is to plug the computer into a grounded power strip that is plugged into the wall socket and then shut the power strip off. That way you are not supplying power to the computer while you are working on it.

The scariest thing about ESD is that it doesn't always kill — sometimes it just maims. A board or a RAM chip that receives a substantial shock may not fail right away. It may not fail for days, weeks, or months. It may start to show intermittent behavioral problems that cannot be isolated. These are absolutely the worst kinds of problems to troubleshoot because they may not occur for long periods and may not be of a type that points to any one component. It may be the CPU itself that received the shock, and no matter how many boards you replace, it does not solve the problem. This is why ESD should be treated so seriously; any time you handle a component or open the computer, you should be very, very careful.

CALLING CUSTOMER SUPPORT

If, after all these precautions and general maintenance, you must still call Avid Support, at least take heart in a very short, best-in-class wait time. To make it go even faster, have certain answers prepared since almost all support calls start with the same basic questions. Know your versions, operating system, CPU model, and Avid model. Be able to describe what you were trying to do when the problem occurred. This is especially crucial for video engineers who have not taken the time to learn the software. The editor describes the problem and the engineer cannot explain it to the support rep in enough detail. What is the exact wording of the error? Some errors are pretty obscure like "missing a quiesce." Write it down and don't fake it with "something was missing, I think." When did the problem begin? Right after you put in the new RAM? Pull the RAM out. Does the problem happen every time you perform a particular operation or is it really random? Can you repeat it? The simplest answer to any error that seems random is to shut down the computer and restart.

The other very important thing you need to give to Customer Support is your system ID. If you don't have a system ID, you may not get any support! That is the way Avid determines whether you or your company has a valid support contract. Not something you want to discover at 3 A.M. The system ID can often be found by just doing a Finder search for "sys" or launching the utility Dongle Dumper, but if you can't launch the computer that won't help much. Write it down or use Dongle Dumper and print it out.

Consider how much more helpful your Customer Support Representative will be if you do not immediately launch into a tirade of abuse. They know you are frustrated or you wouldn't have called. They are trained to help and to help calm you down, but you can make everyone's job easier by being civil and professional. And if you don't hear the hold music, you are not on hold so don't say insulting comments about them to everyone else in the suite. The rep may have his or her headset muted, but he or she is still listening!

We can't deal with all the techniques, potential problems, and error messages in this short space. If you want to know more about your system, there are courses in troubleshooting and a whole curriculum to become an Avid Certified Support Representative. You don't need to be a technician to keep your system running happily most of the time, but you should have some basic knowledge of what is going on under the hood. Good maintenance routines and a healthy dose of caution are two necessary items when dealing with sophisticated and complicated systems. Keep it simple and you will be rewarded with fewer steps and less stress when you need to troubleshoot.

14

Nonlinear Video Assistants

With the adoption of new technology has come the blending of postproduction responsibilities. Producers and writers become offline editors, offline editors become online editors, and the difference between film and video begins to blur. One thing that this shift has created is many new people who can edit well, but who don't have the inclination, ability, or time to get involved in the technical requirements.

This opens up the possibilities for an important position: the nonlinear assistant. There have always been film assistants whose responsibilities are pretty clear. They handle all the day-to-day requirements of film handling, organization, and preparation for the editor. There have also been video assistants in the past, although their roles have changed through the years and at times have been eliminated altogether. It is inconceivable to lack a film assistant on a major feature, but many high-end production companies operate quite well without video assistants.

Assistants are very important if the design of the postproduction facility is focused around a central machine room. The editor initiates the communication via an intercom system and tapes are changed, set up, and dubbed by this voice on the other end. Occasionally, that assistant is in the same room and can speed up the editing process by doubling as a sound engineer or a character generator operator.

The elimination or devaluing of the video assistant makes it harder for young people to break into the business. Since the film assistant on a nonlinear project may work a second shift while the editor works throughout the day, there is less opportunity for the interaction between master and apprentice. With tense clients who are paying large sums per hour for time-critical work, the production company that puts an unknown or untested quantity in the driver's seat is taking a risk. They may lose their client forever to the competition or may have to discount the session to appease them.

How, then, do you break into this business? There are no second video assistants like there would be in film, so what is the entry-level position? Many times it is whatever the facility needs at the time: a tape dubber, a receptionist, even a courier; however, with the advent of nonlinear editing, there is the nonlinear assistant. In large and busy nonlinear postproduction facilities, there may be one nonlinear assistant per shift and three shifts per day. The entry level then becomes the graveyard shift and eventually the day or evening shift, where editors can discretely observe skills that keep them in demand. The job doesn't really require the ambition to become an editor, but the people who gain the most from the nonlinear assistant position are those who need to know these subtle skills to move on to the next level.

What responsibilities should such assistants be expected to perform? Much of the knowledge they need has been covered in this book. In fact, many editors perceive many of the techniques in this book as something only an assistant would perform. Others see it as required knowledge before starting a job! It is when a facility desires such a specialization of labor, either for personnel or billable reasons, that the assistants have the most value. They must perform all of the functions and have all the knowledge required to keep the systems running. The post supervisor instructs the assistant on all the requirements to keep as many jobs running smoothly as possible. The administrator or supervisor sees the big picture and the assistant performs the tasks.

These important daily tasks include capturing, media management, basic maintenance, backup, and output. Anything that is required to prepare the suite for the editor and the smooth transition from one project to another is appropriate for the assistant. Each facility has its own set of requirements, but mastery of all these skills can make someone very valuable to any busy postproduction facility.

CAPTURING

Capturing also implies following logs, creating bins, setting levels, and understanding drives. The logs are handed off with the understanding that the marked takes or possibly everything should be captured and the master clips named according to the description in the logs. Bad logs mean bad bins unless the assistant knows something about the specific job and is given the freedom to create better master clip names.

- Assess each take as to whether only video or only audio should be captured, thus maximizing use of disk space. Don't capture video for the voiceover!
- Check to see if everything, even incomplete takes, should be captured.
- Don't assume that just because there isn't enough drive space that certain shots must be left out. It is the assistant's responsibility to find the drives, connect them, and capture everything as required.

- The bins can be named based on tape name, and the editor will determine where to put the shots based on content later.
- Watch the audio and video levels! Distorted audio and blown-out video can come back to haunt the project at the finishing stage
- Learn how to read a vectorscope and waveform monitor or risk being bypassed by those who can!

DRIVES

Understanding drives is crucial to making sure the captured video can be played back. There is a setting called Drive Filtering under the General Setting. If this is on, only drives capable of playing back the selected resolution are available. Unfortunately, when people use non-Avid drives, they must disable this setting all the time. This is because when the system does not recognize the firmware loaded onto the media drives, it assumes the drive is incapable of higher resolution playback. This is not always true, but it is safer than assuming every drive can play back every resolution. Not paying attention at this stage can mean having to copy huge amounts of material to the proper drives. It is possible to capture audio to a Zip drive by accident with Drive Filtering turned off! Generally, however, if there is a mismatch between resolution and drive capabilities, there will be an error about either the video or audio overrunning its buffers. Usually, however, this error is because you have set the Macintosh RAM cache too high on OS 9.x and earlier.

- If you can, split the audio and video to separate drives, *not* separate partitions of the same drive.
- The drives should be named so there is no confusion between what is a new drive and what is just another partition on the same drive.
- Never overfill the drives. Leave a minimum of 10 to 15 percent free on any partition.
- You can split a single media file across multiple drives after it hits a certain length, like over 30 minutes. This will simplify capturing long clips. You will no longer need to split a master clip to fit on two drives and end up confusing everyone. This is capturing a single master clip to multiple media files. You can also group drives so that when one fills up you have control over which drive gets used next. Just don't separate two drives that are sharing media files for the same clip!
- Have a thorough understanding of SCSI principles (terminators, IDs, etc.).
- Be sure you can recognize the difference between drive types and speeds so that you do not slow down the performance of all the drives

on the SCSI chain by adding a slow drive in the wrong place. Have the right SCSI terminators, cables, and adapters.
- Be comfortable with mounting and dismounting Media Docks or even older RMAGS (the original removables for media playback).
- All drives should have unique names and physical labels so they can be moved to any system and still be identified.

Understanding drives also means knowing how to resuscitate an ailing one and knowing when to call it quits and get another. The number of drives returned to Avid with nothing wrong is astounding. If you can get a drive back to full health in an hour or two, how does that impact the production schedule compared to waiting for the morning rush delivery?

- Know the replacement policies of your non-Avid drives.
- Learn how to use all the Avid drive utilities and make sure you have the latest versions.
- Don't always load the newest firmware until you know all the ramifications for your configuration. Read the Release Notes if in doubt.
- Know how to mount, repartition, and, as a last resort, perform a destructive read/write test.
- Dealing with very large media files still requires different rules for maintenance. Don't assume you can run any drive utility on a media drive.

MEDIA MANAGEMENT

All the media on all the drives is under your jurisdiction so knowing what to keep and what to discard is both incredibly important and commonplace.

- Know how to lock and unlock media files and delete precomputes.
- Know how long it takes to copy media from one drive to another.
- Delete media through the Media Tool, Media Manager, or drag entire projects to the trash in their media folders after using MediaMover. Have a regular plan to delete precomputes.
- Understand the network and how it helps you to move media efficiently.
- Learn Media Mover and the Unity Administration tools if your facility has them.
- Become the network expert if you can and research the possibilities.
- Figure out how to improve network throughput.
- Reduce the number of drives that must be moved.

BASIC MAINTENANCE

If you aspire to become an editor, there are people who will hold your technical expertise against you. They think you cannot be technically proficient and a true artist. You may have to work a little harder to prove them wrong. It takes only a couple of success stories where you save the system *and* the project before employers see the value in an editor who reduces his or her own downtime.

- Take a Macintosh or Windows support class or a basic troubleshooting class.
- Consider becoming an Avid Certified Support Representative (ACSR).
- Always have a floppy disk or some other removable media that you can boot from and run disk-recovery programs.
- Be ready to strip all unneeded extensions from a Macintosh system and know which ones are the absolute minimum to run your specific system.
- Be ready to do a clean reinstallation of the system software.
- Between projects do everything you can to make all the systems as similar as possible.

BACKING UP

Consolidating is the most important feature for backing up. Some people have enough time and tape to back up everything in a project, but more likely, you will be forced to decide what to keep and what to discard.

- Learn all the variations for consolidating the final sequence.
- Know how to back up only the media needed to recreate the sequence.
- Create a database that can both retrieve the project and the individual bins.
- Print out the bin for each tape and include that with the tape. You will have a paper archive when all else fails.

OUTPUT

All forms of output are important and critical to the next step in the project. The Digital Cut may be the master or the approval copy. If it is the master, then the levels must be perfect and if it is just a VHS for approval, then it must contain all the video tracks of graphics and all the audio, mixed or direct.

- Do you have a time code generator that can burn in the sequence time-code to the Digital Cut?
- Spot-check the EDL for accuracy.
- A cut list should be scrutinized *at every cut* because the stakes are so much higher and the chance for adjustment at the next stage more remote.
- Make as many EDL versions and as many disks as you have time for.
- Expect that, when the online assembly finally comes, someone may ask for something more. If the request is for video-only EDLs, make a few with audio, too, just in case.
- Make several printouts to cover yourself, save time, and help the online editors if there is a mistake at their session. Everybody makes mistakes and, if you have an original copy of the EDL on paper, you can isolate the mistake more quickly as something done wrong after your handoff.
- Talk to professional sound studios. What are the most common mistakes you are likely to make when handing off media and projects to an AudioVision or ProTools session? What is correct for one production company may be wrong for another — maybe because the other facility handles your files wrong!
- Document all the steps you take to prepare the files and label everything clearly.
- It really is part of your job to prevent other people from making mistakes!

RECAPTURING

- Does every shot need to be recaptured? It is possible to make an EDL and a cut list with media offline. You can even relink to master clips that are offline just for the correct version of metadata.
- If the changes requested pertain only to the open and close of a sequence, then you might be able to get away with capturing only those sections. You can leave the rest of the media offline.
- If the need for recapturing is because the levels were set wrong, then the saved settings for that tape must be deleted before recapturing.
- If some of the footage needs to be captured in black and white, make sure to set the Capture Tool (Compression Tool on older systems) to monochrome. Even more important, change it back when you're finished! The third monitor shows you a color image all the time because it is monitoring what the signal looks like before it is processed. How many times have people walked into an Avid suite during capturing at low resolution, looked at the client monitor, and said, "Hey, that doesn't look so bad!"? The monitor during capturing does not reflect whether you have left monochrome on or are capturing at the wrong resolution.

BLACKING TAPE

It is desirable to know absolutely everything about connecting and operating video decks. This information is beyond the scope of this book, but it would include knowledge about:

- Reference signals
- Signal termination
- Loop through
- Deck-to-deck editing
- Deck front panel input choices
- Blacking tape (sometimes called "black and coding" in American English or "black and bursting" in the Queen's English)

After spending an afternoon with a well-known, much-decorated documentary editor, I asked about his deck connections. "What's a BNC?" he said. Although he had a great attitude toward the new technology, there was quite a learning curve involved in making him self-sufficient.

If you are not supported by staff engineers, I highly recommend that you learn how to clean the video heads. Oxide flakes off videotape and sticks in the tiny gap that video heads use to read the video signal. If you have a regular staff of technicians, then *don't touch*, but make sure the video heads are serviced on a regular basis. If no one is regularly cleaning the heads on the video deck, you should take on the responsibility as routine. Cleaning the heads with a proper head-cleaning kit once a week during heavy usage is not a bad idea, and if you are using a consumer deck, of course, there are head-cleaning cassettes. Consumer-quality tape loses oxide faster than professional tape (Walter Murch cut *The English Patient* using S-VHS and a Film Composer).

- There is a significant difference between digital and analog cleaning procedures, so make sure you know the deck you are working on.
- Digital decks do some self-cleaning.
- Do *not* clean with alcohol. Alcohol leaves a gummy film when it dries.
- Use a freon substitute and a wipe that is recommended by the manufacturer.
- Some rental facilities cover the edges of the VTR top cover with a seal so they know if it was opened and tampered with. Check with the rental company before breaking the seal.

For connecting Betacam SP decks, or any decks that have both composite and component input, I recommend looping the blackburst signal out of the reference input to the composite input. This serves several purposes, but the most important is that you will never accidentally use the composite input for your Digital Cut. It is always a black signal. The other benefit is that you can change the position of the switch on the front panel of the deck that changes the deck input from component to composite. Then you can begin to black tape without disconnecting or reconnecting any cables. Always monitor the output of any deck that is recording or you may end up recording black when you don't mean to!

If you buy tapes by the case, a good practice is to black them all whenever there is any downtime. By turning down the audio inputs and switching over to composite video input, which has been connected to a source of black, you can black tapes at a moment's notice throughout the day. Be sure to turn the audio inputs back up (or pop them back into the preset position) and throw the front panel switch back to YRB (component) before you start a digital cut!

When blacking tapes, set the four switches under the front panel of the Betacam deck to:

- DF (drop-frame) or NDF (non-drop-frame) if in NTSC
- Internal
- Record run
- Preset

This ensures that the timecode is generated from the internal timecode generator. It is not looking for an external signal. The timecode will only increment (run) when the tape is recording, and the starting point is wherever you preset it to start, no matter what other timecode is on the tape.

The ability to preset a starting number for timecode is achieved by manipulating the buttons on the outside of the front panel. Each deck is slightly different, but on Sony decks other than the UVW and DSR series:

- Press the Hold button.
- Use either the + and − buttons or the search knob to choose hours, minutes, seconds, and frames.
- Press the Set button.

Although the number in the LED does not show it yet, when you begin recording, the timecode starts incrementing from the desired preset number. The UVW and DSR series use a simple menu system to set timecode.

Other switches to check on a Betacam SP deck are on top of the front panel (as opposed to on its face or hidden by it), where the record inhibit switch is also located. The 2/4 field switch can sometimes be the cause of field-based problems when capturing graphics from tape. You should always capture everything with this switch in the four-field position (NTSC).

If you are trying to lay four discreet channels of audio to the deck, you'll need to switch the front panel audio switch that says 1/2 and 3/4. If it's on 1/2, then the audio sent to 1 and 2 is automatically laid to 3 and 4 if those input levels are up. Many people use channels 3 and 4 on the Betacam SP because of the superior signal-to-noise ratio. They may want four discreet channels from the Avid to the deck so the stereo music can go to channels 3 and 4 while the narration and sound bites go to channels 1 and 2.

A final note of caution: Tracks 3 and 4 are recorded helically, alongside the video. If you make a video insert edit on a Betacam SP master, you will wipe out tracks 3 and 4 for the length of the edit!

Even as Betacam SP decks are replaced by DV or other digital decks, these principles will always stay the same. Be familiar with how to set timecode on any deck you work with.

There may be more specific requirements for assistants at your facility, but these should cover the basic skills to make you useful from day one. All markets are slightly different with different terminology and different kinds of clientele. Every chance you get to watch the editors work should be snatched up, not just to see how they use the interface, but also to see what they are doing with it. Observe the way creative ideas are thrown around, accepted, rejected, modified, and experimented. This is always more important than the equipment being used.

You have an advantage over assistants in the past because, if you can get permission, you can take the opportunity to cut your own versions of the scenes, commercials, or segments when the suite would otherwise be empty. You can't mess up the film or add wear to the tapes just dissecting what the editor did and trying a few variations of your own. If you are lucky, you can get the editor to check them out and offer suggestions. A recommendation by the editor may move your career along faster than anything else.

With the more affordable Avid systems available to you to learn with, you should always be practicing when you can. Get time on a system, even if it is at home, to improve your skills as a storyteller or your command of effects. In the past you would have had to come into a production company in the middle of the night to get your hands on the "big iron." Now you can be creative and work on your editing chops on a laptop. Take advantage of this wonderful gift of evolving technology to improve as an artist. Positions change and so does technology. If you stay flexible, always willing to learn something new, you will always stay valuable to your employer. This book outlined some of the more important techniques to keep in mind with the Avid editing systems, but by no means all of them. To keep up with the speed of change takes more effort and more research than you might have expected. Stay focused on the most important aspect of the technology—the storytelling—and you will never be out of date.

Index